Recent Developments in Harmonic Analysis and its Applications

CONTEMPORARY MATHEMATICS

792

Recent Developments in Harmonic Analysis and its Applications

AMS Special Session
Harmonic Analysis
March 26–27, 2022
Virtual

Shaoming Guo
Zane Kun Li
Brian Street
Editors

AMERICAN MATHEMATICAL SOCIETY

2020 *Mathematics Subject Classification.* Primary 11D25, 42B10, 42B20, 42B25, 32–00.

Library of Congress Cataloging-in-Publication Data

Names: AMS Special Session on Harmonic Analysis (2022 : Online) — Guo, Shaoming, 1988- editor. — Li, Zane Kun, 1991- editor. — Street, Brian, 1981- editor.

Title: Recent developments in harmonic analysis and its applications : Virtual AMS Special Session on Harmonic Analysis, March 26-27, 2022 / Shaoming Guo, Zane Kun Li, Brian Street, editors.

Description: Providence, Rhode Island : American Mathematical Society, [2024] — Series: Contemporary mathematics, 0271-4132 ; volume 792 — Includes bibliographical references.

Identifiers: LCCN 2023044954 — ISBN 9781470471408 (paperback) — ISBN 9781470476052 (ebook)

Subjects: LCSH: Harmonic analysis–Congresses. — AMS: Number theory – Diophantine equations – Cubic and quartic equations. — Harmonic analysis on Euclidean spaces – Harmonic analysis in several variables – Fourier and Fourier-Stieltjes transforms and other transforms of Fourier type. — Harmonic analysis on Euclidean spaces – Harmonic analysis in several variables – Singular and oscillatory integrals (Calderón-Zygmund, etc.). — Harmonic analysis on Euclidean spaces – Harmonic analysis in several variables – Maximal functions, Littlewood-Paley theory. — Several complex variables and analytic spaces – General reference works (handbooks, dictionaries, bibliographies, etc.).

Classification: LCC QA403 .A5278 2024 — DDC 515/.2433–dc23/eng/20231106 LC record available at https://lccn.loc.gov/2023044954

---

10 9 8 7 6 5 4 3 2 1 24 23 22 10 20 19

Contents

Preface

Harmonic analysis has gone through rapid developments in the past decade. A large number of important works, in connection with geometric measure theory, analytic number theory, partial differential equations, several complex variables, etc., have appeared in the last few years. We would like to take this opportunity to collect some of these works.

We would like to thank all the authors of the papers collected in this volume. It is their contributions that made this volume possible.

Shaoming Guo
Zane Li
Brian Street

Contemporary Mathematics
Volume **792**, 2024
https://doi.org/10.1090/conm/792/15892

Length of sets under restricted families of projections onto lines

Terence L. J. Harris

ABSTRACT. Let $\gamma : I \to S^2$ be a C^2 curve with $\det(\gamma, \gamma', \gamma'')$ nonvanishing, and for each $\theta \in I$ let ρ_θ be orthogonal projection onto the span of $\gamma(\theta)$. It is shown that if $A \subseteq \mathbb{R}^3$ is a Borel set of Hausdorff dimension strictly greater than 1, then $\rho_\theta(A)$ has positive length for a.e. $\theta \in I$. This answers a question raised by Käenmäki, Orponen and Venieri.

1. Introduction

Let S^2 be the unit sphere in $\mathbb{R}^3$, let $\gamma : I \to S^2$ be a C^2 curve with $\det(\gamma, \gamma', \gamma'')$ nonvanishing on an interval I, and let ρ_θ be orthogonal projection onto the span of $\gamma(\theta)$, given by

$$\rho_\theta(x) = \langle x, \gamma(\theta)\rangle\gamma(\theta), \qquad x \in \mathbb{R}^3.$$

In [**2**, Conjecture 1.6], Fässler and Orponen conjectured that for any analytic set $A \subseteq \mathbb{R}^3$

$$\dim \rho_\theta(A) = \min\{\dim A, 1\}, \qquad \text{a.e. } \theta \in I,$$

where $\dim B$ means the Hausdorff dimension of B. For general C^2 curves this was resolved by Pramanik, Yang and Zahl [**11**], and independently by Gan, Guth and Maldague [**5**]. The case where γ is a non-great circle was resolved earlier by Käenmäki, Orponen and Venieri [**7**], who asked [**7**, p. 4] whether $\dim A > 1$ implies that $\rho_\theta(A)$ has positive length for a.e. $\theta \in I$. This had been shown previously by Fässler and Orponen [**2**, Theorem 1.9] in the special case where A is a self-similar set without rotations. The following theorem resolves the general case.

THEOREM 1.1. *If $A \subseteq \mathbb{R}^3$ is an analytic set with $\dim A > 1$, then*

$$\dim \left\{\theta \in I : \mathcal{H}^1(\rho_\theta(A)) = 0\right\} \leq \frac{4 - \dim A}{3}.$$

Throughout, the symbol $\mathcal{H}^s$ will be used for the s-dimensional Hausdorff measure on Euclidean space. By Frostman's lemma, Theorem 1.1 will follow from Theorem 1.2 below. For the statement, some notation will be defined first. Given a Borel measure μ on $\mathbb{R}^3$ and $\alpha \geq 0$, define

$$c_\alpha(\mu) = \sup_{\substack{x \in \mathbb{R}^3 \\ r > 0}} \frac{\mu(B(x,r))}{r^\alpha}.$$

2020 *Mathematics Subject Classification.* 28A78; 28A80.
Key words and phrases. Hausdorff dimension, orthogonal projection.

For each $\theta \in I$, the pushforward measure $\rho_{\theta\#}\mu$ is defined by

$$(\rho_{\theta\#}\mu)(E) = \mu\left(\rho_\theta^{-1}(E)\right),$$

for any Borel set E.

THEOREM 1.2. *If μ is a Borel measure on $\mathbb{R}^3$ with $c_\alpha(\mu) < \infty$ for some $\alpha > 1$, then*

$$\dim\left\{\theta \in I : \rho_{\theta\#}\mu \not\ll \mathcal{H}^1\right\} \leq \frac{4-\alpha}{3}.$$

The notation $\rho_{\theta\#}\mu \not\ll \mathcal{H}^1$ means that $\rho_{\theta\#}\mu$ is not absolutely continuous with respect to $\mathcal{H}^1$. The proof of Theorem 1.2 is similar to the proof of Theorem 8 in [**4**], which solved the analogous problem of projections onto planes. It uses a variant of the decomposition of a measure into "good" and "bad" parts, which originated in [**6**].

Given $s \geq 0$, a set $A \subseteq \mathbb{R}^3$ is called an s-set if A is $\mathcal{H}^s$-measurable with $0 < \mathcal{H}^s(A) < \infty$. Theorem 1.2 implies Theorem 1.1, but it also implies the following slightly stronger version for s-sets.

THEOREM 1.3. *Suppose that $s > 1$, and that $A \subseteq \mathbb{R}^3$ is an s-set. Let μ be the Borel measure defined by $\mu(F) = \mathcal{H}^s(F \cap A)$ for any Borel set $F \subseteq \mathbb{R}^3$, and let*

$$E = \left\{\theta \in I : \rho_{\theta\#}\mu \not\ll \mathcal{H}^1\right\}.$$

Then

$$\dim E \leq \frac{4-s}{3},$$

and

$$\mathcal{H}^1(\rho_\theta(B)) > 0, \qquad \textit{for all } \theta \in I \setminus E.$$

for any $\mathcal{H}^s$-measurable set $B \subseteq A$ with $\mathcal{H}^s(B) > 0$.

Theorem 1.3 is related to a lemma of Marstrand (see [**9**, Lemma 13]), which states that for $s > 1$, for any s-set A in the plane, there is a measure zero set of exceptional directions, such that any s-set $B \subseteq A$ projects onto a set of positive Lebesgue measure outside of this set of exceptional directions.

Theorem 1.3 also implies Theorem 1.1, since any analytic set of infinite $\mathcal{H}^s$ measure contains a closed set of positive finite $\mathcal{H}^s$ measure [**1**].

2. Proof of Theorem 1.2 and Theorem 1.3

Throughout this section, $\gamma : I \to S^2$ will be a fixed C^2 unit speed curve with $\det(\gamma, \gamma', \gamma'')$ nonvanishing on I, where I is a compact interval. For $A \subseteq \mathbb{R}^3$, $m(A)$ will denote the Lebesgue measure of A.

DEFINITION 2.1. Let

$$\Lambda = \bigcup_{j\geq 1} \Lambda_j,$$

where each Λ_j is a collection of boxes τ of dimensions $1 \times 2^{j/2} \times 2^j$, forming a finitely overlapping cover of the ~ 1-neighbourhood of the truncated light cone Γ_j in the standard way, where

$$\Gamma_j = \left\{t\gamma(\theta) : 2^{j-1} \leq |t| \leq 2^j, \quad \theta \in I\right\}.$$

Each $\tau \in \Lambda_j$ has an angle θ_τ such that the long axis of τ is parallel to $\gamma(\theta_\tau)$, the medium axis of τ is parallel to $\gamma'(\theta_\tau)$, and the short axis of τ is parallel to

$(\gamma \times \gamma')(\theta_\tau)$. Let $\{\psi_\tau\}_{\tau\in\Lambda}$ be a smooth partition of unity subordinate to the cover $\{1.1\tau : \tau \in \Lambda\}$ of $\bigcup_{\tau\in\Lambda} \tau$, such that for any $\xi, v \in \mathbb{R}^3$, $t \in \mathbb{R}$ and any positive integer n,

$$\left|\left(\frac{d}{dt}\right)^n \psi_\tau(\xi + tv)\right| \lesssim_n \left|2^{-j}\langle v, \gamma(\theta_\tau)\rangle\right|^n + \left|2^{-j/2}\langle v, \gamma'(\theta_\tau)\rangle\right|^n + |\langle v, (\gamma \times \gamma')(\theta_\tau)\rangle|^n. \tag{2.1}$$

Fix $\widetilde{\delta} > 0$. For each $\tau \in \Lambda$, let $\mathbb{T}_\tau$ be a collection of boxes (referred to as "planks") of dimensions $2^{j\widetilde{\delta}} \times 2^{j(\widetilde{\delta}-1/2)} \times 2^{j(\widetilde{\delta}-1)}$, forming a finitely overlapping cover of $\mathbb{R}^3$, such that for each $T \in \mathbb{T}_\tau$, the short axis of T is parallel to $\gamma(\theta_\tau)$, the medium axis is parallel to $\gamma'(\theta_\tau)$, and the long axis is parallel to $(\gamma \times \gamma')(\theta_\tau)$ (so that T is "dual" to τ). Let $\{\eta_T\}_{T\in\mathbb{T}_\tau}$ be a smooth partition of unity subordinate to the cover $\mathbb{T}_\tau$ of $\mathbb{R}^3$, such that for any $x, v \in \mathbb{R}^3$, any $t \in \mathbb{R}$ and any positive integer n,

$$\left|\left(\frac{d}{dt}\right)^n \eta_T(x + tv)\right| \lesssim_n \left|2^{j(1-\widetilde{\delta})}\langle v, \gamma(\theta_\tau)\rangle\right|^n + \left|2^{j(1/2-\widetilde{\delta})}\langle v, \gamma'(\theta_\tau)\rangle\right|^n + \left|2^{-j\widetilde{\delta}}\langle v, (\gamma \times \gamma')(\theta_\tau)\rangle\right|^n. \tag{2.2}$$

Given a finite complex Borel measure μ on $\mathbb{R}^3$, define

$$M_T\mu = \eta_T\left(\mu * \widecheck{\psi_\tau}\right).$$

For each $\theta \in I$, let π_θ be orthogonal projection onto the orthogonal complement of $(\gamma \times \gamma')(\theta)$. The following lemma is essentially a special case of Lemma 5 from [**4**], but the proof will be included here for completeness.

LEMMA 2.2. *There exists an $r > 0$, depending only on γ, such that the following holds. Let $j \geq 1$ and let $\tau \in \Lambda_j$. If $\theta \in I$ is such that $2^{j(\widetilde{\delta}-1/2)} \leq |\theta_\tau - \theta| \leq r$, then for any $T \in \mathbb{T}_\tau$, for any positive integer N and for any $f \in L^1(\mathbb{R}^3)$,*

$$\|\pi_{\theta\#}M_T f\|_{L^1(\mathcal{H}^2)} \leq C2^{-j\widetilde{\delta}N}\|f\|_{L^1(\mathbb{R}^3)}, \tag{2.3}$$

where $C = C\left(N, \gamma, \widetilde{\delta}\right)$.

PROOF. For any $x \in (\gamma \times \gamma')(\theta)^\perp$,

$$(\pi_{\theta\#}M_T f)(x) = \int_{\mathbb{R}^3}\int_\tau f(y)\psi_\tau(\xi)e^{-2\pi i\langle \xi, x-y\rangle}\left[\int_{\mathbb{R}} \eta_T(x + t(\gamma \times \gamma')(\theta))e^{-2\pi i t\langle \xi, (\gamma\times\gamma')(\theta)\rangle}\, dt\right] d\xi\, dy. \tag{2.4}$$

Let $\xi \in \tau$. By integrating by parts n times, and using (2.2),

$$\begin{aligned}\left|\int_{\mathbb{R}} \eta_T(x + t(\gamma \times \gamma')(\theta))e^{-2\pi i t\langle \xi, (\gamma\times\gamma')(\theta)\rangle}\, dt\right| &\lesssim \int_{\mathbb{R}} \chi_T(x + t(\gamma \times \gamma')(\theta))\, dt \\ \times \frac{1}{|\langle(\gamma \times \gamma')(\theta), \xi\rangle|^n}\Big(2^{j(1-\widetilde{\delta})}&|\langle(\gamma \times \gamma')(\theta), \gamma(\theta_\tau)\rangle| + 2^{j(1/2-\widetilde{\delta})}|\langle(\gamma \times \gamma')(\theta), \gamma'(\theta_\tau)\rangle| \\ &+ 2^{-j\widetilde{\delta}}|\langle(\gamma \times \gamma')(\theta), (\gamma \times \gamma')(\theta_\tau)\rangle|\Big)^n.\end{aligned} \tag{2.5}$$

By the definition of τ,

$$\begin{aligned}&\langle(\gamma\times\gamma')(\theta),\xi\rangle\\&=\xi_1\langle(\gamma\times\gamma')(\theta),\gamma(\theta_\tau)\rangle+\xi_2\langle(\gamma\times\gamma')(\theta),\gamma'(\theta_\tau)\rangle+\xi_3\langle(\gamma\times\gamma')(\theta),(\gamma\times\gamma')(\theta_\tau)\rangle,\end{aligned}$$

where $2^{j-2}\leq|\xi_1|\leq 2^{j+2}$, $|\xi_2|\leq 2^{j/2}$ and $|\xi_3|\leq 1$. Let $\varepsilon:=|\theta_\tau-\theta|$. By a second order Taylor approximation (using that γ is C^2) and the scalar triple product formula,

$$|\xi_1\langle(\gamma\times\gamma')(\theta),\gamma(\theta_\tau)\rangle|\sim\varepsilon^2 2^j,$$

provided r is small enough. Moreover

$$|\xi_2\langle(\gamma\times\gamma')(\theta),\gamma'(\theta_\tau)\rangle|\lesssim 2^{j/2}\varepsilon,\quad |\xi_3\langle(\gamma\times\gamma')(\theta),(\gamma\times\gamma')(\theta_\tau)\rangle|\leq 1.$$

Since $\varepsilon\geq 2^{j(\widetilde{\delta}-1/2)}$, it follows that

$$|\langle(\gamma\times\gamma')(\theta),\xi\rangle|\sim\varepsilon^2 2^j,$$

provided that r is sufficiently small and j is sufficiently large. Using second order Taylor approximation in a similar way gives that

$$\begin{aligned}&2^{j(1-\widetilde{\delta})}\,|\langle(\gamma\times\gamma')(\theta),\gamma(\theta_\tau)\rangle|\\&+2^{j(1/2-\widetilde{\delta})}\,|\langle(\gamma\times\gamma')(\theta),\gamma'(\theta_\tau)\rangle|+2^{-j\widetilde{\delta}}\,|\langle(\gamma\times\gamma')(\theta),(\gamma\times\gamma')(\theta_\tau)\rangle|\lesssim 2^{j(1-\widetilde{\delta})}\varepsilon^2.\end{aligned}$$

It follows that

$$(2.5)\lesssim_n 2^{-j\widetilde{\delta}n}\int_{\mathbb{R}}\chi_T(x+t(\gamma\times\gamma')(\theta))\,dt.$$

Substituting this into (2.4) gives that

$$|(\pi_{\theta\#}M_T f)(x)|\lesssim_n 2^{-j\widetilde{\delta}n}\,\|f\|_{L^1(\mathbb{R}^3)}\,m(\tau)\int_{\mathbb{R}}\chi_T(x+t(\gamma\times\gamma')(\theta))\,dt,$$

for any $x\in(\gamma\times\gamma')(\theta)^\perp$. Integrating over $t\in\mathbb{R}$ and $x\in(\gamma\times\gamma')(\theta)^\perp$ gives (2.3). □

The following lemma is essentially the same as Lemma 4 from [**4**], but again the proof is included for completeness.

LEMMA 2.3. *Let $j\geq 1$ and let $\tau\in\Lambda_j$. For any finite compactly supported Borel measure μ,*

$$\|M_T\mu\|_{L^1(\mathbb{R}^3)}\leq 2^{3j\widetilde{\delta}}\mu(2T)+C_N 2^{-j\widetilde{\delta}N}\mu(\mathbb{R}^3),$$

for any positive integer N.

PROOF. By definition,

$$\begin{aligned}\|M_T\mu\|_{L^1(\mathbb{R}^3)}&=\int\eta_T(x)\left|\int\widecheck{\psi_\tau}(x-y)\,d\mu(y)\right|dx\\&\leq\int_T\int_{2T}\left|\widecheck{\psi_\tau}(x-y)\right|\,d\mu(y)\,dx\\&\quad+\int_T\int_{\mathbb{R}^3\setminus 2T}\left|\widecheck{\psi_\tau}(x-y)\right|\,d\mu(y)\,dx.\end{aligned}\tag{2.6}$$

The first integral satisfies

$$\int_T\int_{2T}\left|\widecheck{\psi_\tau}(x-y)\right|\,d\mu(y)\,dx\leq 2^{3j\widetilde{\delta}}\mu(2T).\tag{2.7}$$

The second integral satisfies

$$\int_T \int_{\mathbb{R}^3 \setminus 2T} \left| \widecheck{\psi_\tau}(x-y) \right| \, d\mu(y) \, dx \leq \mu(\mathbb{R}^3) \int_{\mathbb{R}^3 \setminus T_0} \left| \widecheck{\psi_\tau} \right|,$$

where T_0 is the translate of the plank T to the origin, parallel to T. Integrating by parts, and using (2.1), gives that for any $k \geq 0$, for any $x \in \mathbb{R}^3 \setminus 2^k T_0$,

$$\left| \widecheck{\psi_\tau}(x) \right| \lesssim_N 2^{-kN - j\tilde{\delta}N} m(\tau).$$

Summing a geometric series over $k \geq 0$ gives (with relabelled N)

$$\int_T \int_{\mathbb{R}^3 \setminus 2T} \left| \widecheck{\psi_\tau}(x-y) \right| \, d\mu(y) \, dx \leq C_N \mu(\mathbb{R}^3) 2^{-j\tilde{\delta}N}. \tag{2.8}$$

Putting (2.7) and (2.8) into (2.6) finishes the proof. □

For a function $f : X \to [0, +\infty]$ on a measure space $(X, \mathcal{A}, \mu)$, let $\int_* f \, d\mu$ denote the lower integral of f, defined by

$$\int_* f \, d\mu = \sup \left\{ \int g \, d\mu : g \text{ is simple and } \mathcal{A}\text{-measurable with } 0 \leq g \leq f \right\},$$

where "simple" means that g takes finitely many values. In the application of Lemma 2.4 below, the integrand will be measurable, so the use of the lower integral is not important and is only a technical convenience to avoid measurability issues. The definition of the lower integral is standard; see e.g. [**10**, p. 13].

LEMMA 2.4. *Let $\beta \in [0,1]$, let $\alpha = 4 - 3\beta$, and let λ be a Borel measure supported on I with $c_\beta(\lambda) \leq 1$. Then for any $\epsilon > 0$, there exists $\delta > 0$ such that*

$$\int_* (\rho_{\theta\#}\mu) \left(\bigcup_{D \in \mathbb{D}_\theta} D \right) d\lambda(\theta) \leq C(\lambda, \delta, \epsilon, \gamma) R^{-\delta} \mu(\mathbb{R}^3), \tag{2.9}$$

for any $R \geq 1$, for any Borel measure μ on $B_3(0,1)$ with $c_\alpha(\mu) \leq 1$, and for any family of sets $\{\mathbb{D}_\theta\}$, where each $\mathbb{D}_\theta$ is a disjoint set of intervals of diameter $2R^{-1}$ in the span of $\gamma(\theta)$, each with cardinality $|\mathbb{D}_\theta| \leq R^{1-\epsilon}\mu(\mathbb{R}^3)$.

REMARK 2.5. Lemma 2.4 roughly says that the pushforward $\rho_{\theta\#}\mu$ of a $(4-3\beta)$-dimensional measure μ satisfies a 1-dimensional Frostman condition on average. By taking $\beta = 1$ (for example), this is enough to conclude that projections of 1-dimensional sets are a.e. 1-dimensional, but the main application of Lemma 2.4 here will be to bound the L^1 norm of the "bad" part of the measure in the proof of Theorem 1.2. The "bad" part of the measure corresponds to intervals of large $\rho_{\theta\#}\mu$-mass, which will automatically satisfy the cardinality assumption of Lemma 2.4.

PROOF OF LEMMA 2.4. It may be assumed that $\alpha \leq 3$, since otherwise the lemma is trivial. Since the constant is allowed to depend on γ, it can be assumed that γ is localised to a smaller interval on which Lemma 2.2 holds.

Let ϕ_R be a nonnegative bump function supported in $B(0, R^{-1})$ which integrates to 1, defined by

$$\phi_R(x) = R^3 \phi(Rx),$$

for some fixed nonnegative bump function ϕ with support in $B(0,1)$ such that $\int \phi = 1$. For any θ and any $D \in \mathbb{D}_\theta$, the 1-Lipschitz property of orthogonal projections implies that

$$[\rho_{\theta\#}(\mu * \phi_R)](2D) \geq (\rho_{\theta\#}\mu)(D),$$

where $2D$ is the interval with the same centre as D, but twice the radius. Moreover,

$$c_\alpha(\mu * \phi_R) \lesssim c_\alpha(\mu),$$

so it suffices to prove (2.9) with $\mu * \phi_R$ in place of μ. To simplify notation the new measure will not be relabelled, but it will be assumed throughout that μ is a nonnegative Schwartz function, and that

$$|\widehat{\mu}(\xi)| \leq C_N (R/\xi)^N, \quad \xi \in \mathbb{R}^3, \tag{2.10}$$

for any positive integer N, where C_N is a constant depending only on N.

Let ϵ_0 be any positive real number which is strictly larger than the the infimum over all positive ϵ for which the conclusion of the lemma is true. It suffices to prove that the lemma holds for any $\epsilon > (2\epsilon_0)/3$, so let such an ϵ be given. Let $R \geq 1$ and choose a nonnegative integer J such that $2^J \sim R^{\epsilon/1000}$. Let $\varepsilon > 0$ be such that $\varepsilon \ll \epsilon - \frac{2\epsilon_0}{3}$. Choose $\widetilde{\delta} > 0$ such that $\widetilde{\delta} \ll \min\{\delta_{\epsilon_0}, \varepsilon\}$, where δ_{ϵ_0} is a δ corresponding to ϵ_0 that satisfies (2.9).

Define the "bad" part of μ by

$$\mu_b = \sum_{j \geq J} \sum_{\tau \in \Lambda_j} \sum_{T \in \mathbb{T}_{\tau,b}} M_T \mu, \tag{2.11}$$

where, for each $\tau \in \Lambda_j$, the set of "bad" planks corresponding to τ is defined by

$$\mathbb{T}_{\tau,b} = \left\{ T \in \mathbb{T}_\tau : \mu(4T) \geq 2^{j(\epsilon_0 - 1)} \right\}, \tag{2.12}$$

where $4T$ is a plank with the same centre as T, but scaled by a factor of 4. Define the "good" part of μ by

$$\mu_g = \mu - \mu_b.$$

The Schwartz decay of μ implies that the sum in (2.11) converges in the Schwartz space $\mathcal{S}(\mathbb{R}^3)$. This implies that μ_b and μ_g are Schwartz functions, and in particular they are finite complex measures. Pushforwards of complex measures are defined just as for positive measures. By[1] Cauchy–Schwarz,

$$\begin{aligned}\int_* (\rho_\theta \# \mu) \left(\bigcup_{D \in \mathbb{D}_\theta} D \right) d\lambda(\theta) &\leq \int \|\rho_\theta \# \mu_b\|_{L^1(\mathcal{H}^1)} \, d\lambda(\theta) \\ &\quad + \sup_{\theta \in I} \mathcal{H}^1 \left(\bigcup_{D \in \mathbb{D}_\theta} D \right)^{1/2} \left(\int \|\rho_\theta \# \mu_g\|_{L^2(\mathcal{H}^1)}^2 \, d\lambda(\theta) \right)^{1/2}.\end{aligned}$$

[1] The particular Cauchy–Schwarz technique used here is from [**8**].

The contribution from the "bad" part will be bounded first. By the triangle inequality,

$$\int \|\rho_\theta\#\mu_b\|_{L^1(\mathcal{H}^1)}\, d\lambda(\theta) \leq \sum_{j\geq J}\int \sum_{\tau\in\Lambda_j}\sum_{T\in\mathbb{T}_{\tau,b}} \|\rho_\theta\# M_T\mu\|_{L^1(\mathcal{H}^1)}\, d\lambda(\theta)$$

$$= \sum_{j\geq J}\int \sum_{\substack{\tau\in\Lambda_j:\\ |\theta_\tau-\theta|\leq 2^{j(\widetilde{\delta}-1/2)}}}\sum_{T\in\mathbb{T}_{\tau,b}} \|\rho_\theta\# M_T\mu\|_{L^1(\mathcal{H}^1)}\, d\lambda(\theta) \tag{2.13}$$

$$+ \sum_{j\geq J}\int \sum_{\substack{\tau\in\Lambda_j:\\ |\theta_\tau-\theta|> 2^{j(\widetilde{\delta}-1/2)}}}\sum_{T\in\mathbb{T}_{\tau,b}} \|\rho_\theta\# M_T\mu\|_{L^1(\mathcal{H}^1)}\, d\lambda(\theta). \tag{2.14}$$

Let π_θ be orthogonal projection onto $(\gamma\times\gamma')(\theta)^\perp$. By the inequality

$$\|\rho_\theta\# f\|_{L^1(\mathcal{H}^1)} \leq \|\pi_\theta\# f\|_{L^1(\mathcal{H}^2)},$$

followed by Lemma 2.2,

$$(2.14) \lesssim 2^{-J}\mu(\mathbb{R}^3) \sim R^{-\epsilon/1000}\mu(\mathbb{R}^3).$$

By the inequality

$$\|\rho_\theta\# f\|_{L^1(\mathcal{H}^1)} \leq \|f\|_{L^1(\mathbb{R}^3)},$$

followed by Lemma 2.3,

$$\begin{aligned}(2.13) &\leq \sum_{j\geq J}\int \sum_{\substack{\tau\in\Lambda_j:\\ |\theta_\tau-\theta|\leq 2^{j(\widetilde{\delta}-1/2)}}}\sum_{T\in\mathbb{T}_{\tau,b}} \|M_T\mu\|_{L^1(\mathbb{R}^3)}\, d\lambda(\theta)\\ &\lesssim 2^{-J}\mu(\mathbb{R}^3) + \sum_{j\geq J} 2^{3j\widetilde{\delta}}\int \sum_{\substack{\tau\in\Lambda_j:\\ |\theta_\tau-\theta|\leq 2^{j(\widetilde{\delta}-1/2)}}}\sum_{T\in\mathbb{T}_{\tau,b}} \mu(2T)\, d\lambda(\theta),\end{aligned}$$

The nontail term satisfies

$$\sum_{j\geq J} 2^{3j\widetilde{\delta}}\int \sum_{\substack{\tau\in\Lambda_j:\\ |\theta_\tau-\theta|\leq 2^{j(\widetilde{\delta}-1/2)}}}\sum_{T\in\mathbb{T}_{\tau,b}} \mu(2T)\, d\lambda(\theta) \lesssim \sum_{j\geq J} 2^{10j\widetilde{\delta}}\int \mu(B_j(\theta))\, d\lambda(\theta), \tag{2.15}$$

where, for each $\theta\in I$ and each j,

$$B_j(\theta) = \bigcup_{\substack{\tau\in\Lambda_j:\\ |\theta_\tau-\theta|\leq 2^{j(\widetilde{\delta}-1/2)}}}\bigcup_{T\in\mathbb{T}_{\tau,b}} 2T.$$

The inequality (2.15) used that for each j and each $\theta\in I$, there are $\lesssim 2^{j\widetilde{\delta}}$ sets $\tau\in\Lambda_j$ with the property that $|\theta_\tau-\theta|\leq 2^{j(\widetilde{\delta}-1/2)}$, which means each of the planks $2T$ in the union defining $B_j(\theta)$ intersects $\lesssim 2^{j\widetilde{\delta}}$ of the others. For each T in the union defining $B_j(\theta)$, the set $(4T)\cap B(0,1)$ is contained in a plank T_θ of dimensions

$$\sim 2^{j(2\widetilde{\delta}-1)}\times 2^{j(\widetilde{\delta}-1/2)}\times 1,$$

with short direction parallel to $\gamma(\theta)$, medium direction parallel to $\gamma'(\theta)$, and long direction parallel to $(\gamma\times\gamma')(\theta)$, where the implicit constant depends only on γ; this follows from the second order Taylor approximation for γ. Therefore, the intervals in the set

$$\left\{\rho_\theta(T_\theta) : T \in \mathbb{T}_{\tau,b}, \quad \tau \in \Lambda_j, \quad |\theta_\tau - \theta| \leq 2^{j(\widetilde{\delta}-1/2)}\right\},$$

all have length $\sim 2^{j(2\widetilde{\delta}-1)}$, and form a cover of $\rho_\theta(B_j(\theta) \cap B(0,1))$. By the Vitali covering lemma, there is a disjoint subcollection

$$\{\rho_\theta(T_\theta) : T \in \mathcal{B}_\theta\},$$

indexed by some set $\mathcal{B}_\theta$, such that

$$\{3\rho_\theta(T_\theta) : T \in \mathcal{B}_\theta\},$$

is a cover of $\rho_\theta(B_j(\theta) \cap B(0,1))$. The set $\mathcal{B}_\theta$ has cardinality $|\mathcal{B}_\theta| \leq \mu(\mathbb{R}^3)2^{j(1-\epsilon_0)}$; by disjointness and the definition of the "bad" planks (see (2.12)). Since the conclusion of the lemma holds for ϵ_0, it follows that for each $j \geq J$,

$$\int \mu(B_j(\theta))\, d\lambda(\theta) \leq \int (\rho_{\theta\#}\mu)\left(\bigcup_{T_\theta \in \mathcal{B}_\theta} 3\rho_\theta(T_\theta)\right) d\lambda(\theta) \lesssim 2^{j\left(-\delta_{\epsilon_0}/2+10\widetilde{\delta}\right)}\mu(\mathbb{R}^3).$$

The set $B_j(\theta)$ is piecewise constant in θ over a partition of I into Borel sets, so it may be assumed that the integrands above are Borel measurable. Since $\widetilde{\delta} \ll \delta_{\epsilon_0}$, the inequality above yields

$$(2.13) \lesssim \mu(\mathbb{R}^3)2^{-(J\delta_{\epsilon_0})/100} \sim \mu(\mathbb{R}^3)R^{-(\epsilon\delta_{\epsilon_0})/10^5}.$$

It remains to bound the contribution from μ_g. By the assumptions in the lemma,

$$\sup_{\theta \in I} \mathcal{H}^1\left(\bigcup_{D \in \mathbb{D}_\theta} D\right) \lesssim R^{-\epsilon}\mu(\mathbb{R}^3).$$

Since $\varepsilon \ll \epsilon - (2\epsilon_0)/3$, it suffices to prove that

$$\int \|\rho_{\theta\#}\mu_g\|^2_{L^2(\mathcal{H}^1)}\, d\lambda(\theta) \lesssim \max\left\{R^{2\epsilon_0/3+1000\varepsilon}, R^{\epsilon/2}\right\}\mu(\mathbb{R}^3),$$

By Plancherel's theorem in 1 dimension,

$$\int \|\rho_{\theta\#}\mu_g\|^2_{L^2(\mathcal{H}^1)}\, d\lambda(\theta) = \int\int_{\mathbb{R}} |\widehat{\mu_g}(t\gamma(\theta))|^2\, dt\, d\lambda(\theta).$$

To formally prove this identity, one approach is to rotate $\gamma(\theta)$ to $(1,0,0)$ (using that $\mathcal{H}^1$ is a rotation invariant measure on $\mathbb{R}^3$), and apply Plancherel's theorem on $\mathbb{R}$. By symmetry, by summing a geometric series, and by the rapid decay of $\widehat{\mu}$ outside $B(0,R)$ (see (2.10)), it will suffice to bound

$$\int\int_{2^{j-1}}^{2^j} |\widehat{\mu_g}(t\gamma(\theta))|^2\, dt\, d\lambda(\theta),$$

for any $j \geq 2J$ with $2^j \leq R^{1+\widetilde{\delta}}$; the contribution from the small frequencies can be bounded trivially by the definition of J. For each $\tau \in \Lambda$, define the set of "good" planks corresponding to τ by

$$\mathbb{T}_{\tau,g} = \mathbb{T}_\tau \setminus \mathbb{T}_{\tau,b}.$$

Then

$$\int\int_{2^{j-1}}^{2^j} \left|\widehat{\mu_g}\left(t\gamma(\theta)\right)\right|^2 \, dt \, d\lambda(\theta)$$
$$\lesssim \int\int_{2^{j-1}}^{2^j} \left| \sum_{\tau \in \bigcup_{|j'-j|\leq 2} \Lambda_{j'}} \sum_{T \in \mathbb{T}_{\tau,g}} \widehat{M_T\mu}\left(t\gamma(\theta)\right)\right|^2 dt\, d\lambda(\theta) + 2^{-j}\mu(\mathbb{R}^3).$$

Since the τ's are finitely overlapping,

$$\int\int_{2^{j-1}}^{2^j} \left| \sum_{\tau \in \bigcup_{|j'-j|\leq 2} \Lambda_{j'}} \sum_{T \in \mathbb{T}_{\tau,g}} \widehat{M_T\mu}\left(t\gamma(\theta)\right)\right|^2 dt\, d\lambda(\theta) \lesssim 2^{-j}\mu(\mathbb{R}^3) + \sum_{\tau \in \bigcup_{|j'-j|\leq 2} \Lambda_{j'}} \int\int_{2^{j-1}}^{2^j} \left| \sum_{T \in \mathbb{T}_{\tau,g}} \widehat{M_T\mu}\left(t\gamma(\theta)\right)\right|^2 dt\, d\lambda(\theta). \tag{2.16}$$

The uncertainty principle implies that each of the integrals in the right-hand side of (2.16) is bounded by $2^{-j\beta}$ times the integral of the same function over $\mathbb{R}^3$. More precisely, for each $\tau \in \bigcup_{|j'-j|\leq 2} \Lambda_{j'}$, the contribution from the planks in the above sum with $T \cap B\left(0, 2^{10j\widetilde{\delta}}\right) = \emptyset$ is negligible since μ is supported in $B(0,1)$. The remaining sum

$$g_\tau = \sum_{T \in \mathbb{T}_{\tau,g} : T \cap B\left(0, 2^{10j\widetilde{\delta}}\right) \neq \emptyset} M_T\mu,$$

is equal to $g_\tau \varphi$ where φ is a smooth bump function on $B\left(0, 2^{1+10j\widetilde{\delta}}\right)$ obtained by rescaling a bump function on the unit ball, and therefore by the Cauchy–Schwarz inequality,

$$\left|\widehat{g_\tau}\right|^2 \lesssim \left|\widehat{g_\tau}\right|^2 * \zeta, \tag{2.17}$$

where

$$\zeta(\xi) = \frac{2^{30j\delta}}{1 + |2^{10j\delta}\xi|^{100}}, \qquad \xi \in \mathbb{R}^3.$$

The function in the right-hand side of (2.17) is essentially constant on balls of radius $2^{-10j\widetilde{\delta}}$ (as it inherits this property from ζ), so by discretising the integral in (2.16) into balls of radius $2^{-10j\widetilde{\delta}}$ and using the condition $c_\beta(\lambda) \leq 1$,

$$\int\int_{2^{j-1}}^{2^j} \left| \sum_{T \in \mathbb{T}_{\tau,g}} \widehat{M_T\mu}\left(t\gamma(\theta)\right)\right|^2 dt\, d\lambda(\theta)$$
$$\lesssim 2^{j\left(100\widetilde{\delta} - \beta\right)} \int_{\mathbb{R}^3} \left| \sum_{T \in \mathbb{T}_{\tau,g} : T \cap B\left(0, 2^{10j\widetilde{\delta}}\right) \neq \emptyset} \widehat{M_T\mu}\right|^2 d\xi + 2^{-j}\mu(\mathbb{R}^3),$$

for each $\tau \in \bigcup_{|j'-j|\leq 2} \Lambda_{j'}$. Let

$$\mathbb{T}_{j,g} = \bigcup_{\tau \in \bigcup_{|j'-j|\leq 2} \Lambda_{j'}} \left\{ T \in \mathbb{T}_{\tau,g} : T \cap B\left(0, 2^{10j\tilde{\delta}}\right) \neq \emptyset \right\}.$$

By Plancherel's theorem in $\mathbb{R}^3$ and the finite overlapping property of the T's, it suffices to prove that for any $j \geq 2J$,

$$\sum_{T \in \mathbb{T}_{j,g}} \int_{\mathbb{R}^3} |M_T \mu|^2 \, dx \lesssim 2^{j\left(\beta + \frac{2\epsilon_0}{3} + 100\varepsilon\right)} \mu(\mathbb{R}^3). \tag{2.18}$$

From the definition $M_T \mu = \eta_T \left(\mu * \widecheck{\psi_{\tau(T)}} \right)$ and by Fubini, the left-hand side of the above is equal to

$$\int \sum_{T \in \mathbb{T}_{j,g}} [\eta_T M_T \mu] * \widetilde{\psi_{\tau(T)}} \, d\mu.$$

If $f_T := [\eta_T M_T \mu] * \widetilde{\psi_{\tau(T)}}$, then by the Cauchy–Schwarz inequality with respect to the measure μ, the square of the above is bounded by

$$\int \left| \sum_{T \in \mathbb{T}_{j,g}} f_T \right|^2 d\mu \cdot \mu(\mathbb{R}^3).$$

By the uncertainty principle,

$$\int \left| \sum_{T \in \mathbb{T}_{j,g}} f_T \right|^2 d\mu \lesssim \int \left| \sum_{T \in \mathbb{T}_{j,g}} f_T \right|^2 d\mu_j,$$

where $\mu_j = \mu * \phi_j$ and $\phi_j(x) = \frac{2^{3j}}{1+|2^j x|^N}$, where $N \sim 1000/\tilde{\delta}^2$. By dyadic pigeonholing, there exists a collection $\mathbb{W}$ of planks $T \in \mathbb{T}_{j,g}$ with $\|f_T\|_p$ constant over $T \in \mathbb{W}$ up to a factor of 2, and a union Y of disjoint 2^{-j}-balls Q such that each Q intersects $\sim M$ planks $2T \in \mathbb{W}$ for some dyadic number M, and such that

$$\int \left| \sum_{T \in \mathbb{T}_{j,g}} f_T \right|^2 d\mu_j \lessapprox j^{10} \int_Y \left| \sum_{T \in \mathbb{W}} f_T \right|^2 d\mu_j + 2^{-j} \mu(\mathbb{R}^3)^2,$$

Let $p = 6$. By Hölder's inequality with respect to the Lebesgue measure,

$$\int_Y \left| \sum_{T \in \mathbb{W}} f_T \right|^2 d\mu_j \leq \left\| \sum_{T \in \mathbb{W}} f_T \right\|_{L^p(Y)}^2 \left(\int_Y \mu_j^{\frac{p}{p-2}} \right)^{\frac{p-2}{p}}, \tag{2.19}$$

By the dimension condition $c_\alpha(\mu) \leq 1$ on μ, the definition of Y, and the definition of the "good" planks,

$$\begin{aligned}
\int_Y \mu_j^{\frac{p}{p-2}} &\lesssim 2^{\frac{2j(3-\alpha)}{p-2}} \int_Y \mu_j \\
&\leq 2^{\frac{2j(3-\alpha)}{p-2}} \sum_{Q \subseteq Y} \int_Q \mu_j \\
&\lesssim \left(\frac{1}{M}\right) \cdot 2^{\frac{2j(3-\alpha)}{p-2}} \sum_{Q \subseteq Y} \sum_{\substack{T \in \mathbb{W} \\ 2T \cap Q \neq \emptyset}} \int_{Q \cap 3T} \mu_j \\
&\lesssim \left(\frac{1}{M}\right) \cdot 2^{\frac{2j(3-\alpha)}{p-2}} \sum_{T \in \mathbb{W}} \int_{3T} \mu_j \\
&\lesssim \left(\frac{1}{M}\right) \cdot 2^{\frac{2j(3-\alpha)}{p-2}} \sum_{T \in \mathbb{W}} \mu(4T) + 2^{-100j} \\
&\lesssim 2^{\frac{2j(3-\alpha)}{p-2} + j(\epsilon_0 - 1)} \left(\frac{|\mathbb{W}|}{M}\right).
\end{aligned} \tag{2.20}$$

This bounds the second factor in (2.19), so it remains to bound the first factor.

By rescaling by 2^j, applying the refined decoupling inequality (see Theorem A.1 of the appendix), and then rescaling back,

$$\left\| \sum_{T \in \mathbb{W}} f_T \right\|_{L^p(Y)} \lesssim 2^{j\varepsilon} \left(\frac{M}{|\mathbb{W}|}\right)^{\frac{1}{2} - \frac{1}{p}} \left(\sum_{T \in \mathbb{W}} \|f_T\|_p^2 \right)^{1/2}.$$

Recall that $f_T = [\eta_T M_T \mu] * \widecheck{\psi_{\tau(T)}}$. By applying the Hausdorff-Young inequality, then Hölder's inequality, and then Plancherel's theorem,

$$\|f_T\|_p \lesssim \|M_T \mu\|_2 \, 2^{\frac{3j}{2}\left(\frac{1}{2} - \frac{1}{p}\right)}.$$

Hence

$$\left\| \sum_{T \in \mathbb{W}} f_T \right\|_{L^p(Y)} \lesssim 2^{\frac{3j}{2}\left(\frac{1}{2} - \frac{1}{p} + \varepsilon\right)} \left(\frac{M}{|\mathbb{W}|}\right)^{\frac{1}{2} - \frac{1}{p}} \left(\sum_{T \in \mathbb{W}} \|M_T \mu\|_2^2 \right)^{1/2}. \tag{2.21}$$

Putting (2.21) and (2.20) into (2.19) gives

$$\sum_{T \in \mathbb{T}_{j,g}} \int_{\mathbb{R}^3} |M_T \mu|^2 \lesssim 2^{j\left[\frac{5-2\alpha}{2p} + \frac{1}{4} + \frac{\epsilon_0 (p-2)}{2p} + \frac{3\varepsilon}{2}\right]} \left(\sum_{T \in \mathbb{T}_{j,g}} \int_{\mathbb{R}^3} |M_T \mu|^2 \right)^{1/2} \mu(\mathbb{R}^3)^{1/2}.$$

By canceling the common factor, this gives

$$\sum_{T \in \mathbb{T}_{j,g}} \int_{\mathbb{R}^3} |M_T \mu|^2 \lesssim 2^{j\left[\frac{5-2\alpha}{p} + \frac{1}{2} + \frac{\epsilon_0 (p-2)}{p} + 3\varepsilon\right]} \mu(\mathbb{R}^3).$$

Since $p = 6$ and $\alpha = 4 - 3\beta$, this simplifies to

$$\sum_{T \in \mathbb{T}_{j,g}} \int_{\mathbb{R}^3} |M_T \mu|^2 \lesssim 2^{j\left(\beta + \frac{2\epsilon_0}{3} + 3\varepsilon\right)} \mu(\mathbb{R}^3),$$

which verifies (2.18), and as explained above, this proves the lemma. $\square$

The proof of Theorem 1.2 will be similar to the proof of the lemma.

PROOF OF THEOREM 1.2. It will first be shown that the set

$$\left\{\theta \in I : \rho_{\theta\#}\mu \not\ll \mathcal{H}^1\right\}$$

is Borel measurable. It may be assumed that μ is compactly supported, since if μ_k is the restriction of μ to $B(0,k)$, then

$$\left\{\theta \in I : \rho_{\theta\#}\mu \not\ll \mathcal{H}^1\right\} = \bigcup_{k=1}^{\infty} \left\{\theta \in I : \rho_{\theta\#}\mu_k \not\ll \mathcal{H}^1\right\}.$$

For any positive integer n, the function

$$(\theta, t) \mapsto (\rho_{\theta\#}\mu)\left(B(t\gamma(\theta), 1/n)\right),$$

is lower semicontinuous on $I \times \mathbb{R}$, and is therefore Borel measurable from $I \times \mathbb{R}$ to $[0, +\infty]$ (here $B(x,r)$ denotes the *open* ball or interval of radius r around x). It follows that the function

$$(\theta, t) \mapsto \limsup_{n\to\infty} n\, (\rho_{\theta\#}\mu)\left(B(t\gamma(\theta), 1/n)\right)$$

is Borel measurable from $I \times \mathbb{R}$ to $[0, +\infty]$. By Tonelli's theorem (see e.g. [**12**, Theorem 1.7.15]), the function

$$\theta \mapsto \int_{\mathbb{R}} \limsup_{n\to\infty} n\, (\rho_{\theta\#}\mu)\left(B(t\gamma(\theta), 1/n)\right) dt$$

is Borel measurable from I to $[0, +\infty]$. It follows that the set

$$\left\{\theta \in I : \int_{\mathbb{R}} \limsup_{n\to\infty} n\, (\rho_{\theta\#}\mu)\left(B(t\gamma(\theta), 1/n)\right) dt < \mu(\mathbb{R}^3)\right\},$$

is a Borel measurable subset of I, and this set is equal to $\left\{\theta \in I : \rho_{\theta\#}\mu \not\ll \mathcal{H}^1\right\}$ by the Lebesgue differentiation theorem (see e.g. [**3**, Theorem 3.22]).

As in the proof of the lemma, it may be assumed that γ is localised to a small interval on which Lemma 2.2 holds. It may also be assumed that $\alpha \leq 3$, $c_\alpha(\mu) \leq 1$ and (by countable stability of the Hausdorff dimension) that μ has support in the unit ball. Let β be such that $0 \leq \beta < (4-\alpha)/3$. Let λ be a Borel measure supported on I with $c_\beta(\lambda) \leq 1$. Let $\epsilon > 0$ be such that $\epsilon \ll \frac{4-\alpha}{3} - \beta$. Choose $\widetilde{\delta} > 0$ such that $\widetilde{\delta} \ll \min\{\epsilon, \delta_\epsilon\}$, where δ_ϵ is an exponent corresponding to ϵ from Lemma 2.4. Using Definition 2.1, define μ_b by

$$\mu_b = \sum_{j\geq 1} \sum_{\tau \in \Lambda_j} \sum_{T \in \mathbb{T}_{\tau,b}} M_T\mu, \tag{2.22}$$

where, for each $j \geq 1$ and $\tau \in \Lambda_j$, the set of "bad" planks corresponding to τ is defined by

$$\mathbb{T}_{\tau,b} = \left\{T \in \mathbb{T}_\tau : \mu(4T) \geq 2^{j(\epsilon-1)}\right\}.$$

Since the frequencies are no longer localised, the sum in (2.22) only converges a priori in the space of tempered distributions, but the individual functions $M_T\mu$ are smooth and compactly supported. For λ-a.e. $\theta \in I$,

$$\rho_{\theta\#}\mu_b := \sum_{j\geq 1} \sum_{\tau \in \Lambda_j} \sum_{T \in \mathbb{T}_{\tau,b}} \rho_{\theta\#}M_T\mu, \tag{2.23}$$

where, for λ-a.e. $\theta \in I$, the series will be shown to be absolutely convergent in $L^1(\mathcal{H}^1)$. Since L^1 is always a Banach space, any absolutely convergent series of L^1 functions is convergent in L^1, so the λ-a.e. absolute convergence of (2.23) in $L^1(\mathcal{H}^1)$ will imply that $\rho_{\theta\#}\mu_b \in L^1(\mathcal{H}^1)$ for λ-a.e. $\theta \in I$, and will imply that the series is well-defined as an $L^1(\mathcal{H}^1)$ limit. Define

$$\rho_{\theta\#}\mu_g = \rho_{\theta\#}\mu - \rho_{\theta\#}\mu_b,$$

for each $\lambda \in I$ such that the sum defining $\rho_{\theta\#}\mu_b$ converges in $L^1(\mathcal{H}^1)$ (which will include λ-a.e. $\theta \in I$). It will be shown that

$$\rho_{\theta\#}\mu_g \in L^2(\mathcal{H}^1),$$

for λ-a.e. $\theta \in I$. Together with $\rho_{\theta\#}\mu_b \in L^1(\mathcal{H}^1)$, this will imply that $\rho_{\theta\#}\mu \in L^1(\mathcal{H}^1)$ (or equivalently $\rho_{\theta\#}\mu \ll \mathcal{H}^1$) for λ-a.e. $\theta \in I$.

It will first be shown that

$$\int \sum_{j\geq 1} \sum_{\tau\in\Lambda_j} \sum_{T\in\mathbb{T}_{\tau,b}} \|\rho_{\theta\#}M_T\mu\|_{L^1(\mathcal{H}^1)}\, d\lambda(\theta) < \infty. \tag{2.24}$$

The proof of this is similar to the proof of Lemma 2.4, but some of the details will be included for readability. The left-hand side of (2.24) can be written as

$$\sum_{j\geq 1} \int \sum_{\tau\in\Lambda_j} \sum_{T\in\mathbb{T}_{\tau,b}} \|\rho_{\theta\#}M_T\mu\|_{L^1(\mathcal{H}^1)}\, d\lambda(\theta)$$

$$= \sum_{j\geq 1} \int \sum_{\substack{\tau\in\Lambda_j:\\ |\theta_\tau-\theta|\leq 2^{j(\widetilde{\delta}-1/2)}}} \sum_{T\in\mathbb{T}_{\tau,b}} \|\rho_{\theta\#}M_T\mu\|_{L^1(\mathcal{H}^1)}\, d\lambda(\theta) \tag{2.25}$$

$$+ \sum_{j\geq 1} \int \sum_{\substack{\tau\in\Lambda_j:\\ |\theta_\tau-\theta|> 2^{j(\widetilde{\delta}-1/2)}}} \sum_{T\in\mathbb{T}_{\tau,b}} \|\rho_{\theta\#}M_T\mu\|_{L^1(\mathcal{H}^1)}\, d\lambda(\theta). \tag{2.26}$$

By Lemma 2.2,

$$(2.26) \lesssim \mu(\mathbb{R}^3).$$

By Lemma 2.3,

$$\begin{aligned}(2.25) &\leq \sum_{j\geq 1} \int \sum_{\substack{\tau\in\Lambda_j:\\ |\theta_\tau-\theta|\leq 2^{j(\widetilde{\delta}-1/2)}}} \sum_{T\in\mathbb{T}_{\tau,b}} \|M_T\mu\|_{L^1(\mathbb{R}^3)}\, d\lambda(\theta) \\ &\lesssim \mu(\mathbb{R}^3) + \sum_{j\geq 1} 2^{3j\widetilde{\delta}} \int \sum_{\substack{\tau\in\Lambda_j:\\ |\theta_\tau-\theta|\leq 2^{j(\widetilde{\delta}-1/2)}}} \sum_{T\in\mathbb{T}_{\tau,b}} \mu(2T)\, d\lambda(\theta).\end{aligned}$$

As in the proof of Lemma 2.4, the nontail term satisfies

$$\sum_{j\geq 1} 2^{3j\widetilde{\delta}} \int \sum_{\substack{\tau\in\Lambda_j:\\ |\theta_\tau-\theta|\leq 2^{j(\widetilde{\delta}-1/2)}}} \sum_{T\in\mathbb{T}_{\tau,b}} \mu(2T)\, d\lambda(\theta) \lesssim \sum_{j\geq 1} 2^{10j\widetilde{\delta}} \int \mu(B_j(\theta))\, d\lambda(\theta),$$

where, for each $\theta \in I$ and each j,

$$B_j(\theta) = \bigcup_{\substack{\tau\in\Lambda_j:\\ |\theta_\tau-\theta|\leq 2^{j(\widetilde{\delta}-1/2)}}} \bigcup_{T\in\mathbb{T}_{\tau,b}} 2T.$$

For each T in the union defining $B_j(\theta)$, the set $(4T) \cap B(0,1)$ is contained in a plank T_θ of dimensions

$$\sim 2^{j(2\widetilde{\delta}-1)} \times 2^{j(\widetilde{\delta}-1/2)} \times 1,$$

with short direction parallel to $\gamma(\theta)$, medium direction parallel to $\gamma'(\theta)$, and long direction parallel to $(\gamma \times \gamma')(\theta)$. The intervals in the set

$$\left\{\rho_\theta(T_\theta) : T \in \mathbb{T}_{\tau,b}, \quad \tau \in \Lambda_j, \quad |\theta_\tau - \theta| \leq 2^{j\left(\widetilde{\delta}-1/2\right)}\right\},$$

all have length $\sim 2^{j(2\widetilde{\delta}-1)}$, and cover $\rho_\theta(B_j(\theta) \cap B(0,1))$. By the Vitali covering lemma, there is a disjoint subcollection

$$\{\rho_\theta(T_\theta) : T \in \mathcal{B}_\theta\},$$

indexed by some set $\mathcal{B}_\theta$, such that

$$\{3\rho_\theta(T_\theta) : T \in \mathcal{B}_\theta\},$$

is a cover of $\rho_\theta(B_j(\theta) \cap B(0,1))$. The set $\mathcal{B}_\theta$ has cardinality $|\mathcal{B}_\theta| \leq \mu(\mathbb{R}^3)2^{j(1-\epsilon)}$; by disjointness and the definition of the "bad" planks. By Lemma 2.4, for each $j \geq 1$,

$$\int \mu(B_j(\theta))\, d\lambda(\theta) \leq \int (\rho_{\theta\#}\mu) \left(\bigcup_{T_\theta \in \mathcal{B}_\theta} 3\rho_\theta(T_\theta) \right) d\lambda(\theta) \lesssim 2^{j\left(-\delta_\epsilon/2+10\widetilde{\delta}\right)}\mu(\mathbb{R}^3).$$

Since $\widetilde{\delta} \ll \delta_\epsilon$, summing the above inequality over j gives

$$(2.25) \lesssim \mu(\mathbb{R}^3).$$

It remains to show that $\rho_{\theta\#}\mu_g \in L^2(\mathcal{H}^1)$ for λ-a.e. $\theta \in I$. To prove this, by Plancherel's theorem in 1 dimension it suffices to show that

$$\int \int_{\mathbb{R}} |\widehat{\mu_g}(t\gamma(\theta))|^2 \, dt\, d\lambda(\theta) < \infty.$$

By symmetry and by summing a geometric series, it is enough to show that for any $j \geq 1$,

$$\int \int_{2^{j-1}}^{2^j} |\widehat{\mu_g}(t\gamma(\theta))|^2 \, dt\, d\lambda(\theta) \lesssim 2^{-j\epsilon}.$$

By similar reasoning to the proof of Lemma 2.4, it suffices to show that

$$\sum_{T \in \mathbb{T}_{j,g}} \int_{\mathbb{R}^3} |M_T\mu|^2 \lesssim 2^{j(\beta-10\epsilon)},$$

where

$$\mathbb{T}_{j,g} = \bigcup_{\tau \in \bigcup_{|j'-j|\leq 2} \Lambda_{j'}} \left\{T \in \mathbb{T}_{\tau,g} : T \cap B\left(0, 2^{10j\widetilde{\delta}}\right) \neq \emptyset\right\}.$$

By a similar argument to the proof of Lemma 2.4,

$$\sum_{T \in \mathbb{T}_{j,g}} \int_{\mathbb{R}^3} |M_T\mu|^2 \lesssim 2^{j\left[\frac{5-2\alpha}{p}+\frac{1}{2}+100\epsilon\right]}.$$

Since $\alpha > 4 - 3\beta$, $p = 6$ and $\epsilon \ll \frac{4-\alpha}{3} - \beta$, this implies that

$$\sum_{T \in \mathbb{T}_{j,g}} \int_{\mathbb{R}^3} |M_T\mu|^2 \lesssim 2^{j(\beta-10\epsilon)},$$

which finishes the proof of the theorem. □

PROOF OF THEOREM 1.3. By the density theorem for Hausdorff measures ([**10**, Theorem 6.2]),

$$\limsup_{r\to 0+} \frac{\mathcal{H}^s(A\cap B(x,r))}{r^s} \leq 2^s \qquad \mathcal{H}^s\text{-a.e. } x\in A.$$

It follows that if, for each positive integer n,

$$A_n := \left\{ x\in A : \sup_{0<r<1/n} \frac{\mathcal{H}^s(A\cap B(x,r))}{r^s} < 2^{s+1} \right\},$$

and μ_n is the Borel measure defined by

$$\mu_n(F) = \mathcal{H}^s\left(F\cap A_n \setminus \bigcup_{k=1}^{n-1} A_k \right),$$

for any Borel set F, then

$$\mu = \sum_{n=1}^{\infty} \mu_n, \tag{2.27}$$

and for any $n\geq 1$

$$c_s(\mu_n) \leq \max\left\{2^{2s+1}, (2n)^s\mathcal{H}^s(A)\right\}.$$

By Theorem 1.2, for any $n\geq 1$,

$$\dim\left\{\theta\in I : \rho_{\theta\#}\mu_n \not\ll \mathcal{H}^1\right\} \leq \frac{4-s}{3},$$

By (2.27),

$$\left\{\theta\in I : \rho_{\theta\#}\mu \not\ll \mathcal{H}^1\right\} = \bigcup_{n=1}^{\infty} \left\{\theta\in I : \rho_{\theta\#}\mu_n \not\ll \mathcal{H}^1\right\}.$$

By countable stability of the Hausdorff dimension, it follows that

$$\dim\left\{\theta\in I : \rho_{\theta\#}\mu \not\ll \mathcal{H}^1\right\} \leq \frac{4-s}{3}.$$

This proves the first half of the theorem. For the second half, let $B\subseteq A$ be an $\mathcal{H}^s$-measurable set with $\mathcal{H}^s(B)>0$. Let ν be the Borel measure

$$\nu(F) = \mathcal{H}^s(F\cap B) \qquad (=\mu(F\cap B)),$$

for any Borel set F. It will be shown that

$$\left\{\theta\in I : \rho_{\theta\#}\nu \ll \mathcal{H}^1\right\} \subseteq \left\{\theta\in I : \mathcal{H}^1(\rho_\theta(B))>0\right\}. \tag{2.28}$$

Let $\theta\in I$ be such that $\rho_{\theta\#}\nu \ll \mathcal{H}^1$. Suppose for a contradiction that $\mathcal{H}^1(\rho_\theta(B)) = 0$. Let $\delta>0$ be such that

$$(\rho_{\theta\#}\nu)(F) < \mathcal{H}^s(B),$$

for any Borel set F with $\mathcal{H}^1(F)<\delta$. Since $\mathcal{H}^1(\rho_\theta(B))=0$, there exists a Borel set F containing $\rho_\theta(B)$ with $\mathcal{H}^1(F)<\delta$. Hence

$$\mathcal{H}^s(B) = \mathcal{H}^s(\rho_\theta^{-1}(F)\cap B) = (\rho_{\theta\#}\nu)(F) < \mathcal{H}^s(B),$$

and this contradiction proves (2.28). Thus

$$\begin{aligned}\left\{\theta\in I : \mathcal{H}^1(\rho_\theta(B))=0\right\} &\subseteq \left\{\theta\in I : \rho_{\theta\#}\nu \not\ll \mathcal{H}^1\right\}\\ &\subseteq \left\{\theta\in I : \rho_{\theta\#}\mu \not\ll \mathcal{H}^1\right\}.\end{aligned}$$

□

Appendix A. Refined decoupling

The following inequality is Theorem 9 from [**4**]; it is a refined version of the decoupling theorem for generalised cones.

THEOREM A.1 ([**4**, Theorem 9]). *Let I be a compact interval, and let $\gamma : I \to S^2$ be a C^2 unit speed curve with $\det(\gamma, \gamma', \gamma'')$ nonvanishing on I. Then if $c > 0$ is sufficiently small (depending only on γ), then for any $\epsilon > 0$, there exists $\delta_0 > 0$ such that the following holds for all $0 < \delta < \delta_0$, and any $R \geq 1$. Let Θ_R be a maximal $cR^{-1/2}$-separated subset of I, and for each $\theta \in \Theta_R$, let*

$$\tau(\theta) :=$$
$$\left\{\lambda_1\gamma(\theta) + \lambda_2\gamma'(\theta) + \lambda_3(\gamma \times \gamma')(\theta) : 1/2 \leq \lambda_1 \leq 1, |\lambda_2| \leq R^{-1/2}, |\lambda_3| \leq R^{-1}\right\}.$$

For each $\tau = \tau(\theta)$, let $\mathbb{T}_\tau$ be a ~ 1-overlapping cover of $\mathbb{R}^3$ by translates of

$$\left\{\lambda_1\gamma(\theta) + \lambda_2\gamma'(\theta) + \lambda_3(\gamma \times \gamma')(\theta) : |\lambda_1| \leq R^{\delta}, |\lambda_2| \leq R^{1/2+\delta}, |\lambda_3| \leq R^{1+\delta}\right\}.$$

If $2 \leq p \leq 6$, and

$$\mathbb{W} \subseteq \bigcup_{\theta \in \Theta_R} \mathbb{T}_{\tau(\theta)},$$

and

$$\sum_{T \in \mathbb{W}} f_T$$

is such that $\|f_T\|_p$ is constant over $T \in \mathbb{W}$ up to a factor of 2, with $\operatorname{supp} \widehat{f_T} \subseteq \tau(T)$ and

$$\|f_T\|_{L^\infty(B(0,R)\setminus T)} \leq AR^{-10000}\|f_T\|_p,$$

and Y is a disjoint union of balls in $B_3(0, R)$ of radius 1, such that each ball $Q \subseteq Y$ intersects at most M planks $2T$ with $T \in \mathbb{W}$, then

$$\left\|\sum_{T \in \mathbb{W}} f_T\right\|_{L^p(Y)} \leq C_{A,\gamma,c,\epsilon,\delta} R^{\epsilon} \left(\frac{M}{|\mathbb{W}|}\right)^{\frac{1}{2}-\frac{1}{p}} \left(\sum_{T \in \mathbb{W}} \|f_T\|_p^2\right)^{1/2}.$$

References

[1] R. O. Davies, *Subsets of finite measure in analytic sets*, Indag. Math. **14** (1952), 488–489. Nederl. Akad. Wetensch. Proc. Ser. A **55**. MR0053184

[2] K. Fässler and T. Orponen, *On restricted families of projections in $\mathbb{R}^3$*, Proc. Lond. Math. Soc. (3) **109** (2014), no. 2, 353–381, DOI 10.1112/plms/pdu004. MR3254928

[3] G. B. Folland, *Real analysis*, 2nd ed., Pure and Applied Mathematics (New York), John Wiley & Sons, Inc., New York, 1999. Modern techniques and their applications; A Wiley-Interscience Publication. MR1681462

[4] S. Gan, S. Guo, L. Guth, T. L. J. Harris, D. Maldague, and H. Wang, *On restricted projections to planes in $\mathbb{R}^3$*, Preprint, `arXiv:2207.13844v1`, (2022).

[5] S. Gan, L. Guth, and D. Maldague, *An exceptional set estimate for restricted projections to lines in $\mathbb{R}^3$*, `arXiv:2209.15152v2`, (2022).

[6] L. Guth, A. Iosevich, Y. Ou, and H. Wang, *On Falconer's distance set problem in the plane*, Invent. Math. **219** (2020), no. 3, 779–830, DOI 10.1007/s00222-019-00917-x. MR4055179

[7] A. Käenmäki, T. Orponen, and L. Venieri, *A Marstrand-type restricted projection theorem in $\mathbb{R}^3$*, To appear in Amer. J. Math., `arXiv:1708.04859v2`, (2017).

[8] B. Liu, *Hausdorff dimension of pinned distance sets and the L^2-method*, Proc. Amer. Math. Soc. **148** (2020), no. 1, 333–341, DOI 10.1090/proc/14740. MR4042855

[9] J. M. Marstrand, *Some fundamental geometrical properties of plane sets of fractional dimensions*, Proc. London Math. Soc. (3) **4** (1954), 257–302, DOI 10.1112/plms/s3-4.1.257. MR63439

[10] P. Mattila, *Geometry of sets and measures in Euclidean spaces*, Cambridge Studies in Advanced Mathematics, vol. 44, Cambridge University Press, Cambridge, 1995. Fractals and rectifiability, DOI 10.1017/CBO9780511623813. MR1333890

[11] M. Pramanik, T. Yang, and J. Zahl, *A Furstenberg-type problem for circles, and a Kaufman-type restricted projection theorem in* $\mathbb{R}^3$, arXiv:2207.02259v2, (2022).

[12] T. Tao, *An introduction to measure theory*, Graduate Studies in Mathematics, vol. 126, American Mathematical Society, Providence, RI, 2011, DOI 10.1090/gsm/126. MR2827917

DEPARTMENT OF MATHEMATICS, CORNELL UNIVERSITY, ITHACA, NY 14853, USA

Current address: Department of Mathematics, University of Wisconsin, Madison, Wisconsin 53706

Email address: terry.harris@wisc.edu

Contemporary Mathematics
Volume **792**, 2024
https://doi.org/10.1090/conm/792/15893

New bounds for Stein's square function in $\mathbb{R}^3$

Shengwen Gan, Yifan Jing, and Shukun Wu

ABSTRACT. We prove an $L^p(\mathbb{R}^3)$ bounds on Stein's square function for $p \geq 3.25$. As an application, it improves the maximal Bochner–Riesz conjecture to the same range of p. The main method we used is polynomial partitioning.

1. Introduction

In this paper, the ambient space is always $\mathbb{R}^3$. The purpose of this paper is to prove some L^p estimate for Stein's square function. To begin with, let us recall the Bochner–Riesz mean of order λ:

$$T_t^\lambda f(x) := \int_{\mathbb{R}^3} \Big(1 - \frac{|\xi|^2}{t^2}\Big)_+^\lambda \widehat{f}(\xi) e^{ix\cdot\xi} d\xi\,, \tag{1.1}$$

where f is a Schwartz function. Stein [**Ste58**] (See also [**SW71**] Chapter 7) introduced a square function $G^\lambda f$ defined as

$$G^\lambda f(x) := \Big(\int_0^\infty \Big|\frac{\partial}{\partial t} T_t^\lambda f(x)\Big|^2 t dt\Big)^{1/2}. \tag{1.2}$$

Our main result is the following.

THEOREM 1.1 (Stein's square function). *Assume $p \geq 3.25$. Then*

$$\|G^\lambda f\|_p \leq C_p \|f\|_p, \qquad \lambda > 3\Big(\frac{1}{2} - \frac{1}{p}\Big). \tag{1.3}$$

An immediate application of Theorem 1.1 is improvements on the maximal Bochner–Riesz problems (See [**Car83**] for instance). Recall that the maximal Bochner–Riesz operator T_*^λ is defined as

$$T_*^\lambda f(x) = \sup_{t>0} |T_t^\lambda f(x)|. \tag{1.4}$$

THEOREM 1.2 (Maximal Bochner–Riesz). *Assume $p \geq 3.25$. Then*

$$\|T_*^\lambda f\|_p \leq C_p \|f\|_p, \qquad \lambda > 1 - 3/p. \tag{1.5}$$

The two dimensional square function problem and the maximal Bochner–Riesz problem were solved by Carbery [**Car83**]. In $\mathbb{R}^3$, the previous best results for Theorem 1.1 and Theorem 1.2 are both $p \geq 10/3$, which were obtained respectively by Lee, Rogers, Seeger in [**LRS12**] and Lee in [**Lee04**]. We remark that the range of λ in both theorem is sharp. See [**LRS12**] for some discussions.

Both the square function $G^\lambda f$ and the maximal function $T_*^\lambda f$ are closely related to the almost everywhere convergence of the Bochner–Riesz mean. Unlike the

Bochner–Riesz operator, the case $p < 2$ is largely different from the case $p > 2$ for the maximal Bochner–Riesz problem. Here we are only interested in the case $p > 2$. For recent results about the maximal operator T_*^λ when $p < 2$, see [**LW20**].

The main method we use to prove Theorem 1.1 is the polynomial partitioning method, which was introduced by Guth and Katz [**GK15**], and later applied by Guth [**Gut16**] to Fourier restriction estimate. The original article [**Gut16**] itself is also a great overview of this method. Roughly speaking, polynomial partitioning allows one to make a uniform partition for the ambient function, and keep some geometric properties for straight lines at the same time.

In the landmark paper [**Gut16**], Guth used polynomial partitioning to improve the restriction conjecture in $\mathbb{R}^3$ to $p > 3.25$. This result was later improved to $p > 13/3$ by Wang [**Wan18**]. Recently in [**Wu23**] and [**GOW$^+$21**], the authors were able to improve the three-dimensional Bochner–Riesz conjecture to the endpoint $p \geq 3.25$, though the methods used in [**Wu23**] and [**GOW$^+$21**] are quite different. In [**GOW$^+$21**], one used the Carleson-Sjölin reduction to convert the Bochner–Riesz problem to an analogue of the restriction problem. In this way, the techniques developed in [**Gut16**] and [**Gut18**] are still applicable.

A key ingredient in [**Wu23**] and in this paper is a backward algorithm. The algorithm allows us to bound a certain broad function using lattice-cube square function (See (1.9)). This is different from the mixed norm $\|\cdot\|_2^\alpha \|\cdot\|_\infty^{1-\alpha}$ that was widely used in the restriction problem, for instance, in [**Gut16**] and [**Wan18**]. In there, the loss when using polynomial partitioning iteratively is partly picked up via the norm $\|\cdot\|_\infty$. Here in this paper, we use the backward algorithm to try to pick up the loss by looking at the norm $\|\cdot\|_2$.

The proof of our main estimate (1.3) is quite technical and involved, so we give a quick sketch of proof to help the readers to digest. The argument here is more intuitive and less rigorous, as we want to avoid messy notations.

We first reduce the square function $G^\lambda f$ to a curved, congruent, Littlewood–Paley-type of square function. Let $R \geq 1$ be a large integer and let $\zeta : \mathbb{R} \to \mathbb{R}$ be a function such that $\widehat{\zeta}$ is a smooth bump function supported in $[-1, 1]$. For each $j = 1, 2, \ldots, R$, set $t_j := 1 + j/R$ and define a Fourier multiplier $\widehat{\zeta_j} : \mathbb{R}^3 \to \mathbb{R}$ as $\widehat{\zeta_j}(\xi) := \widehat{\zeta}(R|\xi| - Rt_j)$. Then each $\widehat{\zeta_j}$ is supported in a thin neighbourhood of the sphere of radius t_j, which we denote by Γ_j. We introduce a spherical square function

$$\mathbf{Sq}f := \Big(\sum_{j=1}^{R} |\zeta_j * f|^2\Big)^{1/2}. \tag{1.6}$$

After several standard reductions, to get (1.3), one only needs to prove for any $\varepsilon > 0$ and $p \geq 3.25$,

$$\|\mathbf{Sq}f\|_{L^p(B_R)} \leq C_\varepsilon R^{\frac{p-3}{p}+\varepsilon}\|f\|_p. \tag{1.7}$$

Now let us focus on (1.7). Set $K := R^{\varepsilon^{10}}$ and let $\mathcal{T} = \{\tau\}$ be a collection of K^{-1} caps in $\mathbb{S}^2$. We use the Gauss maps of the spheres $\Gamma_j := \{|\xi| = t_j\}(1 \leq j \leq R)$ to define a broad part $\mathrm{Br}\mathbf{Sq}f$ and a narrow part $\{\mathbf{Sq}_\tau f\}_{\tau\in\mathcal{T}}$. Each narrow square function $\mathbf{Sq}_\tau f$ is defined as

$$\mathbf{Sq}_\tau f := \Big(\sum_{j=1}^{R} |\zeta_{j,\tau} * f|^2\Big)^{1/2}, \tag{1.8}$$

where the Fourier transform of each kernel $\zeta_{j,\tau}$ is supported in a thin neighbourhood of the portion $\Gamma_j(\tau)$ (Here, $\Gamma_j(\tau)$ is defined to be the set of points on Γ_j where the normal direction lie in τ). Recall that a main step in the broad-narrow argument is parabolic rescaling. More precisely, one can find a linear map $\mathcal{L}$ that maps the K^{-1}-cap $\Gamma_j(\tau)$ to some 1-cap $\tilde{\Gamma}_j$. However in our setting, when we try to run broad-narrow argument as in [**BG11**], there is some technical difficulty: it is hard to find a suitable parabolic rescaling here. As one can see, since we are dealing with many slices $\{\Gamma_j\}$ instead of a single slice, it's hard to find a common linear map $\mathcal{L}$ for all the K^{-1}-caps $\{\Gamma_j(\tau)\}$.

In order to get around this difficulty, we define a second-level broad part and a second-level narrow part associated to each of the first-level narrow square function $\mathbf{Sq}_\tau f$. We iterate this idea until each of the existing narrow square function is restricted to a $R^{-1/2}$-directional cap. Since $K = R^{\varepsilon^{10}}$, there are in total $\varepsilon^{-10}/2$ levels. After this iteration stops, there are many broad parts at different levels. We estimate each remaining broad part directly and add them up eventually.

Suppose that $\mathrm{Br}\mathbf{Sq}_\sigma f$ is a broad function at any intermediate level ℓ, then σ is a $K^{-\ell}$ directional cap. We set $M = K^\ell$. The desired estimate for $\mathrm{Br}\mathbf{Sq}_\sigma f$ is

$$\|\mathrm{Br}\mathbf{Sq}_\sigma f\|_p \le C_\varepsilon R^{\frac{p-3}{p}+\varepsilon} M^{\frac{6-2p}{p}} \Big\| \Big(\sum_q |\Delta_q f|^2 \Big)^{1/2} \Big\|_p. \tag{1.9}$$

The right-hand side is an auxiliary square function, where $\mathbf{q} = \{q\}$ is a tiling in the frequency space $\mathbb{R}^3$ using $R^{-1/2}$-cubes, and $\Delta_q f$ is a smooth Fourier restriction of f on the $R^{-1/2}$-cube q. Once we prove (1.9), we can use Littlewood–Paley theorem for congruent cubes to sum up all broad quantities $\|\mathrm{Br}\mathbf{Sq}_\sigma f\|_p^p$ at all levels.

To obtain (1.9), we use the polynomial partitioning iteration to break $\mathrm{Br}\mathbf{Sq}_\sigma f$ down. Specifically, suppose that there are in total s steps in the iteration. In each intermediate step u, we have a collection of step u cells $\mathcal{O}_u = \{O_u\}$ and step u broad functions $\mathrm{Br}(\mathbf{Sq}_\sigma f)_{O_u}$. All $\|\mathrm{Br}(\mathbf{Sq}_\sigma f)_{O_u}\|_p^p$ are morally the same, and their sum dominate a significant portion of the original one $\|\mathrm{Br}\mathbf{Sq}_\sigma f\|_p^p$. In the final step s, the L^p norm of $\mathrm{Br}(\mathbf{Sq}_\sigma f)_{O_s}$ can be bounded by the L^2 norm of $(\mathbf{Sq}_\sigma f)_{O_s}$ via a bilinear argument. So it remains to sum up $\|(\mathbf{Sq}_\sigma f)_{O_s}\|_2^2$.

Here is the key ingredient of our proof. Assume that there are in total s_t transverse steps and s_c cell steps in the polynomial partitioning iteration. Using the information from the iteration, we can add up all $\|(\mathbf{Sq}_\sigma f)_{O_s}\|_2^2$ as

$$\sum_{O_s \in \mathcal{O}_s} \|(\mathbf{Sq}_\sigma f)_{O_s}\|_2^2 \lesssim R^{O(\varepsilon^2)} R^{-s_t\varepsilon^2} \|\mathbf{Sq}_\sigma f\|_2^2. \tag{1.10}$$

While this estimate is not strong enough. We refine this estimate by proving

$$\sum_{O_s \in \mathcal{O}_s} \|(\mathbf{Sq}_\sigma f)_{O_s}\|_2^2 \lesssim R^{O(\varepsilon^2)} R^{-s_t\delta} \kappa \| \mathit{1}_X \mathbf{Sq}_\sigma f\|_2^2, \tag{1.11}$$

where the factor $\kappa \lesssim 1$ depends heavily on the set X, and X depends on the collection $\mathcal{O}_s$. In (1.11), we either have gain κ in the L^2 estimate, or have gain $\mathit{1}_X$ in the support, which will be useful when converting $\|\mathit{1}_X \mathbf{Sq}_\sigma f\|_2^2$ to the square function in (1.9). The proof of our refinement is based on a backward algorithm built up step by step from the last step s to the first step of the polynomial partitioning iteration. Inside the backward algorithm, the key is an estimate in incidence geometry, which helps us create a relation between tubes at adjacent scales. Since more notations are needed to clarify the backward algorithm, we stop here and leave the details to the main body of the paper, in particular in Section 8.

Organization of the paper. The first part of Section 2 contains some classic result in Harmonic analysis that we will use later. The second part of Section 2, together with Section 3, contain several reductions for our main result—Theorem 1.1. It is reduced to Theorem 3.13, which is also our strongest result. Section 4 contains the wave packet decomposition. In Sections 5 and 6, we build the polynomial partitioning iteration for some vectors, where the iteration is similar to the one in [**Wu23**]. After that, we prove our main result for some special cases in Section 7. In Sections 8 and 9, similar to [**Wu23**], we establish a backward algorithm and prove our main result for the remaining cases. Section 10 is devoted to a bilinear estimate for some square functions.

Notations.

- We let $a \sim b$ mean that $ca \leq b \leq Ca$ for some unimportant constants c and C. We also use $a \lesssim b$ to represent $a \leq Cb$ for an unimportant constant C. These constants may change from line to line.
- We write $A(R) \leq \mathrm{RapDec}(R)B$ to mean that for any power β, there is a constant C_N such that

$$A(R) \leq C_N R^{-N} B \text{ for all } R \geq 1.$$

- For every number $R > 0$ and set S, we denote by $N_R(S)$ the R-neighborhood of the set S.
- The symbol $B^3(x, r)$ represents the open ball centered at x, of radius r, in $\mathbb{R}^3$. The symbol B_r is reserved for $B^3(0, r)$. M, N, C are (big) constants that may change from line to line.
- We use 1_X to denote the characteristic function of a set $X \subset \mathbb{R}^3$.
- By saying a r-tube, we mean a rectangle of dimensions $R^\beta r^{1/2} \times R^\beta r^{1/2} \times r$. Here β is a tiny number to be defined later.
- By saying a r-cap, we mean a cap of radius r in $\mathbb{S}^2$.
- For any cap $\tau \subset \mathbb{S}^2$, we use c_τ to denote the center of τ.

Some numbers. We will encounter many different numbers in the paper. For readers' convenience, we summarize all of them here. Note that here we only give a rough description of these numbers. The precise definition will be given later in the paper.

- ε is a very small number.
- $\delta = \varepsilon^2$.
- $\beta = \varepsilon^{1000}$ is the number used to deal with the rapid-decay tail of wave packet.
- $K = R^{\varepsilon^{10}}$ is the number appearing in the broad-narrow argument.
- $A \sim \log R$ is the number in the subscript of the broad function $\mathrm{Br}_A f$. Also, we would like A to be of form $2^{\mathbb{N}}$.
- $d = R^{\varepsilon^6}$ is the degree of the polynomial in the polynomial partitioning argument.

2. Preliminaries and a first reduction

In this section, we first review some classic results in Fourier analysis that we will repeatedly use later in the paper. Then we make our first reduction to the square function $G^\lambda f$.

2.1. A local L^2-estimate. Let $\Gamma = \{(\bar{\xi}, \Psi(\bar{\xi})) : \bar{\xi} \in \overline{B^2(0,1)}\}$, be the truncated graph of a function $\Psi : \mathbb{R}^2 \to \mathbb{R}$ with bounded Hessian. The following lemma is a local L^2 estimate for functions whose Fourier transforms are supported in a small neighborhood of Γ.

LEMMA 2.1. *Let f be an L^2 function such that $\widehat{f}$ is supported in $N_\rho(\Gamma)$, where ρ is a positive number much smaller than 1. Let $1 \geq \sigma \geq \rho$ and let $\mathcal{P} \subset \mathbb{R}^n$ be a tube of length σ^{-1} and radius $r > 1$ in the physical space, pointing to the vertical direction e_3. Suppose that $\varphi_{\mathcal{P}}$ is a smooth cutoff function with respect to $\mathcal{P}$ that $|\varphi_{\mathcal{P}}(x)| \gtrsim 1$ on $\mathcal{P}$, $\widehat{\varphi}_{\mathcal{P}}$ is supported in a dual slab $B^2(0, 3r^{-1}) \times [-3\sigma, 3\sigma]$ and $|\widehat{\varphi}_{\mathcal{P}}(\xi)| \gtrsim |\mathcal{P}|$ in a smaller slab $B^2(0, r^{-1}) \times [-\sigma, \sigma]$. Then*

$$\|f\varphi_{\mathcal{P}}\|_2 \lesssim \rho^{1/2}\sigma^{-1/2}\|f\|_2. \tag{2.1}$$

PROOF. The proof is similar to the one in [**Wu23**], dealing with the simple case $\sigma = r^{-1}$. We omit details here. □

2.2. A weighted estimate for pseudo-congruent square functions. The weighted estimate we are going to prove in this subsection is similar to the result in [**Cór81**]. Let $T \subset \mathbb{R}^3$ be a fixed rectangular parallelepiped of dimensions $\rho_1 \times \rho_2 \times \rho_3$ centered at the origin, and let $\{e_1', e_2', e_3'\}$ be the corresponding orthonormal basis. For any positive integer N, we define two weights

$$w_{T,N}(x) = (1 + |x_1/\rho_1| + |x_2/\rho_2| + |x_3/\rho_3|)^{-N} \tag{2.2}$$

$$\widetilde{w}_{T,N}(x) = \frac{1}{|T|}(1 + |x_1/\rho_1| + |x_2/\rho_2| + |x_3/\rho_3|)^{-N}. \tag{2.3}$$

Using the weight $\widetilde{w}_{T,N}$ we can introduce the definition for a class of smooth, rapidly decreasing functions associated to T.

DEFINITION 2.2. We say a finite collection of smooth functions $\{\varphi_j\}$ is "adapted to T", if $|\varphi_j(x)| \leq C_N \widetilde{w}_{T,N}(x)$ uniformly in j for any $N \in \mathbb{N}$.

Next, in the frequency space, we consider a family of finitely overlapping "pseudo-congruent" rectangular parallelepipeds $\Omega = \{\omega\}$ with dimensions $\rho_1^{-1} \times \rho_2^{-1} \times \rho_3^{-1}$. By pseudo-congruent we mean that for any two $\omega_1, \omega_2 \in \Omega$, $\omega_1 \subset C\widetilde{\omega}_2$, where $\widetilde{\omega}_2$ is the parallelepiped congruent to ω_2 and $\widetilde{\omega}_2$ has the same center as ω_1 does. Assume that for each $\omega \in \Omega$ there is a smooth function $\widehat{\varphi}_\omega$ such that the collection $\{\varphi_\omega\}$ is adapt to a rectangular parallelepiped T centered at the origin. As usual, we let $\{e_1, e_2, e_3\}$ be the orthonormal basis associated to T. Now define $\Delta_\omega f := (\widehat{\varphi}_\omega \widehat{f})^\vee$ as the smooth Fourier restriction of f on ω. Then, we have

LEMMA 2.3. *For any positive measurable function g,*

$$\sum_{\omega\in\Omega} \int |\Delta_\omega f|^2 g \leq C_N \int |f|^2 (\widetilde{w}_{T,N} * g). \tag{2.4}$$

PROOF. Following the idea in [**Cór81**], we are going to use the Poisson summation formula. First, let us introduce some more notations. Let θ_T the be rectangular parallelepiped dual to T centered at the origin, and let $\theta = C\theta_T$. We choose the absolute constant C large enough so the following is true: If we consider the lattice translations

$$\{\theta_m : \theta_m := \theta + (m_1\rho_1^{-1}, m_2\rho_2^{-1}, m_3\rho_3^{-1}), m \in \mathbb{Z}^3\}, \tag{2.5}$$

then for each $\omega \in \Omega$, there is a lattice point $m_\omega \in \mathbb{Z}^3$ such that $2\omega \subset \theta_{m_\omega}$. Indeed, we can pick $m_\omega \in \rho_1^{-1}\mathbb{Z} \times \rho_2^{-1}\mathbb{Z} \times \rho_3^{-1}\mathbb{Z}$ be an arbitrary lattice point which also belongs to 2ω. Since parallelepipeds in Ω are finitely overlapping, we have for every $\omega_1 \in \Omega$, all but finitely many $\omega_2 \in \Omega$ satisfy $m_{\omega_1} \neq m_{\omega_2}$. Let $\widehat{\phi}$ be a bump function associated to θ so that ϕ is dominated by $w_{T,N}$, and let $\widehat{\phi}_m(\xi) := \widehat{\phi}(\xi + m)$ for every translate $m \in \mathbb{Z}^3$. Hence $\widehat{\phi}_{m_\omega}\widehat{\Delta_\omega f} = \widehat{\Delta}_\omega f$, so

$$|\Delta_\omega f(x)|^2 = |\varphi_\omega * (\phi_{m_\omega} * f)(x)|^2 \lesssim (|(\phi_{m_\omega} * f)|^2 * |\varphi_\omega|)(x). \tag{2.6}$$

Since $|\varphi_\omega|$ is dominated by the weight $\widetilde{w}_{T,N}$, and since $m_{\omega_1} \neq m_{\omega_2}$ for most pairs (ω_1, ω_2), one has

$$\sum_{\omega\in\Omega} \int |\Delta_\omega f|^2 g \lesssim \sum_{\omega\in\Omega} \int |(\phi_{m_\omega} * f)|^2 (\widetilde{w}_{T,N} * g) \lesssim \sum_{m\in\mathbb{Z}^3} \int |(\phi_m * f)|^2 (\widetilde{w}_{T,N} * g). \tag{2.7}$$

Next, we are going to estimate $\sum_m |\phi_m * f|^2$ via the Poisson summation formula. Without loss of generality, we assume $\rho_1 = \rho_2 = \rho_3 = 1$. Clearly,

$$\sum_{m\in\mathbb{Z}^3} |(\phi_m * f)|^2 = \int_{[0,1]^3} \Big| \sum_{m\in\mathbb{Z}^3} (\phi_m * f) e^{2\pi i m\cdot z} \Big|^2 dz. \tag{2.8}$$

Notice that $\phi_m(x) = \phi(x)e^{-2\pi i m\cdot x}\phi(x)$. By Poisson summation formula (See for instance [**Gra14**] Section 3.1.5), one has

$$\begin{aligned} \sum_{m\in\mathbb{Z}^3} (\phi_m * f) e^{2\pi i m\cdot z} &= \sum_{m\in\mathbb{Z}^3} e^{2\pi i m\cdot z} \int \phi(y) f(x-y) e^{-2\pi i m\cdot y} dy \\ &= \sum_{m\in\mathbb{Z}^3} \phi(m+z) f(x-m-z). \end{aligned} \tag{2.9}$$

Plug this back into (2.8) so that

$$\sum_{m\in\mathbb{Z}^3} |(\phi_m * f)|^2 = \int_{[0,1]^3} \Big| \sum_{m\in\mathbb{Z}^3} \phi(m+z) f(x-m-z) \Big|^2 dz, \tag{2.10}$$

which can be bounded using Hölder's inequality as

$$\sum_{m\in\mathbb{Z}^3} |(\phi_m * f)|^2 \lesssim \int_{[0,1]^3} \sum_{m\in\mathbb{Z}^3} |\phi(m+z)| \cdot |f(x-m-z)|^2 dz \lesssim \widetilde{w}_{T,N} * (|f|^2)(x). \tag{2.11}$$

Finally, noting $\widetilde{w}_{T,N}(x) = \widetilde{w}_{T,N}(-x)$ and $\widetilde{w}_{T,N} * \widetilde{w}_{T,N} \lesssim \widetilde{w}_{T,N}$, we combine (2.7) and the above estimate to conclude (2.4). □

2.3. A first reduction for the square function $G^\lambda f$. In the final part of this section, let us make a reduction for $G^\lambda f$. It consists of several smaller steps, each of which is stated as a lemma.

The first step consists of partition in the frequency space. Suppose that $R \geq 1$ is a big number. For simplicity let us assume $R \in \mathbb{N}$. Let $\psi : \mathbb{R} \to \mathbb{R}$ be a smooth bump function supported in $[-1, 1]$. We introduce a spherical function

$$T_{R,t} f(x) := \int_{\mathbb{R}^3} \psi\Big(\frac{1 - |\xi/t|}{R^{-1}}\Big) \widehat{f}(\xi) e^{ix\cdot\xi} d\xi\,, \tag{2.12}$$

and a spherical square function $G_R f$ as

$$G_R f(x) := \Big(\int_0^\infty \Big|T_{R,t} f(x)\Big|^2 \frac{dt}{t}\Big)^{1/2}. \tag{2.13}$$

After partitioning the Bochner–Riesz multiplier $\left(1-\frac{|\xi|^2}{t^2}\right)_+^\lambda$ canonically as in [**Car83**] and using the triangle inequality, we get that to prove (1.3), one only needs to prove

$$\|G_R f\|_p \le C_\varepsilon R^{\frac{1}{2}-\frac{3}{p}+\varepsilon}\|f\|_p \tag{2.14}$$

for any $\varepsilon > 0$ and $p \ge 3.25$. Using [**GRY20**, Proposition 4.2] (which is a generalization of [**See88**]), it suffices to consider the local square function

$$H_R f(x) := \Big(\int_1^2 \big|T_{R,t} f(x)\big|^2 \frac{dt}{t}\Big)^{1/2} \tag{2.15}$$

and prove for any $\varepsilon > 0$ and $p \ge 3.25$ that

$$\|H_R f\|_p \le C_\varepsilon R^{\frac{1}{2}-\frac{3}{p}+\varepsilon}\|f\|_p \tag{2.16}$$

We put the argument above into Lemma 2.4:

LEMMA 2.4 (First step). *To prove* (1.3), *one only needs to show* (2.16).

Our next step is to discretize the square function $H_R f$. This step in fact is not necessary, but we feel that it is better to work on the discrete version of the square function. Let $\Phi(\bar{\xi};t)$ be a real function with $\bar{\xi} \in B^2(0,1)$ and $t \in [1,2]$, and let $\{t_j\}_{1\le j\le R}$ be any collection of R^{-1}-separated points in the interval $[1,2]$. For each j, define the truncated surface Γ_j as:

$$\Gamma_j := \{(\bar{\xi}, \Phi(\bar{\xi};t_j)) : |\bar{\xi}| \le 1/2\}. \tag{2.17}$$

We need to impose certain regularity conditions and curvature condition on the function Φ. Specifically, assume that when $t \in [1,2]$ and $\bar{\xi} \in B^2(0,1/2)$, Φ satisfies:

$$\begin{cases} |\partial_t \Phi| \sim 1, \ |\partial_t \nabla_{\bar{\xi}} \Phi| = O(1), \\ \text{Both two eigenvalues of } \nabla^2_{\bar{\xi}}\Phi \text{are} \sim -1, \ \text{uniform in } t. \end{cases} \tag{2.18}$$

The conditions in the first line basically says that any Γ_j is roughly a translation copy of another $\Gamma_{j'}$ when j and j' are close enough. In particular, if $|j-j'| \le R^{1/2}$ and $|\bar{\xi}_1 - \bar{\xi}_2| \le R^{-1/2}$, we have $|G_j(\bar{\xi}_1) - G_{j'}(\bar{\xi}_2)| \lesssim R^{-1/2}$, where G_j is the Gauss map of Γ_j. The condition in the second line means that the surfaces $\{\Gamma_j\}_{1\le j\le R}$ are all parabolic. Here are two good examples for $\Phi(\bar{\xi};t)$:

(1) $\Phi(\bar{\xi};t) = (t - |\bar{\xi}|^2)^{1/2}$, a family of of spheres;
(2) $\Phi(\bar{\xi},t) = t - |\bar{\xi}|^2$, a family of translated paraboloids.

For each j, let $\eta_j : \mathbb{R}^2 \to \mathbb{R}$ and let $\psi_j : \mathbb{R} \to \mathbb{R}$ be two smooth bump functions with η_j supported in $B^2(0,1/2)$ and ψ_j supported in $[-1,1]$. Suppose that ϕ : $\mathbb{R}^3 \times [1,2] \to \mathbb{R}$ is an arbitrary smooth function that takes value in $[1/2,2]$ and satisfies $|\partial^\alpha \phi| \le C_\alpha$ for any multi-index $\alpha \in \mathbb{N}^4$. Define a general Fourier multiplier

$$\widehat{m}_j(\xi) := \eta_j(\bar{\xi})\psi_j\Big(\frac{\xi_3 - \Phi(\bar{\xi};t_j)}{R^{-1}}\phi(\xi;t_j)\Big) \tag{2.19}$$

and the associated operator

$$S_j f(x) := m_j * f(x). \tag{2.20}$$

REMARK 2.5. Heuristically, one can think of $\phi(\xi;t) = 1$. But for generality, we still keep $\phi(\xi;t)$ in our later argument. Also, $\widehat{m}_j(\xi)$ is morally a bump function to the R^{-1}-neighborhood of the surface $\{\xi : \xi_3 = \Phi(\bar{\xi};t_j), |\bar{\xi}| \le 1/2\}$.

Now we can introduce the discrete square function associated to the general surfaces $\{\Gamma_j\}_{1\leq j\leq R}$ that we defined in (2.17).

DEFINITION 2.6 (Square function). For any Schwartz function f, define

$$\mathrm{Sq}f(x) := \Big(\sum_{j=1}^{R} |S_j f(x)|^2\Big)^{1/2}. \tag{2.21}$$

LEMMA 2.7 (Second step). *Let* $\mathrm{Sq}f$ *be defined in* (2.21). *Suppose that for any* $\varepsilon > 0$ *and* $p \geq 3.25$,

$$\|\mathrm{Sq}f\|_p \leq C_\varepsilon R^{\frac{p-3}{p}+\varepsilon}\|f\|_p. \tag{2.22}$$

Then (2.16) *is true.*

PROOF. Using polar coordinate and a smooth partition of unity for $\mathbb{S}^2$, we can partition the whole space $\mathbb{R}^3$ into smaller conic region, so that any two vectors in a single conic region make an angle $\leq 1/100$. Note that the Fourier multiplier $\psi\big(R(1-|\xi/t|)\big)$ is a radial function. To prove (2.16), we can apply the aforementioned partition to the multiplier so that by rotational symmetry, we only need to consider the operator $S_{R,t}$ defined as

$$S_{R,t}f(x) := \int_{\mathbb{R}^3} \varphi(\xi)\psi\Big(\frac{1-|\xi/t|}{R^{-1}}\Big)\widehat{f}(\xi)e^{ix\cdot\xi}d\xi. \tag{2.23}$$

Here $\varphi(\xi)$ is a smooth function that is supported in the region

$$\{\xi : \xi_3 \geq 100\sqrt{\xi_1^2+\xi_2^2}, 1/2 \leq |\xi| \leq 2\}.$$

Note that in (2.15), the variable t only ranges in $[1,2]$. We can freely add a smooth cutoff functions $\eta(\bar{\xi})$ in the Fourier multiplier of $S_{R,t}$, where η is supported in the ball $B^2(0,1/2)$ and $=1$ on $B^2(0,1/4)$. Namely,

$$S_{R,t}f(x) = \int_{\mathbb{R}^3} \varphi(\xi)\eta(\bar{\xi})\psi\Big(\frac{1-|\xi/t|}{R^{-1}}\Big)\widehat{f}(\xi)e^{ix\cdot\xi}d\xi. \tag{2.24}$$

To express the Fourier multiplier of $S_{R,t}$ via something similar to (2.19), we take

$$\Phi(\bar{\xi};t) = (t^2-|\bar{\xi}|^2)^{1/2}, \qquad \phi(\xi;t) = \frac{\xi_3+(t^2-|\bar{\xi}|^2)^{1/2}}{t(t+|\xi|)}, \tag{2.25}$$

so that one has

$$S_{R,t}f(x) = \int_{\mathbb{R}^3} \varphi(\xi)\eta(\bar{\xi})\psi\Big(\frac{-\xi_3+\Phi(\bar{\xi};t)}{R^{-1}}\phi(\xi;t)\Big)\widehat{f}(\xi)e^{ix\cdot\xi}d\xi. \tag{2.26}$$

When $t\in[1,2]$, the function $\phi(\xi;t)$ is strictly positive on the support of $\varphi(\xi)$, so we may redefine $\phi(\xi;t)$ for ξ outside of $\operatorname{supp}\varphi$ to make $\phi(\xi;t)$ always positive without changing the value of $S_{R,t}f$. In a word, we may assume $\phi(\xi;t)$ takes the value in $[1/2,2]$.

Since $\varphi(\xi)$ is a smooth Fourier multiplier, one can reduce $S_{R,t}$ to a similar operator still denoted by

$$S_{R,t}f(x) = \int_{\mathbb{R}^3} \eta(\bar{\xi})\psi\Big(\frac{-\xi_3+\Phi(\bar{\xi};t)}{R^{-1}}\phi(\xi;t)\Big)\widehat{f}(\xi)e^{ix\cdot\xi}d\xi. \tag{2.27}$$

Hence to show (2.16), it suffices to prove that for any $\varepsilon>0$ and $p\geq 3.25$,

$$\|S_{R,t}f\|_{L^p_x L^2_t([1,2])} \leq C_\varepsilon R^{\frac{1}{2}-\frac{3}{p}+\varepsilon}\|f\|_p. \tag{2.28}$$

One can deduce (2.28) from (2.22) by discretizing the variable t. We omit the details. □

In the final step, we break the discrete square function $\mathrm{Sq}f$ in the physical space. For any point $m \in R^{-1/2}\mathbb{Z}^3$, we choose a $CR^{-1/2}$-cube q_m for some absolute big constant C only depending on the function Φ. The constant C will be determined later in the proof of Lemma 3.9. Denote the collection of these $CR^{-1/2}$-cube by $\mathbf{q}$. Let $\widehat{\varphi} : B^3(0,2) \to [0,1]$ be a bump function that $\widehat{\varphi} = 1$ in the unit ball $B^3(0,1)$. For each $q_m \in \mathbf{q}$, we define $\widehat{\varphi}_{q_m}(\xi) := \widehat{\varphi}(2CR^{1/2}(\xi - m))$ and

$$\Delta_q f := \varphi_q * f. \tag{2.29}$$

The estimate we would like to prove in the rest of the paper is the following.

THEOREM 2.8. *Let* $\mathrm{Sq}f$ *be defined in* (2.21). *For any* $p \geq 3.25$ *and* $\varepsilon > 0$, *one has*

$$\|\mathrm{Sq}f\|_{L^p(B_R)} \leq C_{\varepsilon,N} R^{\frac{p-3}{p}+\varepsilon} \Big\| \Big(\sum_q |\Delta_q f|^2\Big)^{\frac{1}{2}} \Big\|_{L^p(w_{B_{R^{1+\varepsilon}},N})} \tag{2.30}$$

Note that if (2.30) is true for B_R, it is also true for any R-ball in $\mathbb{R}^3$. Hence we can raise both sides in (2.30) to p-th power, sum up all the R-balls in $\mathbb{R}^3$ and use Lemma 2.3 together with duality to concludes (2.22). We also put it into a lemma.

LEMMA 2.9 (Third step). *Let* $\mathrm{Sq}f$ *be defined in* (2.21). *Then* (2.30) *implies* (2.22).

Combining all the lemmas in this subsection, we conclude the first reduction which we state as Proposition 2.10.

PROPOSITION 2.10 (First reduction). *Theorem* 2.8 *implies Theorem* 1.1.

3. A second reduction: The broad-narrow reduction

The second reduction is made for the function $1_{B_R}\mathrm{Sq}f$ in (2.30). The reduction is slightly different from the original broad-narrow argument in [**BG11**]. We begin with the definition of Gauss maps associated to the surfaces $\{\Gamma_j\}$, then we introduce general setups of broad-narrowness for a directional cap at any intermediate scale between 1 and $R^{-1/2}$. After that, we reduce (2.30) to a local estimate for broad part, which is our main result. Because of technical reasons, in the end of this section, we define some auxiliary broad functions and prove some properties for them.

3.1. Gauss maps.

DEFINITION 3.1 (Gauss map). Let Φ and $\{\Gamma_j\}$ be given in (2.17). For each j, we define the Gauss map of the surface Γ_j as:

$$G_j(\bar{\xi}) = \frac{(\nabla_{\bar{\xi}}\Phi(\bar{\xi};t_j),1)}{|(\nabla_{\bar{\xi}}\Phi(\bar{\xi};t_j),1)|}. \tag{3.1}$$

Note that $G_j(\bar{\xi})$ is just the normal vector of Γ_j at $(\bar{\xi},\Phi(\bar{\xi};t_j))$.

Lemmas 3.2 and 3.3 contain some properties about the Gauss map G_j.

LEMMA 3.2. *For any $1 \leq j \leq R$, the Gauss map G_j is smooth and injective. In particular, when $\bar{\xi} \in B^2(0,1/2)$, one has*

$$|\partial^\alpha G_j| \leq C_\alpha, \quad \alpha \in \mathbb{N}^2 \tag{3.2}$$

uniformly for all G_j.

PROOF. Since $\nabla_{\bar{\xi}}\Phi = O(1)$, one can check (3.2) directly. The fact that $\Phi(\bar{\xi};t_j)$ has positive second fundamental form implies that the intersection between the surface $\{\bar{\xi} \in B^2(0,1) : (\bar{\xi};\Phi(\bar{\xi};t_j))\}$ and an arbitrary plane in $\mathbb{R}^3$ is a two dimensional curve with nonvanishing curvature. Thus, we can prove by contradiction that G_j is injective. □

LEMMA 3.3. *Suppose that $\sigma \subset \mathbb{S}^2$ is any M^{-1}-cap with $1 \leq M \leq R^{1/2}$. Then for any $1 \leq j \leq R$, the set $G_j^{-1}(\sigma)$ is morally an M^{-1}-ball. That is, there is an M^{-1}-ball B_j in $\mathbb{R}^2$ such that*

$$cB_j \subset G_j^{-1}(\sigma) \subset CB_j \tag{3.3}$$

for two absolute constants $c < 1$ and $C > 1$.

PROOF. One only need to use the fact that both two eigenvalues of $\nabla_{\bar{\xi}}^2\Phi$ are ~ -1 uniformly in t, as mentioned in (2.18). We omit the details. □

3.2. Broad-narrow reduction. In this subsection, we will do the broad-narrow reduction. First, we fix the number $\kappa = \lceil \log R^{1/2}/\log K \rceil$. We are interested in the caps of scales $K^{-\ell}$ $(0 \leq \ell \leq \kappa)$. Note that $K = R^{\varepsilon^{10}}$, so $\kappa \leq \varepsilon^{-10}$ is bounded.

We claim that for any $0 \leq \ell \leq \kappa$, there is a set of finitely overlapping $K^{-\ell}$-caps that cover $\mathbb{S}^2$. We denote them by $\Sigma_\ell = \{\sigma_\ell\}$. They satisfy the following properties.

Given a scale $M = K^\ell$, for any cap $\sigma \in \Sigma_\ell$ being an M^{-1}-cap and for each $1 \leq j \leq R$, there is a kernel $m_{j,\sigma}$ with the formula

$$\widehat{m}_{j,\sigma}(\xi) = \eta_{j,\sigma}(\bar{\xi})\psi_j\Big(\frac{\xi_3 - \Phi(\bar{\xi};t_j)}{R^{-1}}\phi(\xi;t)\Big), \tag{3.4}$$

where $\eta_{j,\sigma}$ is a smooth bump function satisfying the following properties:

(1) $\eta_{j,\sigma}$ is supported in $G_j^{-1}(2\sigma)$, where G_j is the Gauss map defined in (3.1),
(2) $\sum_{\sigma \in \Sigma_\ell} \eta_{j,\sigma}(\bar{\xi}) = 1$ when $\bar{\xi} \in B^2(0,1/2)$,
(3) For any multi-index $\alpha \in \mathbb{N}^2$ that $|\alpha| = N$, $|\partial^\alpha \eta_{j,\sigma}| \leq C_N M^N$.

Also there is a relation " $<$ " between caps $\sigma \in \Sigma_\ell$ and caps $\tau \in \Sigma_{\ell+1}$, so that for any $\sigma \in \Sigma_\ell$ we have

$$\eta_{j,\sigma} = \sum_{\tau:\tau<\sigma} \eta_{j,\tau}. \tag{3.5}$$

Moreover, we can make $m_{j,\sigma_0}(\xi) = m_j(\xi)$ (see (2.19)).

We will construct the bump functions $\eta_{j,\sigma}$. We begin with $\Sigma_0 = \{\mathbb{S}^2\}$ which consists of a single cap. We simply set $\eta_{j,\mathbb{S}^2}(\bar{\xi}) = \eta_j(\bar{\xi})$, then $m_{j,\sigma_0}(\xi) = m_j(\xi)$. Suppose we have already constructed $\eta_{j,\sigma}$ for a $\sigma \in \Sigma_\ell$ being an M^{-1}-cap (Here we set $M = K^\ell$). Let $\mathcal{T}_\sigma = \{\tau\}$ be a collection of $K^{-1}M^{-1}$-caps in $\mathbb{S}^2$ that form a cover of σ. Define $\widetilde{\varphi}_\tau : \mathbb{S}^2 \to \mathbb{R}^+$ as a smooth partition of unity associated to this cover so that $\widetilde{\varphi}_\tau$ is supported in 2τ. The smooth cutoff functions $\widetilde{\varphi}_\tau$ satisfies the following derivatives estimates: for any multi-index $\alpha \in \mathbb{N}^2$ that $|\alpha| = N$, $|\partial^\alpha \widetilde{\varphi}_\tau| \leq C_N(MK)^N$. From Lemma 3.2 we know the Gauss map G_j is injective, so the pullback $\varphi_{j,\tau} = \widetilde{\varphi}_\tau \circ G_j$ forms a partition of unity of $1_{\text{supp}(\eta_{j,\sigma})}(\bar{\xi})$. Therefore, by setting $\eta_{j,\tau} = \varphi_{j,\tau}\eta_{j,\sigma}$ we have

$$\eta_{j,\sigma} = \sum_{\tau \in \mathcal{T}_\sigma} \eta_{j,\tau}. \tag{3.6}$$

We set $\Sigma_{\ell+1} := \cup_{\sigma \in \Sigma_\ell} \mathcal{T}_\sigma$, and set $\tau < \sigma$ if $\tau \in \mathcal{T}_\sigma$. We can verify (3.5). Also, the derivative estimates of $\eta_{j,\tau}$ is by chain rule and noting $\ell \le \kappa$ is finite.

Define $S_{j,\sigma}f$ and the square function $\mathrm{Sq}_\sigma f$ associated to the cap σ as

$$S_{j,\sigma}f := m_{j,\sigma} * f, \qquad \mathrm{Sq}_\sigma f(x) := \Big(\sum_{j=1}^{R} |S_{j,\sigma}f(x)|^2\Big)^{1/2}. \tag{3.7}$$

Now we can introduce the broad function $\mathrm{Br}_A\mathrm{Sq}_\sigma f$.

Definition 3.4 (Broadness). Given $A \in \mathbb{N}$ and any $x \in B_R$, we define $\mathrm{Br}_A\mathrm{Sq}_\sigma f(x)$ as the $A+1$-th largest number in $\{\mathrm{Sq}_\tau f(x)\}_{\tau \in \mathcal{T}_\sigma}$. That is,

$$\mathrm{Br}_A\mathrm{Sq}_\sigma f(x) = \min_{\tau_1,\dots,\tau_A \in \mathcal{T}_\sigma} \max_{\substack{\tau \ne \tau_l, \\ 1 \le l \le A}} \mathrm{Sq}_\tau f(x). \tag{3.8}$$

It is clear from the definition that for any $x \in B_R$

$$\mathrm{Sq}_\sigma f(x) \le K^2 \mathrm{Br}_A\mathrm{Sq}_\sigma f(x) + A \sup_{\tau \in \mathcal{T}_\sigma} \mathrm{Sq}_\tau f(x). \tag{3.9}$$

The first term is called the "broad part" and the second term is called the "narrow part". Taking the L^p-norm on both sides, we have

$$\int_{B_R} |\mathrm{Sq}_\sigma f|^p \lesssim K^2 \int_{B_R} |\mathrm{Br}_A\mathrm{Sq}_\sigma f|^p + A \sum_{\tau \in \mathcal{T}_\sigma} \int_{B_R} |\mathrm{Sq}_\tau f|^p. \tag{3.10}$$

We first set $\sigma_0 = \mathbb{S}^2$, so by (3.10) we get

$$\int_{B_R} |\mathrm{Sq}_{\sigma_0} f|^p \lesssim K^2 \int_{B_R} |\mathrm{Br}_A\mathrm{Sq}_{\sigma_0} f|^p + A \sum_{\sigma_1 \in \mathcal{T}_{\sigma_0}} \int_{B_R} |\mathrm{Sq}_{\sigma_1} f|^p. \tag{3.11}$$

We keep the broad part in (3.11), and run the broad-narrow argument for each narrow part to get

$$\int_{B_R} |\mathrm{Sq}_{\sigma_1} f|^p \lesssim K^2 \int_{B_R} |\mathrm{Br}_A\mathrm{Sq}_{\sigma_1} f|^p + A \sum_{\sigma_2 \in \mathcal{T}_{\sigma_1}} \int_{B_R} |\mathrm{Sq}_{\sigma_2} f|^p. \tag{3.12}$$

Note that $\mathrm{Sq}f = \mathrm{Sq}_{\sigma_0} f$. One can repeat this argument for $\kappa \sim \log R^{1/2}/\log K$ steps to break down all the narrow parts, and obtain the caps of scale $K^{-\kappa} \sim R^{-1/2}$ at the final step. Let us conclude this broad-narrow decomposition into a proposition.

Proposition 3.5 (Broad-narrow decomposition). *Let $K (K < R)$ be a big number and let $A = [\log R]$. Define $\kappa := \lceil \log R^{1/2}/\log K \rceil$. Then, we can break down $\|1_{B_R}\mathrm{Sq}f\|_p^p$ into (here θ has diameter $R^{-1/2}$)*

$$\int_{B_R} |\mathrm{Sq}f|^p \lesssim \sum_{\ell=1}^{\kappa-1} A^\ell K^2 \sum_{\sigma_\ell \in \Sigma_\ell} \int_{B_R} |\mathrm{Br}_A\mathrm{Sq}_{\sigma_\ell} f|^p + A^\kappa \sum_{\theta \in \Theta} \int_{B_R} |\mathrm{Sq}_\theta f|^p. \tag{3.13}$$

The kernel $m_{j,\theta}$ defined in (3.7) decays rapidly outside a $R^{1/2} \times R^{1/2} \times R$ tube with direction c_θ. In fact, we have:

LEMMA 3.6. *For any $\beta > 0$, let T be a $R^{1/2+\beta} \times R^{1/2+\beta} \times R^{1+\beta}$ tube with direction c_θ centered at the origin. Then recalling* (2.3), *one has*

$$|m_{j,\theta}(x)| \lesssim \widetilde{w}_{T,N}(x). \tag{3.14}$$

In particular, the family of smooth functions $\{R^{-O(\beta)} m_{j,\theta}\}$ is adapted to T (Recall Definition 2.2).

PROOF. In what follows everything is taken with respect to the standard basis $\{e_1, e_2, e_3\}$. One can write $\eta_{j,\theta}(\bar{\xi}) = \eta(R^{1/2}(\bar{\xi} - c_{j,\theta}))$ for some smooth function η supported in $B^2(0, 3/4)$, where the vector $(c_{j,\theta}, \Phi(c_{j,\theta}))$ is parallel to c_θ. Hence

$$m_{j,\theta}(x) = \int_{\mathbb{R}^3} e^{ix\cdot\xi} \eta(R^{1/2}(\bar{\xi} - c_{j,\theta}))\psi_j\Big(\frac{\xi_3 - \Phi(\bar{\xi}; t_j)}{R^{-1}}\phi(\xi; t)\Big) d\xi. \tag{3.15}$$

Consider the following change of variables:

$$\mathcal{L}_\tau : (\bar{\omega}, \omega_3) = \Big(\frac{\bar{\xi} - c_{j,\theta}}{R^{-1/2}}, \frac{\xi_3 - \nabla\Phi(c_{j,\theta}) \cdot \bar{\xi} - \Phi(c_{j,\theta}) + \nabla\Phi(c_{j,\theta}) \cdot c_{j,\theta}}{R^{-1}}\Big). \tag{3.16}$$

which leads to the linear transform

$$\begin{cases} \bar{\xi} = R^{-1/2}\bar{\omega} + c_{j,\theta} \\ \xi_3 = R^{-1}\omega_3 + R^{-1/2}\nabla\Phi(c_{j,\theta}) \cdot \bar{\omega} + \Phi(c_{j,\theta}). \end{cases} \tag{3.17}$$

Now let us introduce a new function $\Psi(\bar{\omega})$ defined as

$$\Psi(\bar{\omega}) := R[\Phi(c_{j,\theta} + R^{-1/2}\bar{\omega}) - \Phi(c_{j,\theta}) - R^{-1/2}\nabla\Phi(c_{j,\theta}) \cdot \bar{\omega}], \tag{3.18}$$

so that we can rewrite $|m_{j,\theta}(x)|$ as

$$|m_{j,\theta}(x)| = R^{-2}\Big| \int_{\mathbb{R}^3} e^{iR^{-1/2}(x_3\nabla\Phi(c_{j,\theta}) + \bar{x})\cdot\bar{\omega}} e^{iR^{-1}x_3\omega_3} a(\omega; t_j) d\omega \Big|. \tag{3.19}$$

Here $a(\omega; t_j) = \eta(\bar{\omega})\psi_j((\omega_3 - \Psi(\bar{\omega}; t_j))\widetilde{\phi}(\omega; t_j))$ is a smooth function, with $\widetilde{\phi}(\omega; t_j) = \phi(\xi; t_j)$. Finally, one can obtain (3.14) by the method of stationary phase. We leave out the details. □

In order to get (2.30), we only need to bound the right-hand side of (3.13). Before going even further, let us introduce one more notation and one more lemma.

DEFINITION 3.7. Let $\mathbf{q} = \{q\}$ be the collection of finitely overlapping $CR^{-1/2}$-cubes in the frequency space that was introduced in the paragraph above (2.29). For any cap $\tau \subset \mathbb{S}^2$ and any $1 \le j \le R$, let $\Gamma_j(\tau)$ be the subset of Γ_j where the normal directions lie in 2τ:

$$\Gamma_j(\tau) := \{\xi \in \Gamma_j : G_j(\bar{\xi}) \in 2\tau\}, \tag{3.20}$$

where G_j is the Gauss map introduced in (3.1). Define $\mathbf{q}(\tau) \subset \mathbf{q}$ as

$$\mathbf{q}(\tau) := \{q \in \mathbf{q} : \text{thereisa } j,\ 1 \le j \le R,\ q \cap \Gamma_j(\tau) \ne \varnothing\}. \tag{3.21}$$

When $\Phi(\bar{\xi}; t) = t - |\bar{\xi}|^2/2$ or $\Phi(\bar{\xi}; t) = (t - |\bar{\xi}|^2)^{1/2}$ which means $\{\Gamma_j\}$ is a family of paraboloids or spheres, cubes in $\mathbf{q}(\tau)$ are roughly contained in a straight tube, whose radius is the same as the radius of the cap τ. For general Φ satisfying (2.18), cubes in $\mathbf{q}(\tau)$ are morally contained in a curved tube instead. The exact distribution of cubes in $\mathbf{q}(\tau)$ does not concern us. What we need for $\mathbf{q}(\tau)$ is Lemma 3.8.

LEMMA 3.8. *Suppose that $\mathcal{T}$ is a collection of finitely overlapping E^{-1}-caps in $\mathbb{S}^2$ with $1 < E \le R^{1/2}$. Then the collection $\{\mathbf{q}(\tau)\}_{\tau\in\mathcal{T}}$ is finitely overlapped.*

PROOF. Fix a cap τ_1 and an arbitrary $CR^{-1/2}$-cube $q_1 \in \mathbf{q}(\tau_1)$. We claim that if $|\tau_1 - \tau_2| \geq C'E^{-1}$ for some large enough constant C', then there does not have a $CR^{-1/2}$ cube $q_2 \in \mathbf{q}(\tau_2)$ such that $q_2 \cap q_1 \neq \varnothing$. This suffices to prove our lemma.

To prove the our claim, we first note that by the condition $|\partial_t \Phi| \sim 1$ in (2.18), there are $O(R^{1/2})$ many j such that $q_1 \cap \Gamma_j \neq \varnothing$. Let us denote by J the collection of these j. For a fixed $j \in J$, from Lemma 3.3 we know that the pullbacks $\{G_j^{-1}(\tau)\}_{\tau \in \mathcal{T}}$ are finitely overlapped. Also, via (2.18) we know that $\text{dist}(G_j^{-1}(\tau), G_{j'}^{-1}(\tau)) \lesssim R^{-1/2}$ if $|j - j'| \lesssim R^{-1/2}$. These two arguments in particular implies that if $\text{dist}(\tau_1, \tau_2) \geq C'R^{-1/2}$ for some large enough constant C', $\text{dist}(G_j^{-1}(\tau_1), G_{j'}^{-1}(\tau_2)) \geq 100CR^{-1/2}$ for any $j' \in J$. It gives our claim. $\square$

The second part of (3.13) is easy to handle by Lemma 3.8 and the next lemma.

LEMMA 3.9. *Let* $\text{Sq}_\theta f$ *be defined in* (3.7) *and let* $\Delta_q f$ *be defined in* (2.29). *Suppose that* $\theta \in \Theta$ *is a* $R^{-1/2}$*-cap. Then for any* $\varepsilon > 0$ *and* $2 \leq p < \infty$,

$$\|\text{Sq}_\theta f\|_{L^p(B_R)} \leq C_\varepsilon R^\varepsilon \Big\| \Big(\sum_{q \in \mathbf{q}(\theta)} |\Delta_q f|^2 \Big)^{\frac{1}{2}} \Big\|_{L^p(w_{B_{R^{1+\varepsilon}}, N})}. \tag{3.22}$$

PROOF. We let $\beta = \varepsilon^{1000}$ and let T be the $R^{1/2+\beta} \times R^{1/2+\beta} \times R^{1+\beta}$ rectangular tube centered at the origin with direction c_θ, where $c_\theta \in \mathbb{S}^2$ is the center of θ. Then by Definition 2.2 and (3.14), the kernels $\{R^{-O(\beta)} m_{j,\sigma}\}_j$ are all adapted to the tube T. Also, recalling that q is a $CR^{-1/2}$-cube, we choose C large enough such that if $\Gamma_j(\theta) \cap \frac{1}{10} q \neq \varnothing$, then $\Gamma_j(\theta) \subset q$. Hence for each $CR^{-1/2}$ cube $q \in \mathbf{q}(\theta)$, by Lemma 2.3 one has

$$\sum_{j: \Gamma_j(\theta) \cap \frac{1}{10} q \neq \varnothing} \int |m_{j,\theta} * f|^2 g \leq C_{N,\beta} R^{O(\beta)} \int |\Delta_q f|^2 \widetilde{w}_{T,N} * g \tag{3.23}$$

Note that there is a 1-bounded, positive function $g \in L^{p'/2}(B_R)$ with $\|g\|_{p'/2} = 1$, such that

$$\|\text{Sq}_\theta f\|_{L^p(B_R)}^2 = \sum_j \int |m_{j,\theta} * f|^2 g. \tag{3.24}$$

We sum up all $j \in \{1, 2, \dots R\}$ in (3.23) to have

$$\|\text{Sq}_\theta f\|_{L^p(B_R)}^2 \leq C_{N,\beta} R^{O(\beta)} \int \sum_{q \in \mathbf{q}(\theta)} |\Delta_q f|^2 \widetilde{w}_{T,N} * g. \tag{3.25}$$

Note that $\|\widetilde{w}_{T,N} * g\|_{L^{p'/2}} \lesssim \|\widetilde{w}_{T,N}\|_{L^1} \|g\|_{L^{p'/2}} \lesssim \|g\|_{L^{p'/2}}$, so we have

$$\|\text{Sq}_\theta f\|_{L^p(B_R)} \leq C_\beta R^{O(\beta)} \Big\| \Big(\sum_{q \in \mathbf{q}(\theta)} |\Delta_q f|^2 \Big)^{\frac{1}{2}} \Big\|_{L^p(w_{B_{R^{1+\beta}}, N})}. \tag{3.26}$$

Since $\beta = \varepsilon^{1000} < \varepsilon$, it proves (3.22) $\square$

The first part of (3.13) is our main focus. In fact, we will prove the following result.

LEMMA 3.10. *Let* $K = R^{\varepsilon^{10}}$ *be a big number and let* $A = [\log R]$. *Recall the definition of* $\text{Sq}_\sigma f$ *in* (3.7). *Suppose that* $M = K^\ell$ *for some* $1 \leq \ell < \log R^{1/2} / \log K$

and $\sigma \in \Sigma_\ell$ is an M^{-1}-cap. Then for any $\varepsilon > 0$ and $p \geq 3.25$,

$$\|\mathrm{Br}_A \mathrm{Sq}_\sigma f\|_{L^p(B_R)} \leq C_\varepsilon R^{\frac{p-3}{p}+\varepsilon} M^{\frac{6-2p}{p}} \Big\| \Big(\sum_{q \in \mathbf{q}(\sigma)} |\Delta_q f|^2 \Big)^{\frac{1}{2}} \Big\|_{L^p(w_{B_{R^{1+\varepsilon}},N})}. \tag{3.27}$$

The function $\Delta_q f$ was introduced in (2.29).

Via Lemma 3.8, in (3.13), we can sum up the first part using Lemma 3.10, and sum up the second part using Lemma 3.9 to conclude (2.30). Since Lemma 3.8 and Lemma 3.9 are already verified, we have the following lemma

LEMMA 3.11. *Lemma* 3.10 *implies Theorem* 2.8.

From now on, let us fix M and σ. By rotating the coordinates we may also assume σ is the M^{-1}-cap centered at the north pole of $\mathbb{S}^2$. Since we only care about $\mathrm{Sq}_\sigma f$, we may assume $\widehat{f}$ is supported in $\cup_{q \in \mathbf{q}(\sigma)} q$, so we can replace $\mathbf{q}(\sigma)$ by $\mathbf{q}$ to rewrite (3.27) as

$$\|\mathrm{Br}_A \mathrm{Sq}_\sigma f\|_{L^p(B_R)} \leq C_\varepsilon R^{\frac{p-3}{p}+\varepsilon} M^{\frac{6-2p}{p}} \Big\| \Big(\sum_{q \in \mathbf{q}} |\Delta_q f|^2 \Big)^{\frac{1}{2}} \Big\|_{L^p(w_{B_{R^{1+\varepsilon}},N})}. \tag{3.28}$$

Let us introduce one more definition to further reduce (3.28) to a more local estimate, which is our main result.

DEFINITION 3.12 (Rescaled balls). Let $\sigma \subset \mathbb{S}^2$, $M > 1$ be fixed as above. For any radius $r > 1$, we say a geometric object in $\mathbb{R}^3$ is a "rescaled r-ball", if it is a tube of length r and radius $M^{-1}r$, pointing to the direction c_σ (c_σ is the center of the cap σ). We use $\mathcal{P}(x, r)$ to denote a rescaled r-ball centered at $x \in \mathbb{R}^3$. For simplicity, we use $\mathcal{P}_r$ to denote $\mathcal{P}(0, r)$.

Note that the kernel $m_{j,\sigma}$ (see (3.4)) is essentially supported in the rescaled ball $\mathcal{P}_R$ (see for instance Lemma 3.6). This is our motivation for introducing the term "rescaled ball". Under the orthonormal coordinate $\{e_1, e_2, e_3\}$ with $e_3 = c_\sigma$, we define an associated weight $w_{\mathcal{P}_R}$ as

$$w_{\mathcal{P}_R}(x) := \Big(1 + \frac{M|x_1|}{R} + \frac{M|x_2|}{R} + \frac{|x_3|}{R}\Big)^{-N_0}. \tag{3.29}$$

Here N_0 is a big number.

THEOREM 3.13 (The main result for broad functions). *Let $A = [\log R]$. Then for any $\varepsilon > 0$ and $p \geq 3.25$,*

$$\int_{\mathcal{P}_R} |\mathrm{Br}_A \mathrm{Sq}_\sigma f|^p \leq C_\varepsilon R^{p-3+p\varepsilon} M^{6-2p} \int \Big(\sum_{q \in \mathbf{q}} |\Delta_q f|^2 \Big)^{p/2} w_{\mathcal{P}_R}. \tag{3.30}$$

By summing up certain translations, Theorem 3.13 implies (3.28) and hence Lemma 3.10. Thus, via Lemma 3.11, we can conclude our second reduction in the next proposition.

PROPOSITION 3.14 (Second reduction). *Theorem* 3.13 *implies Theorem* 2.8.

Combining Proposition 2.10 and 3.14, we know that Theorem 3.13 implies our main result Theorem 1.1. From now on, let us focus on the broad function $\mathrm{Br}_A \mathrm{Sq}_\sigma f$ and the estimate (3.30).

3.3. Auxiliary broad functions. For some technical reasons, we need to define the broad part for vector-valued functions. We begin with the definition of a square function for vectors.

DEFINITION 3.15 (Square function for vector-valued functions). Suppose that we are given a vector-valued function $\boldsymbol{g} = \{g_1, \cdots, g_R\}$. We define the square function of $\boldsymbol{g}$ as:

$$\mathrm{Sq}\boldsymbol{g}(x) := \Big(\sum_{j=1}^{R} |g_j(x)|^2\Big)^{1/2}. \tag{3.31}$$

Now we can introduce the broad part for vector-valued function $\boldsymbol{g} = \{g_1, \ldots, g_R\}$. We will not need to define broad function for an arbitrary vector $\boldsymbol{g}$, but only for those appear in the one-step polynomial partitioning algorithm in Section 5. As a result, we will make some assumption on $\boldsymbol{g}$. We assume each component g_j has Fourier support in $N_{CM^{-2}}\Gamma_j(\sigma)$ (note that M, σ are already fixed after Lemma 3.11). We also remind the reader that $\mathcal{T}_\sigma = \{\tau\}$ is the set of $(MK)^{-1}$-caps such that $\tau < \sigma$. For any $\tau \in \mathcal{T}_\sigma$, we define

$$g_{j,\tau} := \check{\eta}_{j,\tau} * g_j, \qquad \boldsymbol{g}_\tau := \{g_{1,\tau}, \ldots, g_{R,\tau}\}. \tag{3.32}$$

DEFINITION 3.16. Given $A \in \mathbb{N}$ and any $x \in \mathcal{P}_R$, we define $\mathrm{Br}_A\mathrm{Sq}\boldsymbol{g}(x)$ as the $A+1$ largest number in $\{\mathrm{Sq}(\boldsymbol{g}_\tau)(x)\}_{\tau \in \mathcal{T}_\sigma}$. That is,

$$\mathrm{Br}_A\mathrm{Sq}\boldsymbol{g}(x) := \min_{\tau_1, \ldots, \tau_A \in \mathcal{T}_\sigma} \max_{\substack{\tau \neq \tau_l, \\ 1 \le l \le A}} \mathrm{Sq}(\boldsymbol{g}_\tau)(x). \tag{3.33}$$

To see its relationship with $\mathrm{Br}_A\mathrm{Sq}_\sigma f(x)$, we set

$$f_j := S_{j,\sigma} f, \qquad \boldsymbol{f} := \{f_1, \ldots, f_R\}, \tag{3.34}$$

then one has $\mathrm{Br}_A\mathrm{Sq}_\sigma f(x) = \mathrm{Br}_A\mathrm{Sq}\boldsymbol{f}(x)$.

Probably the most important property we will use for the broad function is the following weak version of triangle inequality. It allows us to keep the "broad" property when decomposing the original function.

LEMMA 3.17 (Triangle inequality). *If $\boldsymbol{f} = \boldsymbol{g} + \boldsymbol{h}$, then*

$$\mathrm{Br}_A\mathrm{Sq}\boldsymbol{f}(x) \le \mathrm{Br}_{A/2}\mathrm{Sq}\boldsymbol{g}(x) + \mathrm{Br}_{A/2}\mathrm{Sq}\boldsymbol{h}(x) \tag{3.35}$$

PROOF. By the definition of the broad function, suppose that there are two collection of caps $\{\tau_i\}_{i=1}^{A/2}, \{\tau_i'\}_{i=1}^{A/2}$ such that

$$\mathrm{Br}_{A/2}\mathrm{Sq}\boldsymbol{g}(x) = \max_{\substack{\tau \neq \tau_k, \\ 1 \le k \le A/2}} \mathrm{Sq}(\boldsymbol{g}_\tau)(x), \tag{3.36}$$

$$\mathrm{Br}_{A/2}\mathrm{Sq}\boldsymbol{h}(x) = \max_{\substack{\tau \neq \tau_k', \\ 1 \le k \le A/2}} \mathrm{Sq}(\boldsymbol{h}_\tau)(x). \tag{3.37}$$

Since $\mathrm{Sq}(\boldsymbol{f}_\tau) \le \mathrm{Sq}(\boldsymbol{g}_\tau) + \mathrm{Sq}(\boldsymbol{h}_\tau)$, we have

$$\mathrm{Br}_A\mathrm{Sq}\boldsymbol{f}(x) \le \max_{\substack{\tau \neq \tau_k, \tau \neq \tau_k', \\ 1 \le k \le A/2}} \mathrm{Sq}(\boldsymbol{f}_\tau)(x) \le \mathrm{Br}_{A/2}\mathrm{Sq}\boldsymbol{g}(x) + \mathrm{Br}_{A/2}\mathrm{Sq}\boldsymbol{h}(x) \tag{3.38}$$

as desired. □

The broad function is dominated by a certain bilinear function. First, we define

$$\mathrm{Bil}\boldsymbol{g}(x) := \sum_{\tau_1,\tau_2\in\mathcal{T}_\sigma} |\mathrm{Sq}(\boldsymbol{g}_{\tau_1})(x)|^{1/2}|\mathrm{Sq}(\boldsymbol{g}_{\tau_2})(x)|^{1/2}, \tag{3.39}$$

where in the summation τ_1, τ_2 are non adjacent caps.

LEMMA 3.18. *For any* $A \geq 2$, *we have* $\mathrm{Br}_A\mathrm{Sq}\boldsymbol{g}(x) \leq \mathrm{Bil}\boldsymbol{g}(x)$.

PROOF. The proof is just by definition. □

4. Wave packet decomposition

In this section, we discuss the wave packet decomposition. Fix a scale $R \geq \rho \geq R^\varepsilon$. Let $\boldsymbol{g} = \{g_1, \cdots, g_R\}$ be a vector-valued function with $\mathrm{supp}(\widehat{g}_j) \subset N_{R^\beta\rho^{-1}}\Gamma_j$. We build the wave packet decomposition for $\boldsymbol{g}$ at scale ρ as follows.

Let $\{\tau\}$ be a set of $\rho^{-1/2}$-caps that form a finitely overlapping cover of $\mathbb{S}^2$. Then for each $1 \leq j \leq R$, $\{\Gamma_j(\tau)\}$ form a finitely overlapping cover of Γ_j. Let $\{\widehat{\varphi}_{j,\tau}(\xi)\}$ be a partition of unity associated to $N_{R^\beta\rho^{-1}}\Gamma_j$ so that each $\widehat{\varphi}_{j,\tau}$ is a smooth bump function for the slab $N_{R^\beta\rho^{-1}}\Gamma_j(\tau)$ of dimensions $\rho^{-1/2} \times \rho^{-1/2} \times R^\beta\rho^{-1}$. In this way, $\sum_\tau \widehat{\varphi}_{j,\tau}$ forms a smooth approximation to $1_{N_{R^\beta\rho^{-1}}\Gamma_j}$, and $\sum_\tau \widehat{\varphi}_{j,\tau} = 1$ on $N_{R^\beta\rho^{-1}}\Gamma_j$. We define

$$g_{j,\tau} := \varphi_{j,\tau} * g_j, \quad \boldsymbol{g}_\tau := \{g_{1,\tau}, \cdots, g_{R,\tau}\}. \tag{4.1}$$

So, we can partition each g_j of $\boldsymbol{g}$ in the frequency space as

$$g_j = \sum_\tau g_{j,\tau}. \tag{4.2}$$

Note this is slightly different from (3.32), but they both play the role of restricting the Fourier support to the cap $N_{R^\beta\rho^{-1}}\Gamma_j$. From now on we adopt this new definition, as (3.32) is only used in the broad narrow reduction.

Next, we partition the physical space. Fix a cap τ. After rotating c_τ, without loss of generality we assume $c_\tau = e_3$. We can choose a partition of unity $\{\zeta_v\}_{v\in\mathbb{Z}^3}$ such that for any $v \in \mathbb{Z}^3$, $\zeta_v(x) = \zeta(x - v)$, where $\zeta : \mathbb{R}^3 \to \mathbb{R}^+$ is a smooth function whose Fourier transform is supported in the unit ball in the frequency space, and ζ decays rapidly outside the unit ball in the physical space.

To handle the rapidly decaying tail, we modify the partition of unity $\{\zeta_v\}$ a little bit. Recall the tiny number $\beta = \delta^{1000}$. For each point $u \in R^\beta\mathbb{Z}^3$, we define

$$\eta_u(x) = \sum_{v\in E(u)} \zeta_v(x), \text{ for } E_u := \{v : -R^\beta/2 < v_l - u_l \leq R^\beta/2,\ l = 1, \cdots, n\}. \tag{4.3}$$

Hence $\{\eta_u\}_{u\in R^\beta\mathbb{Z}^3}$ also forms a smooth partition of unity. The advantage of using this new partition of unity is that the supports of functions $\{\eta_u\}$ are essentially disjoint in the following sense: if $u, u' \in R^\beta\mathbb{Z}^3$ satisfy $|u - u'| \geq 10R^\beta$, then $\min_x(|\eta_u(x)|, |\eta_{u'}(x)|) = \mathrm{RapDec}(R)$. This means that each η_u essentially interacts with a bounded number (say at most 100) of other $\eta_{u'}$.

For each $u \in R^\beta\mathbb{Z}^3$, we consider the rescaled function

$$\widetilde{\eta}_u(x) := \eta_u(\rho^{-1/2}x_1, \rho^{-1/2}x_{n-1}, R^\beta\rho^{-1}x_n).$$

From the discussion above, we see $\widetilde{\eta}_u$ is essentially supported in a rectangle of dimensions $R^\beta\rho^{1/2} \times R^\beta\rho^{1/2} \times \rho$. We also call it a ρ-tube. This tube is defined by

$$T_u := \{x \in \mathbb{R}^3, |x_1 - u_1| \leq R^\beta\rho^{1/2},\ |x_2 - u_2| \leq R^\beta\rho^{1/2},\ |x_3 - u_3| \leq \rho\}. \tag{4.4}$$

For any $x \in \mathbb{R}^3 \setminus 2T_u$, we have

$$\widetilde{\eta}_u(x) \lesssim \mathrm{RapDec}(R)\Big(1 + \frac{|x_1 - u_1|}{\rho^{1/2}} + \frac{|x_2 - u_2|}{\rho^{1/2}} + \frac{|x_3 - u_3|}{\rho R^{-\beta}}\Big)^{-4}. \tag{4.5}$$

We also have that the Fourier transform of $\widetilde{\eta}_u$ is supported in a $\rho^{-1/2} \times \rho^{-1/2} \times R^\beta \rho^{-1}$-slab. This is the reason that we assume the Fourier support of the function is in $R^\beta\rho^{-1}$-neighborhood of the surface.

Note that we assumed $c_\tau = e_3$. For general τ, we also defined the partition of unity $\{\widetilde{\eta}_u\}$ and the tubes $\{T_u\}$, where the coreline of each tube T_u is parallel to the direction c_τ. We denote by

$$\mathbb{T}_\tau[\rho] := \{T_u\} \tag{4.6}$$

the collection of these tubes. In order to reveal the essential support of each $\widetilde{\eta}_u$, we use the notation

$$\mathit{1}^*_{T_u}(x) := \widetilde{\eta}_u(x).$$

Let us continue working on the right-hand side of (4.2). We further partition each $g_{j,\tau}$ into

$$g_{j,\tau} = \sum_{T \in \mathbb{T}_\tau[\rho]} g_{j,\tau}\mathit{1}^*_T.$$

From now on, we just set $g_{j,T} := g_{j,\tau}\mathit{1}^*_T$. We obtain the wave packet decomposition:

$$g_j = \sum_\tau \sum_{T \in \mathbb{T}_\tau[\rho]} g_{j,T}, \quad \text{where } g_{j,T} := g_{j,\tau}\mathit{1}^*_T \text{ for } T \in \mathbb{T}_\tau[\rho]. \tag{4.7}$$

If $T \in \mathbb{T}_\tau[\rho]$, then the Fourier transform of $\mathit{1}^*_T$ is supported in a slab centered at the origin which has dimensions $\rho^{-1/2} \times \rho^{-1/2} \times \rho^{-1}$, and the normal direction of this slab is c_τ. So, $\widehat{g_{j,T}}$ is supported in $N_{R^\beta\rho^{-1}}\Gamma_j(2\tau)$

This is the scale ρ wave packet decomposition. Each $g_{j,T}$ is a single wave packet. Since τ and T are both independent to the factor j, we just write $\boldsymbol{g}_\tau := \{g_{1,\tau}, \dots, g_{R,\tau}\}$ and $\boldsymbol{g}_T = \boldsymbol{g}_\tau \mathit{1}^*_T := \{g_{1,T_\tau}, \dots, g_{R,T_\tau}\}$ (when $T \in \mathbb{T}_\tau[\rho]$). We also call $\boldsymbol{g}_T$ a single wave packet.

We have the following L^2-orthogonality for our wave packets.

LEMMA 4.1 (L^2-orthogonality). *For an arbitrary collection $\mathbb{T}' \subset \cup_\tau \mathbb{T}_\tau[\rho]$, we have*

$$\Big\| \sum_{T \in \mathbb{T}'} g_{j,T} \Big\|_2^2 \lesssim \sum_{T \in \mathbb{T}'} \|g_{j,T}\|_2^2 \lesssim \|g_j\|_2^2. \tag{4.8}$$

This estimate is uniform for all $1 \le j \le R$.

We have another useful lemma.

LEMMA 4.2. *Let $\boldsymbol{g} = \{g_1, \cdots, g_R\}$ be a vector-valued function with $\mathrm{supp}(\widehat{g}_j) \subset N_{R^\beta\rho^{-1}}\Gamma_j$. Let T be a ρ-tube and $\boldsymbol{g}_T$ be the wave packet at scale ρ. We have*

$$\|\mathrm{Sq}\boldsymbol{g}_T\|_2^2 \lesssim \|\mathit{1}_{10T}\mathrm{Sq}\boldsymbol{g}\|_2^2 + \mathrm{RapDec}(\rho)\|\mathrm{Sq}\boldsymbol{g}\|_2^2.$$

PROOF. We just need to prove for each j we have

$$\|g_{j,T}\|_2^2 \lesssim \|\mathit{1}_{10T}g_j\|_2^2 + \mathrm{RapDec}(\rho)\|g_j\|_2^2.$$

By definition, we have $g_{j,T} = (\varphi_{j,\tau} * g_j) 1_T^*$, and $|\varphi_{j,\tau}(x)| \lesssim 1_{T_0}^*(x)$ (T_0 is a translation of T that centered at the origin). We can rewrite as

$$g_{j,T} = \big(\varphi_{j,\tau} * (1_{10T}\ g_j)\big) 1_T^* + \big(\varphi_{j,\tau} * (1_{\mathbb{R}^3 \setminus 10T}\ g_j)\big) 1_T^*.$$

The second term is negligible, since it is essentially supported in $\big(T_0 + (\mathbb{R}^3 \setminus 10T)\big) \cap T = \emptyset$. So, we have

$$\|g_{j,T}\|_2^2 \le \Big\| \big(\varphi_{j,\tau} * (1_{10T}\ g_j)\big) 1_T^* \Big\|_2^2 + \mathrm{RapDec}(\rho)\|g_j\|_2^2 \lesssim \|1_{10T} g_j\|_2^2 + \mathrm{RapDec}(\rho)\|g_j\|_2^2.$$

□

We also have the following lemma comparing wave packets at different scales. The argument is standard, so we omit the details.

DEFINITION 4.3. For a tube T of dimensions $\rho \times \rho \times r$ with $\rho \le r$, we define $\omega(T) \subset \mathbb{S}^2$ to be a cap of radius ρ/r whose center is the direction of T. We call $\omega(T)$ the direction cap of T. For two tubes T_i of dimensions $\rho_i \times \rho_i \times r_i$ with $\rho_i \le r_i$ ($i = 1, 2$). We say $T_1 < T_2$ if

$$T_1 \subset 50T_2 \quad \text{and} \quad \omega(T_2) \subset 10\omega(T_1).$$

LEMMA 4.4. *Fix two scales $R^\varepsilon \le \rho \le r$. Let T_1 be a ρ-tube and T_2 be a r-tube such that $T_1 \not< T_2$. Then for a wave packet g_{T_2}, we have*

$$(g_{T_2})_{T_1} = \mathrm{RapDec}(R)\|g\|_2.$$

Similarly, for a wave packet g_{T_1}, we also have

$$(g_{T_1})_{T_2} = \mathrm{RapDec}(R)\|g\|_2.$$

5. Modified polynomial partitioning

Starting in this section, we set $\delta := \varepsilon^2$, $d := R^{\varepsilon^6}$, $\beta := \varepsilon^{1000}$. Let us introduce some definitions regarding to the zero set of polynomials.

DEFINITION 5.1. Suppose $Q_1, \dots, Q_k$ are polynomials in $\mathbb{R}^n$. We say $Z(Q_1, \dots, Q_k)$ is a transverse complete intersection if for any $x \in Z(Q_1, \dots, Q_k)$, the vectors $\nabla Q_1(x), \dots, \nabla Q_k(x)$ are linearly independent.

DEFINITION 5.2. We say a polynomial P in $\mathbb{R}^n$ is nonsingular, if $\nabla P(x) \ne 0$ for any $x \in Z(P)$.

The modified polynomial partitioning we need is the following.

PROPOSITION 5.3. *Let $\rho > 0$. Suppose that g is a nonnegative L^1 function in $\mathbb{R}^n$ supported in a rescaled ρ-ball $\mathcal{P}$. Then for any $d \in \mathbb{Z}^+$, there exists a polynomial P with degree $O(d)$, such that*

(1) *There are $\sim d^n$ many cells O contained in $\mathcal{P} \setminus Z(P)$, satisfying*

$$\int_O g \sim d^{-n} \int_{\mathbb{R}^n} g. \tag{5.1}$$

(2) *Each of these cells O in (1) lie in a smaller rescaled ρd^{-1}-ball.*

We omit the proof of this proposition. In fact, by a horizontal nonisotropic scaling, we can reduce Proposition 5.3 to the case where the support of g is the usual ball B_ρ. In this case, the corresponding polynomial argument was shown in [**Wan18**].

Next, we state the one-step polynomial partitioning algorithm based on the idea in [**Gut16**]. This may be less intuitive, but one of its advantage is that when building up our iteration in next section, we can apply this algorithm directly.

ALGORITHM 1. **Inputs:** $(\rho, U, \boldsymbol{g}_U, A)$. More precisely our inputs are:

(1) A scale ρ ($M^2 R^{\varepsilon/100} \le \rho \le R$);
(2) A set U which is contained in a rescaled ρ-ball $\mathcal{P}_U$;
(3) A vector-valued function $\boldsymbol{g}_U = \{g_{1,U}, \cdots, g_{R,U}\}$. Each component $g_{j,U}$ has Fourier support in $N_{R^\beta \rho^{-1}}(\Gamma_j(\sigma))$;
(4) An integer $A \ge 10$.

Outputs: We obtain a polynomial P of degree $O(d)$ where $d := R^{\varepsilon^6}$, and a wall $W = U \cap N_{\rho^{1/2} R^\delta} Z(P)$. Also, one of the following three cases holds: cellular case, transverse case and tangent case.

Cellular case:

(1) We obtain a collection of cells $\mathcal{O} = \{O\}$ which we call cellular cells. They satisfies: each O is contained in a rescaled ρd^{-1}-ball $\mathcal{P}_O$ and

$$|\mathcal{O}| \sim d^3. \tag{5.2}$$

(2) We obtain tube sets $\{\mathbb{T}_O\}_{O\in\mathcal{O}}$ and a set of functions $\{\boldsymbol{g}_O\}_{O\in\mathcal{O}}$ which are indexed by $\mathcal{O}$. They satisfy an L^2-estimate

$$\sum_{O\in\mathcal{O}} \|\mathrm{Sq}\boldsymbol{g}_O\|^2 \le \|\mathrm{Sq}\boldsymbol{g}_U\|_2^2; \tag{5.3}$$

and a broad estimate

$$\int_U |\mathrm{Br}_A \mathrm{Sq}\boldsymbol{g}_U|^p \sim d^3 \int_O |\mathrm{Br}_A \mathrm{Sq}\boldsymbol{g}_O|^p \text{ for each } O \in \mathcal{O}. \tag{5.4}$$

(3) The Fourier transform of each component $g_{j,O}$ of $\boldsymbol{g}_O$ satisfies:

$$\operatorname{supp} \widehat{g_{j,O}} \subset N_{R^\beta(\rho d^{-1})^{-1}}(\Gamma_j(\sigma)). \tag{5.5}$$

Transverse case:

(1) We obtain a collection of cells $\mathcal{B} = \{B\}$ which we call transverse cells. They satisfy: each B is a subset of W and each B is contained in a rescaled $\rho R^{-\delta}$-ball $\mathcal{P}_B$ with $\delta = \varepsilon^2$.
(2) We obtain tube sets $\{\mathbb{T}_{B,trans}\}_{B\in\mathcal{B}}$ and a set of functions $\{\boldsymbol{g}_{B,trans}\}_{B\in\mathcal{B}}$ which are indexed by $\mathcal{B}$. They satisfy an L^2-estimate

$$\sum_{B\in\mathcal{B}} \|\mathrm{Sq}\boldsymbol{g}_{B,trans}\|_2^2 \le \mathrm{Poly}(d) R^{-\delta} \|\mathrm{Sq}\boldsymbol{g}_U\|_2^2; \tag{5.6}$$

and a broad estimate

$$\int_U |\mathrm{Br}_A \mathrm{Sq}\boldsymbol{g}_U|^p \lesssim \log R \sum_{B\in\mathcal{B}} \int_B |\mathrm{Br}_{A/2} \mathrm{Sq}\boldsymbol{g}_{B,trans}|^p. \tag{5.7}$$

Also, for each $B \in \mathcal{B}$, the quantity

$$\int_B |\mathrm{Br}_{A/2} \mathrm{Sq}\boldsymbol{g}_{B,trans}|^p \tag{5.8}$$

are the same up to a constant factor.

(3) The Fourier transform of each component $g_{j,B,trans}$ of $\boldsymbol{g}_{B,trans}$ satisfies:

$$\operatorname{supp}\widehat{g}_{j,B,trans} \subset N_{R^\beta(\rho R^{-\delta})^{-1}}(\Gamma_j(\sigma)). \tag{5.9}$$

Tangent case:

(1) We obtain a collection of cells $\mathcal{B} = \{B\}$ which we call tangent cells. They satisfy: each $B \subset W$; each B is contained in a rescaled $\rho R^{-\delta}$-ball $\mathcal{P}_B$.

(2) We obtain tube sets $\{\mathbb{T}_{B,tang}\}_{B\in\mathcal{B}}$ and a set of functions $\{\boldsymbol{g}_{B,tang}\}_{B\in\mathcal{B}}$ which are indexed by $\mathcal{B}$. They satisfy an L^2-estimate

$$\sum_{B\in\mathcal{B}} \|\mathrm{Sq}\boldsymbol{g}_{B,tang}\|_2^2 \le R^{3\delta}\|\mathrm{Sq}\boldsymbol{g}_U\|_2^2; \tag{5.10}$$

and a broad estimate

$$\int_U |\mathrm{Br}_A\mathrm{Sq}\boldsymbol{g}_U|^p \lesssim \log R \sum_{B\in\mathcal{B}} \int_B |\mathrm{Br}_{A/2}\mathrm{Sq}\boldsymbol{g}_{B,tang}|^p. \tag{5.11}$$

(3) The Fourier transform of each component $g_{j,B,tang}$ of $\boldsymbol{g}_{B,tang}$ satisfies:

$$\operatorname{supp}\widehat{g}_{j,B,tang} \subset N_{R^\beta(\rho R^{-\delta})^{-1}}(\Gamma_j(\sigma)). \tag{5.12}$$

PROOF. The rest of this section is devoted to the proof of the Algorithm 1. We apply Proposition 5.3 to the function $1_U|\mathrm{Br}_A\mathrm{Sq}\boldsymbol{g}_U|^p$ to obtain a polynomial P. Also, we obtain a collection of cells $\widetilde{\mathcal{O}} = \{\widetilde{O}\}$ such that: $|\widetilde{\mathcal{O}}| \sim d^3$, each $\widetilde{O}$ is contained in a rescaled ρd^{-1}-ball, and

$$\int_{\widetilde{O}} |\mathrm{Br}_A\mathrm{Sq}\boldsymbol{g}_U|^p \sim d^{-3}\int_U |\mathrm{Br}_A\mathrm{Sq}\boldsymbol{g}_U|^p \tag{5.13}$$

for each cell $\widetilde{O}$.

Let the wall be $W = U \bigcap N_{\rho^{1/2}R^\delta}Z(P)$. We define the shrunken cell $O := \widetilde{O}\backslash W$ and the collection of them $\mathcal{O} := \{O\}$. Since $U = (\sqcup O) \sqcup W$, we have the following inequality:

$$\int_U |\mathrm{Br}_A\mathrm{Sq}\boldsymbol{g}_U|^p \lesssim \sum_{O\in\mathcal{O}} \int_O |\mathrm{Br}_A\mathrm{Sq}\boldsymbol{g}_U|^p + \int_W |\mathrm{Br}_A\mathrm{Sq}\boldsymbol{g}_U|^p. \tag{5.14}$$

Invoking the wave packet decomposition at scale ρ, we can write

$$\boldsymbol{g}_U = \sum_{T\in\mathbb{T}} (\boldsymbol{g}_U)_T, \tag{5.15}$$

where $\mathbb{T}$ is a set of ρ-tubes coming from the wave packet decomposition at scale ρ. Since we are only interested in the value of $\boldsymbol{g}_U(x)$ at $x \in \mathcal{P}_U$, we can assume $\mathbb{T}$ is contained in $\mathcal{P}_U$ by ignoring the rapid decaying term. In the following discussion, we will define $\mathbb{T}'_O, \mathbb{T}'_{B,trans}$ and $\mathbb{T}'_{B,tang}$ which are subsets of $\mathbb{T}$.

First, we analyze the first term on the right-hand side of (5.14). For each cell $O \in \mathcal{O}$, define $\mathbb{T}'_O$ as the collection of tubes such that $O \cap T \ne \varnothing$. It is proved in [**Gut16**] that

LEMMA 5.4. *Each $T \in \mathbb{T}$ belongs to at most $deg(P)+1 = O(d)$ many sets $\mathbb{T}'_O$.*

We define the vector-valued function $\boldsymbol{g}'_O$ associated to the cell O as

$$\boldsymbol{g}'_O := \sum_{T\in\mathbb{T}'_O} (\boldsymbol{g}_U)_T. \tag{5.16}$$

Let $\mathcal{P}_O$ be the rescaled ρd^{-1}-ball containing the cell O. We do wave packet decomposition for $\boldsymbol{g}'_O$ at scale ρd^{-1}, and let $\mathbb{T}_O$ be the set of ρd^{-1}-tubes that are contained in $\mathcal{P}_O$. We define

$$\boldsymbol{g}_O := \sum_{T\in\mathbb{T}_O} (\boldsymbol{g}'_O)_T. \tag{5.17}$$

For any $x \in \mathcal{P}_O$, one has

$$\boldsymbol{g}_U(x) = \boldsymbol{g}'_O(x) + \mathrm{RapDec}(R)\|\mathrm{Sq}\boldsymbol{g}\|_2 = \boldsymbol{g}_O(x) + \mathrm{RapDec}(R)\|\mathrm{Sq}\boldsymbol{g}\|_2. \tag{5.18}$$

Replacing $\boldsymbol{g}'_O$ by $\boldsymbol{g}_U$ in (5.17), we have

$$\boldsymbol{g}_O = \sum_{T\in\mathbb{T}_O} (\boldsymbol{g}_U)_T + \mathrm{RapDec}(R)\|\boldsymbol{g}\|_2. \tag{5.19}$$

By Lemma 4.4, we see that some wave packets on the right-hand side of (5.17) are negligible. After deleting some of them from $\mathbb{T}_O$, we can assume: for any $T \in \mathbb{T}_O$, there exists $T' \in \mathbb{T}'_O$ so that $T < T'$ (see Definition 4.3).

By (5.18), we have

$$\sum_{O\in\mathcal{O}} \int_O |\mathrm{Br}_A \mathrm{Sq}\boldsymbol{g}_U|^p \lesssim \sum_{O\in\mathcal{O}} \int_O |\mathrm{Br}_A \mathrm{Sq}\boldsymbol{g}_O|^p + \mathrm{RapDec}(R)\|\mathrm{Sq}\boldsymbol{g}\|_2^p. \tag{5.20}$$

Next, let us analyze the second term on the right-hand side of (5.14). We choose a collection of rescaled $\rho R^{-\delta}$-balls $\widetilde{\mathcal{B}} = \{\widetilde{B}\}$ that form a finitely overlapping cover of U. Define

$$B := \widetilde{B} \cap W \text{ and } \mathcal{B} := \{B\}. \tag{5.21}$$

For each B, we define $\mathbb{T}'_{B,tang}$ and $\mathbb{T}'_{B,trans}$ which are subsets of $\mathbb{T}$ as follows.

DEFINITION 5.5. $\mathbb{T}'_{B,tang}$ is the set of ρ-tubes $T \in \mathbb{T}$ obeying the following two conditions:

- $T \cap B \neq \varnothing$,
- If z is any nonsingular point of Z lying in $10\widetilde{B} \cap 10T$, then

$$|\mathrm{Angle}(v(T), T_z Z(P))| \leq \rho^{-1/2} R^{\delta}. \tag{5.22}$$

DEFINITION 5.6. $\mathbb{T}'_{B,trans}$ is the set of ρ-tubes $T \in \mathbb{T}$ obeying the following two conditions:

- $T \cap B \neq \varnothing$.
- There exists a nonsingular point z of Z lying in $10\widetilde{B} \cap 10T$, such that

$$|\mathrm{Angle}(v(T), T_z Z(P))| > \rho^{-1/2} R^{\delta}. \tag{5.23}$$

The following lemma for the transverse tubes in proved in [**Gut16**].

LEMMA 5.7. *Each $T \in \mathbb{T}$ belongs to at most* $\mathrm{Poly}(deg(P)) = \mathrm{Poly}(d)$ *many sets* $\mathbb{T}'_{B,trans}$.

We will also include Lemma 5.8 which will be used in Section 10. First, we cover B using finitely overlapping rescaled $M\rho^{1/2+\beta}$-balls $\{Q\}$ such that $Q \cap B \neq \varnothing$. Then we define

$$\mathbb{T}_{B,tang,Q} := \{T \in \mathbb{T}'_{B,tang}, T \cap 2Q \neq \varnothing\}. \tag{5.24}$$

Then for any Q, all tangent tubes intersecting Q morally lie in a thin neighborhood of a plane. This is proved in the next lemma.

LEMMA 5.8. *There is a plane $V \subset \mathbb{R}^3$ such that every tubes in $\mathbb{T}_{B,tang,Q}$ is contained in $N_{100\rho^{1/2}R^\delta}(V)$.*

PROOF. A crucial fact we use in the proof is that the directions of all tangent tubes are roughly c_σ, which is also the direction of the rescaled ball Q. We pick an arbitrary tube $T \in \mathbb{T}_{B,tang,Q}$. Since $Q \cap W \neq \varnothing$, there is a nonsingular point $z \in Z(P) \cap 10\widetilde{B}$ such that $z \in 2Q$. Let $V = T_z Z(P)$ be the tangent plane at z, so from the angular condition (5.22), we have $T \subset N_{100\rho^{1/2}R^\delta}(V)$.

It remains to verify $T' \subset N_{100\rho^{1/2}R^\delta}(V)$ for every other tube $T' \in \mathbb{T}_{B,tang,Q}$. Indeed, since $T' \cap Q$ is not empty, and since the direction of T' and the direction of Q make an angle less than M^{-1}, the dilated set $2Q$ is contained in $6T'$. Hence for the same nonsingular point $z \in Z(P) \cap 10\widetilde{B}$ chosen above, $z \in 6T'$. Employing the angular condition (5.22) again, we see that $T' \subset N_{\rho^{1/2}R^{O(\delta)}}(V)$ as desired. □

Similar to the cellular case, for each B we define $\boldsymbol{g}'_{B,tang}$ and $\boldsymbol{g}'_{B,trans}$ as

$$\boldsymbol{g}'_{B,trans} := \sum_{T \in \mathbb{T}'_{B,trans}} (\boldsymbol{g}_U)_T, \quad \boldsymbol{g}'_{B,tang} := \sum_{T \in \mathbb{T}'_{B,tang}} (\boldsymbol{g}_U)_T. \tag{5.25}$$

Let $\mathcal{P}_B$ be the rescaled $\rho R^{-\delta}$-ball containing the cell B. We do wave packet decomposition for $\boldsymbol{g}'_{B,trans}$ (resp. $\boldsymbol{g}'_{B,tang}$) at scale $\rho R^{-\delta}$. Let $\mathbb{T}_{B,trans}$ be the set of $\rho R^{-\delta}$-tubes T such that $T \subset \mathcal{P}_B$ and $T \not< T'$ for any $T' \in \mathbb{T}'_{B,tang}$. Let $\mathbb{T}_{B,tang}$ be the set of $\rho R^{-\delta}$-tubes T such that $T \subset \mathcal{P}_B$ and $T < T'$ for some $T' \in \mathbb{T}'_{B,tang}$. We define

$$\boldsymbol{g}_{B,trans} := \sum_{T \in \mathbb{T}_{B,trans}} (\boldsymbol{g}_U)_T, \quad \boldsymbol{g}_{B,tang} := \sum_{T \in \mathbb{T}_{B,tang}} (\boldsymbol{g}_U)_T. \tag{5.26}$$

For any $x \in \mathcal{P}_B$, one has

$$\boldsymbol{g}_U(x) = \boldsymbol{g}_{B,trans}(x) + \boldsymbol{g}_{B,tang}(x) + \mathrm{RapDec}(R)\|\mathrm{Sq}\boldsymbol{g}\|_2. \tag{5.27}$$

By the triangle inequality for broad function (3.35), for any $x \in B$, we have

$$\mathrm{Br}_A\mathrm{Sq}\boldsymbol{g}_U(x) \le \mathrm{Br}_{A/2}\mathrm{Sq}\boldsymbol{g}_{B,trans}(x) + \mathrm{Br}_{A/2}\mathrm{Sq}\boldsymbol{g}_{B,tang}(x) + \mathrm{RapDec}(R)\|\mathrm{Sq}\boldsymbol{g}\|_2. \tag{5.28}$$

Consequently, we have

$$\int_W |\mathrm{BrSq}\boldsymbol{g}_U|^p \lesssim \sum_B \int_B |\mathrm{BrSq}\boldsymbol{g}_{B,trans}|^p + \sum_B \int_B |\mathrm{BrSq}\boldsymbol{g}_{B,tang}|^p + \mathrm{RapDec}(R)\|\mathrm{Sq}\boldsymbol{g}\|_2^p. \tag{5.29}$$

Combining (5.14), (5.20) and (5.29), we get

$$\int_U |\mathrm{Br}_A\mathrm{Sq}\boldsymbol{g}_U|^p \lesssim \sum_{O \in \mathcal{O}} \int_O |\mathrm{Br}_A\mathrm{Sq}\boldsymbol{g}_O|^p + \sum_{B \in \mathcal{B}} \int_B |\mathrm{Br}_{A/2}\mathrm{Sq}\boldsymbol{g}_{B,trans}|^p + \sum_{B \in \mathcal{B}} \int_B |\mathrm{Br}_{A/2}\mathrm{Sq}\boldsymbol{g}_{B,tang}|^p + \mathrm{RapDec}(R)\|\mathrm{Sq}\boldsymbol{g}\|_2^p. \tag{5.30}$$

For simplicity, we just omit the rapid decaying term from now on. We determine which one of the three cases we are in according to which term on the right-hand side of (5.30) dominates.

Cellular case: If the first term on the right-hand side of (5.30) dominates, we say "we are in the cellular case". Together with (5.13), one has

$$\int_U |\mathrm{Br}_A \mathrm{Sq}\boldsymbol{g}_U|^p \lesssim \sum_{O\in\mathcal{O}} \int_O |\mathrm{Br}_A \mathrm{Sq}\boldsymbol{g}_O|^p \le \sum_{\widetilde{O}\in\widetilde{\mathcal{O}}} \int_{\widetilde{O}} |\mathrm{Br}_A \mathrm{Sq}\boldsymbol{g}_O|^p \sim \int_U |\mathrm{Br}_A \mathrm{Sq}\boldsymbol{g}_U|^p,$$

which implies

$$\int_U |\mathrm{Br}_A \mathrm{Sq}(\boldsymbol{g}_U)|^p \sim \sum_{O\in\mathcal{O}} \int_O |\mathrm{Br}_A \mathrm{Sq}(\boldsymbol{g}_O)|^p. \tag{5.31}$$

Also noting that $\int_O |\mathrm{Br}_A \mathrm{Sq}(\boldsymbol{g}_O)|^p \le \int_{\widetilde{O}} |\mathrm{Br}_A \mathrm{Sq}(\boldsymbol{g}_O)|^p$ and $|\mathcal{O}| = |\widetilde{\mathcal{O}}| \sim d^3$, by pigeonholing, we can choose a subset of $\mathcal{O}$ which we still denote by $\mathcal{O}$ such that (5.4) holds for every $O \in \mathcal{O}$, and (5.2) also holds. To show that (5.3) is true, we first use the local L^2 estimate, Lemma 2.1, to obtain

$$\begin{aligned}\sum_{O\in\mathcal{O}} \|\mathrm{Sq}\boldsymbol{g}_O\|_2^2 \sim \sum_{O\in\mathcal{O}} \|\mathrm{Sq}(1_{\mathcal{P}_O}\boldsymbol{g}_{O'})\|_2^2 &\lesssim d^{-1} \sum_{O\in\mathcal{O}} \big\|\mathrm{Sq}\boldsymbol{g}_{O'}\big\|_2^2 \\ &= d^{-1} \sum_{O\in\mathcal{O}} \Big\|\mathrm{Sq}\big(\sum_{T\in\mathbb{T}'_O} (\boldsymbol{g}_U)_T\big)\Big\|_2^2.\end{aligned}$$

Then, from Lemma 5.4 we note that each tube T belongs to $O(d)$ many sets $\mathbb{T}'_O$. This implies

$$\sum_{O\in\mathcal{O}} \Big\|\mathrm{Sq}\big(\sum_{T\in\mathbb{T}'_O} (\boldsymbol{g}_U)_T\big)\Big\|_2^2 \lesssim d\Big\|\mathrm{Sq}\big(\sum_{T\in\mathbb{T}} (\boldsymbol{g}_U)_T\big)\Big\|_2^2 = d\|\mathrm{Sq}\boldsymbol{g}_U\|_2^2. \tag{5.32}$$

Combining the above two inequalities, we prove (5.3).

Transverse case: If the second term on the right-hand side of (5.30) dominates, we say "we are in the transverse case". In this case, we have

$$\int_U |\mathrm{Br}_A \mathrm{Sq}(\boldsymbol{g}_U)|^p \lesssim \sum_{B\in\mathcal{B}} \int_B |\mathrm{Br}_{A/2} \mathrm{Sq}(\boldsymbol{g}_{B,trans})|^p. \tag{5.33}$$

In order to satisfy (5.7), we use pigeonhole principle on $\int_B |\mathrm{Br}_{A/2}\mathrm{Sq}(\boldsymbol{g}_{B,trans})|^p$ to find a subset of $\mathcal{B}$, which we still denote by $\mathcal{B}$, such that

$$\int_B |\mathrm{Br}_{A/2} \mathrm{Sq}(\boldsymbol{g}_{B,trans})|^p \text{ are comparable for } B \in \mathcal{B}, \tag{5.34}$$

and

$$\int_U |\mathrm{Br}_A \mathrm{Sq}(\boldsymbol{g}_U)|^p \lesssim \log R \sum_{B\in\mathcal{B}} \int_B |\mathrm{Br}_{A/2} \mathrm{Sq}(\boldsymbol{g}_{B,trans})|^p. \tag{5.35}$$

This verifies (5.7).

By Lemma 4.4 and the definition of $\mathbb{T}_{B,trans}$, we have for $x \in \mathcal{P}_B$

$$\boldsymbol{g}_{B,trans}(x) = \sum_{T\in\mathbb{T}_{B,trans}} (\boldsymbol{g}'_{B,trans})_T(x) + \mathrm{RapDec}(R)\|\mathrm{Sq}\boldsymbol{g}\|_2. \tag{5.36}$$

So, we have

$$\|\mathrm{Sq}\boldsymbol{g}_{B,trans}\|_2^2 \lesssim \|\mathrm{Sq}(1_{\mathcal{P}_B}\boldsymbol{g}'_{B,trans})\|_2^2.$$

To show (5.6), from Lemma 5.7 we note that each T belongs to at most Poly(d) many sets $\mathbb{T}'_{B,trans}$. Combining this fact with Lemma 2.1 and using the same reasoning as in the cellular case, one can show

$$\sum_{B\in\mathcal{B}}\|\mathrm{Sq}\boldsymbol{g}_{B,trans}\|_2^2\lesssim R^{-\delta}\sum_{B\in\mathcal{B}}\|\mathrm{Sq}(\boldsymbol{g}'_{B,trans})\|=R^{-\delta}\sum_{B\in\mathcal{B}}\|\mathrm{Sq}(\sum_{T\in\mathbb{T}'_{B,trans}}(\boldsymbol{g}_U)_T)\|_2^2$$

and

$$\sum_{B\in\mathcal{B}}\|\mathrm{Sq}(\sum_{T\in\mathbb{T}'_{B,trans}}(\boldsymbol{g}_U)_T)\|_2^2\lesssim \mathrm{Poly}(d)\|\mathrm{Sq}(\sum_{T\in\mathbb{T}}(\boldsymbol{g}_U)_T)\|_2^2=\mathrm{Poly}(d)\|\mathrm{Sq}\boldsymbol{g}_U\|_2^2.$$

Combining the above two inequalities, we prove (5.6).

Tangent case: If the third term on the right-hand side of (5.30) dominates, we say "we are in the tangent case". The tangent case is easier than cellular case and transverse case, because we don't need to prove a very strong L^2-relation like (5.3) or (5.6). The proof of (5.10) is the same as in the transverse case, so we omit the details here. □

REMARK 5.9. Actually our iteration process will immediately stop when we encounter the tangent case, which means we only have one tangent case in the iteration. So, the crude estimate for the tangent case is enough.

6. Polynomial partitioning iteration

In this section, we repeatedly use Algorithm 1 to build our iteration. At each step of the iteration, we endow one of the states: cellular state, transverse state and tangent state. The iteration ends when we arrive in the tangent state, or the scale is very small (slightly larger than M^2). We will discuss more carefully about these two scenarios later.

6.1. Iteration outputs.

Recall that $\boldsymbol{f}$ is defined in (3.34). Let us do scale R wave packet decomposition for $\boldsymbol{f}$, and make the following definition.

DEFINITION 6.1. Define $\boldsymbol{f}_{\mathcal{P}_R}$ as the sum of the wave packets $\boldsymbol{f}_T$ that $T\subset\mathcal{P}_R$.

By Definition 3.16, the broad norm only concerns the wave packets in direction σ, so $\mathrm{Br}_A\mathrm{Sq}\boldsymbol{f}(x)\sim\mathrm{Br}_A\mathrm{Sq}\boldsymbol{f}_{\mathcal{P}_R}(x)$ when $x\in\mathcal{P}_R$. Also we have $\mathrm{Br}_A\mathrm{Sq}_\sigma f(x)=\mathrm{Br}_A\mathrm{Sq}\boldsymbol{f}(x)$ if $x\in\mathcal{P}_R$, which was derived just after Definition 3.16. Therefore,

$$\int_{\mathcal{P}_R}|\mathrm{Br}_A\mathrm{Sq}_\sigma f|^p=\int_{\mathcal{P}_R}|\mathrm{Br}_A\mathrm{Sq}\boldsymbol{f}|^p\lesssim\int_{\mathcal{P}_R}|\mathrm{Br}_A\mathrm{Sq}\boldsymbol{f}_{\mathcal{P}_R}|^p. \tag{6.1}$$

THEOREM 6.2. Fix an integer A. Let $\boldsymbol{f}_{\mathcal{P}_R}=\{f_{1,\mathcal{P}_R},\cdots,f_{R,\mathcal{P}_R}\}$ be given above. We see that the Fourier support of each $f_{j,\mathcal{P}_R}$ is contained in $N_{R^\beta R^{-1}}(\Gamma_j(\sigma))$. Then we have the following outputs:

- There exists an integer s $(0\le s\le\varepsilon^{-10})$ which is the total number of iteration steps. There is a function STATE which we use to record the state of each step:

$$\mathrm{STATE}:\{1,2,\cdots,s\}\to\{\text{cell, trans, tang}\}. \tag{6.2}$$

We require the tangent case appear at most once, and if it appears, it should only appear at the last step. That is: $\mathrm{STATE}(u)=$ tang implies $u=s$. If the tangent case does not appear, then we end with a small radius $r_s\sim R^{\varepsilon/100}M^2$ (r_s is as below).

• At each step u, $u \in \{1, \cdots, s\}$, we have:

(1) A scale r_u for which the explicit formula is

$$r_u = Rd^{-s_c(u)}R^{-\delta s_t(u)}, \tag{6.3}$$

where $r_0 = R$ and the two parameters $s_c(u)$ and $s_t(u)$ are defined as

$$s_c(u) := \#\{1 \le i \le u : \text{ STATE}(i) = \text{cell}\},$$
$$s_t(u) := \#\{1 \le i \le u : \text{ STATE}(i) = \text{trans}\}.$$

(2) A number $a(u)$ defined as

$$a(u) = \#\{1 \le i \le u : \text{ STATE}(i) = \text{trans or tang}\}. \tag{6.4}$$

For convenience we also set $a(0) = 0$. We will see later that we consider the broad operator $\text{Br}_{A/2^{a(u)}}$ at step u.

(3) A set of cells $\mathcal{O}_u = \{O_u\}$ such that each O_u is contained in a rescaled r_u-ball $\mathcal{P}_{O_u}$. For convenience, we set $O_0 = \mathcal{P}_R$. Each O_u has a unique parent $O_{u-1} \in \mathcal{O}_{u-1}$, which we denote by

$$O_u < O_{u-1}. \tag{6.5}$$

Moreover we have the nested property for these cells. That is, for any cell $O_s \in \mathcal{O}_s$, there exist unique $O_u \in \mathcal{O}_u$ $(u = 1, 2, \cdots, s)$ such that

$$O_s < O_s < \cdots < O_1 < O_0 = \mathcal{P}_R. \tag{6.6}$$

(4) A set of r_u-tubes $\mathbb{T}_{O_u}$ and a set of functions $\{\boldsymbol{f}_{O_u}\}_{O_u \in \mathcal{O}_u}$ satisfying

$$\boldsymbol{f}_{O_u} = \sum_{T \in \mathbb{T}_{O_u}} (\boldsymbol{f}_{O_{u-1}})_T + \text{RapDec}(R)\|\text{Sq}\boldsymbol{f}\|_2. \tag{6.7}$$

Here $(\boldsymbol{f}_{O_{u-1}})_T$ is a scale r_u wave packet.

(5) For each step u, one of the following three states will happen: cellular state, transverse state and tangent state. The outputs for each state are the following:

Cellular state: If $\text{STATE}(u) = \text{cell}$, we have the following outputs.

(i) The cells $\mathcal{O}_u$ at step u and the preceding cells $\mathcal{O}_{u-1}$ satisfy the following quantitative relation:

$$|\mathcal{O}_u| \sim d^3|\mathcal{O}_{u-1}|. \tag{6.8}$$

(ii) We have the following L^2-relation between two nearby steps:

$$\sum_{O_u} \|\text{Sq}\boldsymbol{f}_{O_u}\|_2^2 \lesssim \sum_{O_{u-1}} \|\text{Sq}\boldsymbol{f}_{O_{u-1}}\|_2^2. \tag{6.9}$$

(iii) All $\int_{O_u} |\text{Br}_{A/2^{a(u)}}\text{Sq}(\boldsymbol{f}_{O_u})|^p$ $(O_u \in \mathcal{O}_u)$ are the same up to a constant factor, and

$$\sum_{O_{u-1}} \int_{O_{u-1}} |\text{Br}_{A/2^{a(u-1)}}\text{Sq}\boldsymbol{f}_{O_{u-1}}|^p \lesssim \sum_{O_u} \int_{O_u} |\text{Br}_{A/2^{a(u)}}\text{Sq}\boldsymbol{f}_{O_u}|^p \tag{6.10}$$

(iv) The Fourier transform of each component f_{j,O_u} of $\boldsymbol{f}_{O_u}$ satisfies:

$$\text{supp}\,\widehat{f}_{j,O_u} \subset N_{R^\beta r_u^{-1}}(\Gamma_j(\sigma)). \tag{6.11}$$

Transverse state: If $\text{STATE}(u) = \text{trans}$, we have the following outputs.

(i) There exists a number μ_u such that for any $O_{u-1} \in \mathcal{O}_{u-1}$, the quantity $\#\{O_u \in \mathcal{O}_u : O_u < O_{u-1}\}$ is either 0 or $\sim \mu_u$. Also, we have

$$(\log R)^{-2}\mu_u|\mathcal{O}_{u-1}| \lesssim |\mathcal{O}_u| \lesssim \mu_u|\mathcal{O}_{u-1}|. \tag{6.12}$$

(ii) We have the following L^2-relation between nearby steps:

$$\sum_{O_u} \|\mathrm{Sq}\boldsymbol{f}_{O_u}\|_2^2 \lesssim \mathrm{Poly}(d)R^{-\delta} \sum_{O_{u-1}} \|\mathrm{Sq}\boldsymbol{f}_{O_{u-1}}\|_2^2. \tag{6.13}$$

(iii) All the $\int_{O_u} |\mathrm{Br}_{A/2^{a(u)}}\mathrm{Sq}(\boldsymbol{f}_{O_u})|^p$ $(O_u \in \mathcal{O}_u)$ are same up to a constant factor, and

$$\sum_{O_{u-1}} \int_{O_{u-1}} |\mathrm{Br}_{A/2^{a(u-1)}}\mathrm{Sq}\boldsymbol{f}_{O_{u-1}}|^p \lesssim (\log R)^3 \sum_{O_u} \int_{O_u} |\mathrm{Br}_{A/2^{a(u)}}\mathrm{Sq}\boldsymbol{f}_{O_u}|^p. \tag{6.14}$$

(iv) The Fourier transform of each component f_{j,O_u} of $\boldsymbol{f}_{O_u}$ satisfies:

$$\mathrm{supp}\,\widehat{f}_{j,O_u} \subset N_{R^\beta r_u^{-1}}(\Gamma_j(\sigma)). \tag{6.15}$$

Tangent state: If STATE(u) = tang, which means $u = s$, then we have the following outputs.

(i) The cells $\mathcal{O}_s$ satisfies

$$|\mathcal{O}_s| \lesssim R^{3\delta}|\mathcal{O}_{s-1}|. \tag{6.16}$$

(ii) We have the following L^2-relation:

$$\sum_{O_s} \|\mathrm{Sq}\boldsymbol{f}_{O_s}\|_2^2 \lesssim R^{3\delta} \sum_{O_{s-1}} \|\mathrm{Sq}\boldsymbol{f}_{O_{s-1}}\|_2^2. \tag{6.17}$$

(iii) All the $\int_{O_s} |\mathrm{Br}_{A/2^{a(s)}}\mathrm{Sq}\boldsymbol{f}_{O_s}|^p$ $(O_s \in \mathcal{O}_s)$ are same up to a constant factor, and

$$\sum_{O_{s-1}} \int_{O_{s-1}} |\mathrm{Br}_{A/2^{a(s-1)}}\mathrm{Sq}\boldsymbol{f}_{O_{s-1}}|^p \lesssim (\log R)^3 \sum_{O_s} \int_{O_s} |\mathrm{Br}_{A/2^{a(s)}}\mathrm{Sq}\boldsymbol{f}_{O_s}|^p. \tag{6.18}$$

PROOF. We are going to iteratively apply the Algorithm 1 from the last section. Let us begin with the initial step $u = 1$.

Initial step: For convenience, we set $r_0 = R$, $O_0 = \mathcal{P}_R$, $\boldsymbol{f}_{O_0} = \boldsymbol{f}_{\mathcal{P}_R}$. Note that we already have a wave packet decomposition on $\boldsymbol{f}_{\mathcal{P}_R}$ (See Definition 6.1). Apply Algorithm 1 to the tuple $(R, O_0, \boldsymbol{f}_{O_0}, A)$. Then there are three possible cases: cellular case, transverse case and tangent case. We discuss them separately.

Cellular state: If Algorithm 1 results in cellular case, we set

$$\mathrm{STATE}(1) = \mathrm{cell}, \quad r_1 = Rd^{-1}. \tag{6.19}$$

From the cellular case in Algorithm 1, we obtain a collection of cells $\mathcal{O}$, tube sets $\{\mathbb{T}_O\}_{O\in\mathcal{O}}$ and a set of functions $\{\boldsymbol{g}_O\}_{O\in\mathcal{O}}$ that satisfy (5.2), (5.3), (5.4) and (5.5). We define

$$\mathcal{O}_1 := \mathcal{O}, \quad \{\mathbb{T}_{O_1}\}_{O_1\in\mathcal{O}_1} := \{\mathbb{T}_O\}_{O\in\mathcal{O}}, \quad \{\boldsymbol{f}_{O_1}\}_{O_1\in\mathcal{O}_1} := \{\boldsymbol{g}_O\}_{O\in\mathcal{O}}.$$

Also, note that $a(1) = 0$ by (6.4). We can check that (6.8), (6.9), (6.10) and (6.11) all holds for $u = 1$, since they are exactly (5.2), (5.3), (5.4) and (5.5) respectively. We can also verify (6.7), since it is exactly (5.19).

Transverse state: If Algorithm results in transverse case, we set

$$\mathrm{STATE}(1) = \mathrm{trans}, \quad r_1 = RR^{-\delta}. \tag{6.20}$$

From the transverse case in Algorithm 1, we obtain a collection of cells $\mathcal{B}$, tube sets $\{\mathbb{T}_{B,trans}\}_{B\in\mathcal{B}}$ and a set of functions $\{\boldsymbol{g}_{B,trans}\}_{B\in\mathcal{B}}$ that satisfy (5.6), (5.7) and (5.9). We define

$$\mathcal{O}_1 := \mathcal{B},\quad \{\mathbb{T}_{O_1}\}_{O_1\in\mathcal{O}_1} := \{\mathbb{T}_{B,trans}\}_{B\in\mathcal{B}},\quad \{\boldsymbol{f}_{O_1}\}_{O_1\in\mathcal{O}_1} := \{\boldsymbol{g}_{B,trans}\}_{B\in\mathcal{B}}.$$

Also, note that $a(1) = 1$ by (6.4). We can check that (6.13), (6.14) and (6.15) all holds for $u = 1$, since they are exactly (5.6), (5.7) and (5.9) respectively. To check (6.12), we just set

$$\mu_1 = |\mathcal{O}_1|, \tag{6.21}$$

and note that there is only one cell in $\mathcal{O}_0$: $\mathcal{O}_0 = \{\mathcal{P}_R\}$. We can also verify (6.7), since it is exactly (5.26).

Tangent state: If Algorithm 1 results in tangent case, the iteration stop, so $s = 1$. We set

$$\text{STATE}(1) = \text{tang},\quad r_1 = RR^{-\delta}. \tag{6.22}$$

The proof is essentially the same as for the transverse state. From the tangent case in Algorithm 1, we obtain a collection of cells $\mathcal{B}$, tube sets $\{\mathbb{T}_{B,tang}\}_{B\in\mathcal{B}}$ and a set of functions $\{\boldsymbol{g}_{B,tang}\}_{B\in\mathcal{B}}$ that satisfy (5.10), (5.11). We set

$$\mathcal{O}_1 := \mathcal{B},\quad \{\mathbb{T}_{O_1}\}_{O_1\in\mathcal{O}_1} := \{\mathbb{T}_{B,tang}\}_{B\in\mathcal{B}},\quad \{\boldsymbol{f}_{O_1}\}_{O_1\in\mathcal{O}_1} := \{\boldsymbol{g}_{B,tang}\}_{B\in\mathcal{B}}.$$

Also, note that $a(1) = 1$ by (6.4). We can check that (6.17) and (6.18) all hold for $u = 1$, since they are exactly (5.10), (5.11) and (5.12) respectively. To check (6.16), we just note that the number of tangent cells is $\lesssim R^{3\delta}$. We can also verify (6.7), since it is exactly (5.26).

The discussion of **Initial step** is finished.

Next let us move onto any intermediate step. We are going to see how to pass from step u to step $u+1$.

Iteration step: Suppose the iteration is done for step u, so we have a scale r_u, a set of cells $\mathcal{O}_u$, tube sets $\{\mathbb{T}_{O_u}\}_{O_u\in\mathcal{O}_u}$ and a set of functions $\{\boldsymbol{f}_{O_u}\}_{O_u\in\mathcal{O}_u}$. For each O_u, we apply Algorithm 1 to the group $(r_u, O_u, \boldsymbol{f}_{O_u}, A/2^{a(u)})$. To verify this is a valid input, we note that O_u is contained in a rescaled r_u-ball and also (6.11) and (6.15) verify the condition on the Fourier support of $\boldsymbol{f}_{O_u}$.

There are three possible states: cellular state, transverse state and tangent state, depending on the outputs of Algorithm on each O_u. We discuss them separately.

Cellular state: We say "the step $u+1$ is in the cellular state", if at least 1/3 of the cells $O_u \in \mathcal{O}_u$ are in the cellular case. Denote these cells by $\mathcal{O}'_u$ so we have $|\mathcal{O}'_u| \geq \frac{1}{3}|\mathcal{O}_u|$. We set

$$\text{STATE}(u+1) = \text{cell},\quad r_{u+1} = r_u d^{-1}. \tag{6.23}$$

Also note that

$$a(u+1) = a(u). \tag{6.24}$$

For each cell $O_u \in \mathcal{O}'_u$, we have the following outputs due to the Algorithm 1.

(1) We obtain a collection of cells denoted by $\mathcal{O}_{u+1}(O_u) = \{O\}$. They satisfy: $|\mathcal{O}_{u+1}(O_u)| \sim d^3$; each $O \in \mathcal{O}_{u+1}(O_u)$ is contained in a r_{u+1}- rescaled ball $\mathcal{P}_O$.

(2) We obtain tube sets $\{\mathbb{T}_O\}_{O\in\mathcal{O}_{u+1}(O_u)}$ and function sets $\{\boldsymbol{g}_O\}_{O\in\mathcal{O}_{u+1}(O_u)}$ that are indexed by cells in $\mathcal{O}_{u+1}(O_u)$. They satisfy an L^2 estimate

$$\sum_{O\in\mathcal{O}_{u+1}(O_u)} \|\mathrm{Sq}\boldsymbol{g}_O\|^2 \le \|\mathrm{Sq}\boldsymbol{f}_{O_u}\|_2^2, \tag{6.25}$$

and for each $O\in\mathcal{O}_{u+1}(O_u)$, a broad estimate

$$\int_{O_u} |\mathrm{Br}_{A/2^{a(u)}}\mathrm{Sq}\boldsymbol{f}_{O_u}|^p \sim d^3 \int_O |\mathrm{Br}_{A/2^{a(u+1)}}\mathrm{Sq}\boldsymbol{g}_O|^p. \tag{6.26}$$

(3) The Fourier support of each component $g_{j,O}$ of $\boldsymbol{g}_O$ is contained in the set $N_{R^\beta r_{u+1}^{-1}}(\Gamma_j(\sigma))$.

Now we define

$$\mathcal{O}_{u+1} := \bigcup_{O_u\in\mathcal{O}'_u} \mathcal{O}_{u+1}(O_u), \tag{6.27}$$

$$\{\mathbb{T}_{O_{u+1}}\}_{O_{u+1}\in\mathcal{O}_{u+1}} := \{\mathbb{T}_O\}_{O\in\mathcal{O}_{u+1}}, \tag{6.28}$$

$$\{\boldsymbol{f}_{O_{u+1}}\}_{O_{u+1}\in\mathcal{O}_{u+1}} := \{\boldsymbol{g}_O\}_{O\in\mathcal{O}_{u+1}}. \tag{6.29}$$

With all the properties above, we can prove the following results.

(1) Note that $|\mathcal{O}'_u|\ge\frac{1}{3}|\mathcal{O}_u|$ and $|\mathcal{O}_{u+1}(O_u)|\sim d^3$. We have

$$|\mathcal{O}_{u+1}| = \sum_{O_u\in\mathcal{O}'_u} |\mathcal{O}_{u+1}(O_u)| \sim d^3|\mathcal{O}'_u| \sim d^3|\mathcal{O}_u|, \tag{6.30}$$

which verifies (6.8).

(2) By inequality (6.25) and the definition in (6.29), we have

$$\begin{aligned}\sum_{O_{u+1}\in\mathcal{O}_{u+1}} \|\mathrm{Sq}\boldsymbol{f}_{O_{u+1}}\|_2^2 &= \sum_{O_u\in\mathcal{O}'_u}\ \sum_{O_{u+1}\in\mathcal{O}_{u+1}(O_u)} \|\mathrm{Sq}\boldsymbol{f}_{O_{u+1}}\|_2^2 \\ &\lesssim \sum_{O_u\in\mathcal{O}'_u} \|\mathrm{Sq}\boldsymbol{f}_{O_u}\|_2^2 \le \sum_{O_u\in\mathcal{O}_u} \|\mathrm{Sq}\boldsymbol{f}_{O_u}\|_2^2,\end{aligned} \tag{6.31}$$

which verifies (6.9).

(3) By (6.26) and noting that all the $\int_{O_u}|\mathrm{Br}_{A/2^{a(u)}}\mathrm{Sq}(\boldsymbol{f}_{O_u})|^p$ $(O_u\in\mathcal{O}_u)$ are same up to a constant factor, we have that all the

$$\int_{O_{u+1}} |\mathrm{Br}_{A/2^{a(u+1)}}\mathrm{Sq}(\boldsymbol{f}_{O_{u+1}})|^p$$

$(O_{u+1}\in\mathcal{O}_{u+1})$ are the same up to a constant. Also note that $|\mathcal{O}'_u|\ge\frac{1}{3}|\mathcal{O}_u|$. Thus, we have

$$\begin{aligned}\sum_{O_u\in\mathcal{O}_u}\int_{O_u} |\mathrm{Br}_{A/2^{a(u)}}\mathrm{Sq}\boldsymbol{f}_{O_u}|^p &\le 3\sum_{O_u\in\mathcal{O}'_u}\int_{O_u} |\mathrm{Br}_{A/2^{a(u)}}\mathrm{Sq}\boldsymbol{f}_{O_u}|^p \\ &\lesssim \sum_{O_{u+1}\in\mathcal{O}_{u+1}}\int_{O_{u+1}} |\mathrm{Br}_{A/2^{a(u-1)}}\mathrm{Sq}\boldsymbol{f}_{O_{u+1}}|^p.\end{aligned} \tag{6.32}$$

This verifies (6.10).

(4) One also sees that (6.11) holds for $u+1$.

At this point we finish the proof for *Cellular state.*

Transverse state: We say "the step $u+1$ is in the transverse state", if at least 1/3 of the cells $O_u \in \mathcal{O}_u$ are in the transverse case. Denote these cells by $\mathcal{O}'_u$ so we have $|\mathcal{O}'_u| \geq \frac{1}{3}|\mathcal{O}_u|$. We set

$$\text{STATE}(u+1) = \text{trans}, \quad r_{u+1} = r_u R^{-\delta}, \tag{6.33}$$

and note that

$$a(u+1) = a(u) + 1. \tag{6.34}$$

For each cell $O_u \in \mathcal{O}'_u$, we have the following outputs due to Algorithm 1.

(1) We obtain a collection of cells denoted by $\mathcal{O}_{u+1}(O_u) = \{B\}$. Each B is contained in a rescaled r_{u+1}-ball.

(2) We obtain tube sets $\{\mathbb{T}_{B,trans}\}_{B\in\mathcal{O}_{u+1}(O_u)}$ and corresponding function sets $\{\boldsymbol{g}_{B,trans}\}_{B\in\mathcal{O}_{u+1}(O_u)}$, which are indexed by cells in $\mathcal{O}_{u+1}(O_u)$. They satisfy an L^2 estimate

$$\sum_{B\in\mathcal{O}_{u+1}(O_u)} \|\text{Sq}\boldsymbol{g}_{B,trans}\|_2^2 \leq \text{Poly}(d)R^{-\delta}\|\text{Sq}\boldsymbol{f}_{O_u}\|_2^2, \tag{6.35}$$

and for each $B \in \mathcal{O}_{u+1}(O_u)$, a broad estimate

$$\int_{O_u} |\text{Br}_{A/2^{a(u)}}\text{Sq}\boldsymbol{f}_{O_u}|^p \lesssim \log R \sum_{B\in\mathcal{O}_{u+1}(O_u)} \int_B |\text{Br}_{A/2^{a(u+1)}}\text{Sq}\boldsymbol{g}_{B,trans}|^p. \tag{6.36}$$

Also, the quantity

$$\int_B |\text{Br}_{A/2^{a(u+1)}}\text{Sq}\boldsymbol{g}_{B,trans}|^p \tag{6.37}$$

are comparable for all B.

(3) The Fourier transform of each component $g_{j,B,trans}$ of $\boldsymbol{g}_{B,trans}$ is contained in $N_{R^\beta r_{u+1}^{-1/2}}\Gamma_j(\sigma)$.

To derive our outputs, we need to work a bit harder than in the cellular state. We use pigeonhole principle twice to guarantee the uniformity properties (6.12), (6.14).

By dyadic pigeonholing, we can find a subset $\mathcal{O}''_u \subset \mathcal{O}'_u$ with $|\mathcal{O}''_u| \gtrsim (\log)^{-2}|\mathcal{O}'_u|$, such that $\int_B |\text{Br}_{A/2^{a(u+1)}}\text{Sq}(\boldsymbol{g}_{B,trans})|^p$ are comparable for all

$$B \in \bigcup_{O_u\in\mathcal{O}''_u} \mathcal{O}_{u+1}(O_u),$$

and there is a dyadic number μ_{u+1} such that

$$|\mathcal{O}_{u+1}(O_u)| \sim \mu_{u+1} \text{ for every } O_u \in \mathcal{O}''_u, \tag{6.38}$$

We define respectively

$$\mathcal{O}_{u+1} := \bigcup_{O_u\in\mathcal{O}''_u} \mathcal{O}_{u+1}(O_u), \tag{6.39}$$

$$\{\mathbb{T}_{O_{u+1}}\}_{O_{u+1}\in\mathcal{O}_{u+1}} := \{\mathbb{T}_{B,trans}\}_{B\in\mathcal{O}_{u+1}}, \tag{6.40}$$

$$\{\boldsymbol{f}_{O_{u+1}}\}_{O_{u+1}\in\mathcal{O}_{u+1}} := \{\boldsymbol{g}_{B,trans}\}_{B\in\mathcal{O}_{u+1}}. \tag{6.41}$$

With all the properties above, we can show the following results.

(1) By definition, we have $|\mathcal{O}_{u+1}| \sim \mu_{u+1}|\mathcal{O}''_u|$. Since $(\log R)^{-2}|\mathcal{O}'_u| \lesssim |\mathcal{O}''_u| \lesssim |\mathcal{O}'_u|$ and $|\mathcal{O}_u| \sim |\mathcal{O}'_u|$, we can verify (6.12).

(2) By inequality (6.35) and the definition in (6.41), we have

$$\sum_{O_{u+1}\in\mathcal{O}_{u+1}} \|\mathrm{Sq}\boldsymbol{f}_{O_{u+1}}\|_2^2 = \sum_{O_u\in\mathcal{O}''_u} \sum_{O\in\mathcal{O}_{u+1}(O_u)} \|\mathrm{Sq}\boldsymbol{f}_{O_{u+1}}\|_2^2$$
$$\lesssim \mathrm{Poly}(d)R^{-\delta} \sum_{O_u\in\mathcal{O}''_u} \|\mathrm{Sq}\boldsymbol{f}_{O_u}\|_2^2 \le \mathrm{Poly}(d)R^{-\delta} \sum_{O_u\in\mathcal{O}_u} \|\mathrm{Sq}\boldsymbol{f}_{O_u}\|_2^2, \tag{6.42}$$

which verifies (6.13).

(3) By (6.36) and noting that all $\int_{O_u} |\mathrm{Br}_{A/2^{a(u)}}\mathrm{Sq}(\boldsymbol{f}_{O_u})|^p$ $(O_u \in \mathcal{O}_u)$ are same up to a constant factor, we have

$$\sum_{O_u\in\mathcal{O}_u} \int_{O_u} |\mathrm{Br}_{A/2^{a(u)}}\mathrm{Sq}\boldsymbol{f}_{O_u}|^p \lesssim (\log R)^2 \sum_{O_u\in\mathcal{O}''_u} \int_{O_u} |\mathrm{Br}_{A/2^{a(u)}}\mathrm{Sq}\boldsymbol{f}_{O_u}|^p$$
$$\lesssim (\log R)^3 \sum_{O_{u+1}\in\mathcal{O}_{u+1}} \int_{O_{u+1}} |\mathrm{Br}_{A/2^{a(u+1)}}\mathrm{Sq}\boldsymbol{f}_{O_{u+1}}|^p. \tag{6.43}$$

This verifies (6.14).

(4) One also sees that (6.15) holds for $u+1$.

We finish the proof for *Transverse state.*

Tangent state: We say "the step $u+1$ is in the tangent state", if at least 1/3 of the cells $O_u \in \mathcal{O}_u$ are in the tangent case. Actually, one may not encounter the tangent state throughout the iteration. However, once the tangent state appears, the iteration stops and so we have $u = s$.

We can proceed in exactly the same way as we did for the transverse state. We define $\mathcal{O}_s$, $\{\boldsymbol{f}_{O_s}\}_{O_s\in\mathcal{O}_s}$ and $\{\mathbb{T}_{O_s}\}$ in the same way as we did for the transverse state (see (6.39), (6.40) and (6.41)). The proofs for the properties (6.17) and (6.18) are the same as in the transverse case, so we omit the details. For the proof of (6.16), we just note that each cell O_{s-1} has at most $R^{3\delta}$ children O_s from $\mathcal{O}_s$. □

6.2. The main results from the iteration. Let us make a conclusion of the polynomial partitioning iteration. There are two scenarios. The first one is when STATE(s)=tang, which means the iteration ends up in the tangent state. The second one is when STATE(s) ≠tang, which means the iteration ends up with a tiny scale $r_s \le R^{\varepsilon/10}M^2$.

We define two numbers

$$s_c := \#\{1 \le u \le s : \mathrm{STATE}(u) = \mathrm{cell}\}, \tag{6.44}$$

$$s_t := \#\{1 \le u \le s : \mathrm{STATE}(u) = \mathrm{trans}\}. \tag{6.45}$$

It implies that

$$r_s = Rd^{-s_c}R^{-\delta s_t}. \tag{6.46}$$

From now on we write $r = r_s$ for simplicity. Henceforth, recalling (6.1), we can make the conclusion:

If we are in the *first scenario*, we have the theorem:

THEOREM 6.3 (Tangent case). *If* STATE(s) = tang, *and* s_c, s_t, r *are defined as above, we have*

$$\int_{\mathcal{P}_R} |\mathrm{Br}_A \mathrm{Sq} f|^p \lessapprox (\log R)^{3s_t} \sum_{O_s \in \mathcal{O}_s} \int_{O_s} |\mathrm{Br}_{A/2^{a(s)}} \mathrm{Sq} \boldsymbol{f}_{O_s}|^p. \tag{6.47}$$

$$\int_{O_s} |\mathrm{Br}_{A/2^{a(s)}} \mathrm{Sq} \boldsymbol{f}_{O_s}|^p \ (O_s \in \mathcal{O}_s) \text{ are the same up to a constant.} \tag{6.48}$$

$$\sum_{O_s \in \mathcal{O}_s} \|\mathrm{Sq} \boldsymbol{f}_{O_s}\|_2^2 \lessapprox \mathrm{Poly}(d)^{s_t} R^{-\delta s_t} \|\mathrm{Sq} \boldsymbol{f}_{\mathcal{P}_R}\|_2^2. \tag{6.49}$$

$$R^{-\beta} d^{3s_c} \prod_{\mathrm{STATE}(u)=\mathrm{trans}} \mu_u \lessapprox |\mathcal{O}_s| \lessapprox d^{3s_c} \prod_{\mathrm{STATE}(u)=\mathrm{trans}} \mu_u. \tag{6.50}$$

If we are in the *second scenario*, we have the theorem:

THEOREM 6.4 (Small-r case). *If* $\mathrm{STATE}(u) \neq \mathrm{tang}$ *for all* $u \in \{1, \cdots, s\}$, *and* s_c, s_t, r *are defined as above, we have*

$$\int_{\mathcal{P}_R} |\mathrm{Br}_A \mathrm{Sq} f|^p \lessapprox (\log R)^{3s_t} \sum_{O_s \in \mathcal{O}_s} \int_{O_s} |\mathrm{Br}_{A/2^{a(s)}} \mathrm{Sq} \boldsymbol{f}_{O_s}|^p. \tag{6.51}$$

$$\int_{O_s} |\mathrm{Br}_{A/2^{a(s)}} \mathrm{Sq} \boldsymbol{f}_{O_s}|^p \ (O_s \in \mathcal{O}_s) \text{ are the same up to a constant.} \tag{6.52}$$

$$\sum_{O_s \in \mathcal{O}_s} \|\mathrm{Sq} \boldsymbol{f}_{O_s}\|_2^2 \lessapprox \mathrm{Poly}(d)^{s_t} R^{-\delta s_t} \|\mathrm{Sq} \boldsymbol{f}_{\mathcal{P}_R}\|_2^2. \tag{6.53}$$

$$R^{-\beta} d^{3s_c} \prod_{\mathrm{STATE}(u)=\mathrm{trans}} \mu_u \lessapprox |\mathcal{O}_s| \lessapprox d^{3s_c} \prod_{\mathrm{STATE}(u)=\mathrm{trans}} \mu_u. \tag{6.54}$$

$$r \lessapprox R^{\varepsilon/10} M^2. \tag{6.55}$$

7. End when the radius r_s is small

In this section we discuss the second scenario which is much easier than the first scenario. We will use the ineqaulities stated in Theorem 6.4. First, we consider (6.51). Since terms on the right-hand side of (6.51) are the same up to a constant factor, by pigeonholing, there exists a cell O_s with minimal $\|\mathrm{Sq} \boldsymbol{f}_{O_s}\|_2^2$ such that

$$\int_{\mathcal{P}_R} |\mathrm{Br}_A \mathrm{Sq}_\sigma f|^p \lessapprox R^{O(\delta)} |\mathcal{O}_s| \int_{O_s} |\mathrm{Br}_{A/2^{a(s)}} \mathrm{Sq} \boldsymbol{f}_{O_s}|^p \lessapprox R^{O(\delta)} |\mathcal{O}_s| \int_{O_s} |\mathrm{Sq} \boldsymbol{f}_{O_s}|^p. \tag{7.1}$$

The minimality of $\|\mathrm{Sq} \boldsymbol{f}_{O_s}\|_2^2$ and (6.53) yield that

$$\|\mathrm{Sq} \boldsymbol{f}_{O_s}\|_2^2 \lessapprox R^{O(\delta)} R^{-s_t \delta} |\mathcal{O}_s|^{-1} \|\mathrm{Sq} \boldsymbol{f}_{\mathcal{P}_R}\|_2^2. \tag{7.2}$$

To estimate each term on the right-hand side of (7.1), we need the following lemma.

LEMMA 7.1. *For* $2 \le p \le \infty$,

$$\|1_{O_s} \mathrm{Sq} \boldsymbol{f}_{O_s}\|_p \lessapprox M^{4(\frac{1}{p} - \frac{1}{2})} \|\mathrm{Sq} \boldsymbol{f}_{O_s}\|_2 \tag{7.3}$$

PROOF. Recalling the definition of $\mathrm{Sq}\boldsymbol{f}_{O_s}$ in Definition 3.15, we have

$$\|1_{O_s}\mathrm{Sq}\boldsymbol{f}_{O_s}\|_p=\Big(\int_{O_s}\Big(\sum_j|f_{j,O_s}|^2\Big)^{p/2}\Big)^{1/p}. \tag{7.4}$$

By Hölder's inequality, it suffices to prove (7.3) when $p=2$ and $p=\infty$. The case $p=2$ follows easily from Placherel.

Let us consider the case for $p=\infty$. We note that $\widehat{f}_{j,O_s}$ is supported in a slab of dimensions $M^{-1}\times M^{-1}\times M^{-2}$, so via Bernstein's inequality,

$$\|f_{j,O_s}\|_\infty\lesssim M^{-2}\|f_{j,O_s}\|_2. \tag{7.5}$$

Now we can estimate $\|\mathrm{Sq}(\boldsymbol{f}_{O_s})1_{O_s}\|_p$ as

$$\|\mathrm{Sq}(\boldsymbol{f}_{O_s})1_{O_s}\|_\infty=\sup_{x\in O_s}\Big(\sum_j|f_{j,O_s}(x)|^2\Big)^{1/2}\le\Big(\sum_j\sup_{x\in O_s}|f_{j,O_s}(x)|^2\Big)^{1/2}, \tag{7.6}$$

which, using the Berstein's estimate (7.5), is bounded above by

$$M^{-2}\Big(\sum_j\|f_{j,O_s}\|_2^2\Big)^{1/2}=M^{-2}\|\mathrm{Sq}\boldsymbol{f}_{O_s}\|_2. \tag{7.7}$$

This is the desired estimate when $p=\infty$. □

Combining (7.1), (7.2) and (7.3), we have

$$\int_{\mathcal{P}_R}|\mathrm{Br}_A\mathrm{Sq}_\sigma f|^p\lesssim R^{O(\delta)}R^{-\frac{ps_t\delta}{2}}|\mathcal{O}_s|^{1-\frac{p}{2}}M^{4-2p}\|\mathrm{Sq}\boldsymbol{f}_{\mathcal{P}_R}\|_2^{\frac{p}{2}}. \tag{7.8}$$

Now recalling $d^{s_c}R^{s_t\delta}=Rr^{-1}\ge M^{-2}R^{1-\varepsilon/10}$ and $|\mathcal{O}_s|\gtrsim d^{3s_c}$ in (6.54), simple calculations yield that for $p>3$,

$$\int_{\mathcal{P}_R}|\mathrm{Br}_A\mathrm{Sq}f|^p\lesssim R^{p\varepsilon/3}R^{-p/2}M^{4-p}\|\mathrm{Sq}\boldsymbol{f}_{\mathcal{P}_R}\|_2^{\frac{p}{2}}. \tag{7.9}$$

Recall that $\boldsymbol{f}_{\mathcal{P}_R}$ is the sum of the wave packets $\boldsymbol{f}_T$ such that $T\subset\mathcal{P}_R$. Hence one can choose a function $\varphi_{\mathcal{P}_R}$ such that

(1) $\varphi_{\mathcal{P}_R}\lesssim w_{\mathcal{P}_R}$, recalling (3.29),
(2) $\widehat{\varphi}_{\mathcal{P}_R}$ is supported a the dual slab of $\mathcal{P}_R$ centered at the origin,
(3) $\|\mathrm{Sq}\boldsymbol{f}_{\mathcal{P}_R}\|_2^2\lesssim\|\varphi_{\mathcal{P}_R}\mathrm{Sq}\boldsymbol{f}\|_2^2$.

By L^2 orthogonality,

$$\|\varphi_{\mathcal{P}_R}\mathrm{Sq}\boldsymbol{f}\|_2^2\lesssim\int\sum_{\theta\in\Theta}\sum_{j=1}^R|m_{j,\theta}*f|^2\varphi_{\mathcal{P}_R}\lesssim\int\sum_{\theta\in\Theta}\sum_{j=1}^R|m_{j,\theta}*f|^2w_{\mathcal{P}_R}. \tag{7.10}$$

We let $\widetilde{T}_\theta$ be the tube in $\mathbb{T}_\theta$ that is centered at the origin. Then the kernel estimate Lemma 3.6 yields that for fixed θ, the collection of smooth functions $\{R^{-O(\beta)}m_{j,\theta}\}_{1\le j\le R}$ is adapted to $\widetilde{T}_\theta$. By Lemma 2.3 and recalling (2.29), (3.21), we can argue similarly as in Lemma 3.9 to get

$$\int\sum_{\theta\in\Theta}\sum_{j=1}^R|m_{j,\theta}*f|^2w_{\mathcal{P}_R}\lesssim R^{O(\beta)}\int\sum_{\theta\in\Theta}\sum_{q\in\mathbf{q}(\theta)}|\Delta_qf|^2\widetilde{w}_{\widetilde{T}_\theta,N}*w_{\mathcal{P}_R}. \tag{7.11}$$

Note that tube $\widetilde{T}_\theta$ is contained in the rescaled ball $\mathcal{P}_R$. Hence $\sup_\theta \widetilde{w}_{\widetilde{T}_\theta,N} * w_{\mathcal{P}_R} \lesssim w_{\mathcal{P}_R}$, which, via Lemma 3.8, implies

$$\int \sum_{\theta\in\Theta} \sum_{q\in\mathbf{q}(\theta)} |\Delta_q f|^2 \widetilde{w}_{\widetilde{T}_\theta,N} * w_{\mathcal{P}_R} \lesssim \int \sum_{q\in\mathbf{q}} |\Delta_q f|^2 w_{\mathcal{P}_R}. \tag{7.12}$$

Combining (7.9), (7.10), (7.11), (7.12) and using Hölder's inequality, we obtain

$$\int_{\mathcal{P}_R} |\mathrm{Br}_A \mathrm{Sq}_\sigma f|^p \le C_\varepsilon R^{p\varepsilon} R^{p-3} M^{6-2p} \int \Big(\sum_{q\in\mathbf{q}} |\Delta_q f|^2\Big)^{p/2} w_{\mathcal{P}_R},$$

which is just Theorem 3.13. □

8. Backward algorithm

In this section, we discuss the first scenario. That is, the iteration ends with STATE(s)=tang. We use the corresponding estimates Theorem 6.3. However, when doing so, we lose some global information among cells in $\mathcal{O}_s$. The new ingredient is to build up a backward algorithm as in [**Wu23**] to sum up $\|\mathrm{Sq}\boldsymbol{f}_{O_s}\|_2^2$ for $O_s \in \mathcal{O}_s$ efficiently.

For any $a \in [-R, R]$, define L_a as the horizontal plane

$$L_a := \{x : x_3 = a\}. \tag{8.1}$$

For any tube T, we define $\mathring{T}$ as a stretch of T: the tube $\mathring{T}$ has the same coreline and cross section as T, but has infinite length.

In the backward algorithm, we are going to find a refinement $\bar{\mathcal{O}}_u \subset \mathcal{O}_u$ for each step u. And for each $O_u \in \bar{\mathcal{O}}_u$, we will build up an auxiliary tube set $\bar{\mathbb{T}}_{O_u}$ and vector-valued functions $\boldsymbol{h}_{O_u}$. This is discussed in the next theorem.

THEOREM 8.1. *For every $u \in \{1, \cdots, s-1\}$, we can find a refinement of cell $\bar{\mathcal{O}}_u \subset \mathcal{O}_u$ and a refinement of tubes $\{\bar{\mathbb{T}}_{O_u}\}_{O_u\in\bar{\mathcal{O}}_u}$ with $\bar{\mathbb{T}}_{O_u} \subset \mathbb{T}_{O_u}$. Define the vector-valued functions $\{\boldsymbol{h}_{O_u}\}_{O_u\in\bar{\mathcal{O}}_u}$ as*

$$\boldsymbol{h}_{O_u} := \sum_{T\in\bar{\mathbb{T}}_{O_u}} (\boldsymbol{f}_{O_{u-1}})_T; \tag{8.2}$$

These $\boldsymbol{h}_{O_u}$ satisfy the following properties depending on whether we are in cellular state or transverse state at step $u+1$, where $1 \le u \le s-1$.

Cellular state*: If* STATE$(u+1)$ = cell, *we can also find an integer v_{u+1}, so that the following two properties are satisfied:*

$$\sum_{O\in\bar{\mathcal{O}}_{u+1}} \|\mathrm{Sq}\boldsymbol{h}_O\|_2^2 \lesssim R^{O(\beta)} d^{-1} v_{u+1} \sum_{O\in\bar{\mathcal{O}}_u} \|\mathrm{Sq}\boldsymbol{h}_O\|_2^2. \tag{8.3}$$

Define $L_{a,O_u} := L_a \bigcap \{\cup_{T\in\bar{\mathbb{T}}_{O_u}} \mathring{T}\}$. Then uniformly for $a \in [-R, R]$,

$$\max_{O_u\in\bar{\mathcal{O}}_u} \big|N_{Rr_u^{-1/2}} L_{a,O_u}\big| \lesssim d^3 v_{u+1}^{-1} \max_{O_{u+1}\in\bar{\mathcal{O}}_{u+1}} \big|N_{Rr_{u+1}^{-1/2}} L_{a,O_{u+1}}\big|. \tag{8.4}$$

Transverse state*: If* STATE$(u+1)$ = trans, *the following two properties are satisfied:*

$$\sum_{O\in\bar{\mathcal{O}}_{u+1}} \|\mathrm{Sq}\boldsymbol{h}_O\|_2^2 \lesssim \mathrm{Poly}(d) R^{-\delta} \sum_{O\in\bar{\mathcal{O}}_u} \|\mathrm{Sq}\boldsymbol{h}_O\|_2^2. \tag{8.5}$$

Define $L_{a,O_u} := L_a \bigcap \{\cup_{T\in\bar{\mathbb{T}}_{O_u}} \mathring{T}\}$. *Then uniformly for* $a \in [-R, R]$,

$$\max_{O_u\in\bar{\mathcal{O}}_u} \left|N_{Rr_u^{-1/2}} L_{a,O_u}\right| \lesssim \mu_{u+1} \max_{O_{u+1}\in\bar{\mathcal{O}}_{u+1}} \left|N_{Rr_{u+1}^{-1/2}} L_{a,O_{u+1}}\right|. \tag{8.6}$$

Tangent state*: If* STATE$(u + 1)$ = tang *which means* $u + 1 = s$*, the following two properties are satisfied:*

$$\sum_{O\in\mathcal{O}_s} \|\mathrm{Sq}\boldsymbol{f}_O\|_2^2 \lesssim R^{O(\delta)} \sum_{O\in\bar{\mathcal{O}}_{s-1}} \|\mathrm{Sq}\boldsymbol{h}_O\|_2^2. \tag{8.7}$$

Define $L_{a,O_s} := L_a \bigcap \{\cup_{T\in\bar{\mathbb{T}}_{O_s}} \mathring{T}\}$. *Then uniformly for* $a \in [-R, R]$,

$$\max_{O_{s-1}\in\bar{\mathcal{O}}_{s-1}} \left|N_{Rr_{s-1}^{-1/2}} L_{a,O_{s-1}}\right| \lesssim \max_{O_s\in\mathcal{O}_s} \left|N_{Rr_s^{-1/2}} L_{a,O_s}\right|. \tag{8.8}$$

We divide the proof of Theorem 8.1 into two main cases: cellular case and transverse case. The proof relies on a backward induction from the last step s to the first step.

8.1. The base case at step $s-1$**.** In step $s - 1$, which is the first step of the backward algorithm, we set $\bar{\mathcal{O}}_{s-1} := \mathcal{O}_{s-1}$. For each $O_{s-1} \in \bar{\mathcal{O}}_{s-1}$, we set

$$\bar{\mathbb{T}}_{O_{s-1}} := \{T' \in \mathbb{T}_{O_{s-1}} : \exists T \in \mathbb{T}_{O_s}, T < T', \text{ for some } O_s < O_{s-1}\}. \tag{8.9}$$

We use the set $\bar{\mathbb{T}}_{O_{s-1}}$ to define the vector $\boldsymbol{h}_{O_{s-1}}$ as in (8.2). For $O_s < O_{s-1}$, by (6.7) we have

$$\boldsymbol{f}_{O_s} = \sum_{T\in\mathbb{T}_{O_s}} (\boldsymbol{f}_{O_{s-1}})_T + \mathrm{RapDec}(R)\|\mathrm{Sq}\boldsymbol{f}\|_2. \tag{8.10}$$

By (6.7) again, we have

$$\boldsymbol{f}_{O_{s-1}} = \sum_{T'\in\mathbb{T}_{O_{s-1}}} (\boldsymbol{f}_{O_{s-2}})_{T'} + \mathrm{RapDec}(R)\|\mathrm{Sq}\boldsymbol{f}\|_2.$$

Plugging into (8.10) and by Lemma 4.4, we have

$$\begin{aligned} \boldsymbol{f}_{O_s} &= \sum_{T\in\mathbb{T}_{O_s}} \Big(\sum_{T'\in\bar{\mathbb{T}}_{O_{s-1}}} (\boldsymbol{f}_{O_{s-2}})_{T'}\Big)_T + \mathrm{RapDec}(R)\|\mathrm{Sq}\boldsymbol{f}\|_2^2 \\ &= \sum_{T\in\mathbb{T}_{O_s}} (\boldsymbol{h}_{O_{s-1}})_T + \mathrm{RapDec}(R)\|\mathrm{Sq}\boldsymbol{f}\|_2. \end{aligned} \tag{8.11}$$

By L^2 orthogonality, one gets

$$\|\mathrm{Sq}\boldsymbol{f}_{O_s}\|_2^2 \le \|\mathrm{Sq}\boldsymbol{h}_{O_{s-1}}\|_2^2 + \mathrm{RapDec}(R)\|\mathrm{Sq}\boldsymbol{f}\|_2. \tag{8.12}$$

Note that for each fixed $O_{s-1} \in \mathcal{O}_{s-1}$ there are at most $R^{O(\delta)}$ cells $O_s \in \mathcal{O}_s$ with $O_s < O_{s-1}$. Summing over O_s, we obtain

$$\sum_{O\in\mathcal{O}_s} \|\mathrm{Sq}\boldsymbol{f}_O\|_2^2 \lesssim R^{O(\delta)} \sum_{O\in\bar{\mathcal{O}}_{s-1}} \|\mathrm{Sq}\boldsymbol{h}_O\|_2^2 + \mathrm{RapDec}(R)\|\mathrm{Sq}\boldsymbol{f}\|_2^2, \tag{8.13}$$

which is (8.7).

To prove (8.8), we note the fact: for $T \in \mathbb{T}_{O_s}, T' \in \mathbb{T}_{O_{s-1}}$ such that $T < T'$, we have $N_{Rr_{s-1}^{-1/2}} T' \cap \{-R \le x_3 \le R\} \subset N_{Rr_s^{-1/2}} T \cap \{-R \le x_3 \le R\}$. This immediately gives that

$$\max_{O_{s-1}\in\bar{\mathcal{O}}_{s-1}} \left|N_{Rr_{s-1}^{-1/2}} L_{a,O_{s-1}}\right| \lesssim \max_{O_s\in\mathcal{O}_s} \left|N_{Rr_s^{-1/2}} L_{a,O_s}\right|. \tag{8.14}$$

Suppose we have done for step $u+1$. Now we discuss step u. There are two cases.

8.2. Cellular state at step $u+1$. Suppose we are in the situation that STATE$(u+1)=$ cell. Fix a cell $O_u \in \mathcal{O}_u$. To build $\bar{\mathbb{T}}_{O_u}$, our idea is to study the relations between bigger tubes $\mathbb{T}_{O_u}$ and smaller tubes $\{\bar{\mathbb{T}}_{O_{u+1}}\}_{O_{u+1}<O_u}$. Define

$$\bar{\mathcal{O}}_{u+1}(O_u) := \{O_{u+1} \in \bar{\mathcal{O}}_{u+1} : O_{u+1} < O_u\}. \tag{8.15}$$

By induction hypothesis, we have

$$\boldsymbol{h}_{O_{u+1}} = \sum_{T \in \bar{\mathbb{T}}_{O_{u+1}}} (\boldsymbol{f}_{O_u})_T. \tag{8.16}$$

By (6.7), we have

$$\boldsymbol{h}_{O_{u+1}} = \sum_{T \in \bar{\mathbb{T}}_{O_{u+1}}} \Big(\sum_{T' \in \mathbb{T}_{O_u}} (\boldsymbol{f}_{O_{u-1}})_{T'} \Big)_T + \mathrm{RapDec}(R)\|\mathrm{Sq}\boldsymbol{f}\|_2^2. \tag{8.17}$$

For simplicity, let's just ignore the rapidly decaying term. By orthogonality, we have

$$\|\boldsymbol{h}_{O_{u+1}}\|_2^2 \lesssim \sum_{T \in \bar{\mathbb{T}}_{O_{u+1}}} \Big\| \Big(\sum_{T' \in \mathbb{T}_{O_u}} (\boldsymbol{f}_{O_{u-1}})_{T'} \Big)_T \Big\|_2^2. \tag{8.18}$$

Now we sort the bigger tubes in $\mathbb{T}_{O_u}$ via the following definition.

Definition 8.2. For a dyadic integer v, $1 \le v \le O(d)$, we define a subcollection $\mathbb{T}_{O_u}^v \subset \mathbb{T}_{O_u}$ as

$$\mathbb{T}_{O_u}^v := \{T' : \#\{O_{u+1} : T' > T \text{forsome } T \in \mathbb{T}_{O_{u+1}}, O_{u+1} \in \bar{\mathcal{O}}_{u+1}(O_u)\} \sim v\}.$$

Let us return to (8.18). We first sum over all $O_{u+1} \in \mathcal{O}_{u+1}(O_u)$ and then use pigeonholing and the triangle inequality. There is a dyadic number $v = v_{O_u}$ such that

$$\sum_{O_{u+1} \in \bar{\mathcal{O}}_{u+1}(O_u)} \|\boldsymbol{h}_{O_{u+1}}\|_2^2 \lesssim \log R \sum_{O_{u+1} \in \bar{\mathcal{O}}_{u+1}(O_u)} \sum_{T \in \bar{\mathbb{T}}_{O_{u+1}}} \Big\| \Big(\sum_{T' \in \mathbb{T}_{O_u}^v} (\boldsymbol{f}_{O_{u-1}})_{T'} \Big)_T \Big\|_2^2. \tag{8.19}$$

We need some other notations. Let $\{\tau\}$ be a set of $r_{u+1}^{-1/2}$-caps that we chose in the wave packet decomposition. The direction of each r_{u+1}-tube $T \in \bar{\mathbb{T}}_{O_{u+1}}$ corresponds to one τ, which we denote by $\omega(T) = \tau$. We set $\bar{\mathbb{T}}_{O_{u+1},\tau} := \{T \in \bar{\mathbb{T}}_{O_{u+1}} : \omega(T) = \tau\}$. For O_{u+1}, τ and $T' \in \mathbb{T}_{O_u}$, we set $\chi(O_{u+1}, \tau, T') = 1$ if there exists $T \in \bar{\mathbb{T}}_{O_{u+1},\tau}$ such that $T < T'$; otherwise we set $\chi(O_{u+1}, \tau, T') = 0$. Our goal is to show

$$\sum_{T \in \bar{\mathbb{T}}_{O_{u+1}}} \Big\| \Big(\sum_{T' \in \mathbb{T}_{O_u}^v} (\boldsymbol{f}_{O_{u-1}})_{T'} \Big)_T \Big\|_2^2 \lesssim R^\beta d^{-1} \sum_{T' \in \mathbb{T}_{O_u}^v} \sum_{\tau} \chi(O_{u+1}, \tau, T') \|(\boldsymbol{f}_{O_{u-1}})_{T'}\|_2^2. \tag{8.20}$$

The left-hand side of (8.20) equals

$$\sum_{\tau} \sum_{T \in \bar{\mathbb{T}}_{O_{u+1},\tau}} \Big\| \Big(\sum_{T' \in \mathbb{T}_{O_u}^v} (\boldsymbol{f}_{O_{u-1}})_{T'} \Big)_T \Big\|_2^2.$$

By Lemma 4.4, we see it is morally

$$\sim \sum_{\tau} \sum_{T\in\bar{\mathbb{T}}_{O_{u+1},\tau}} \Big\|\Big(\sum_{T'\in\mathbb{T}^v_{O_u}} \chi(O_{u+1},\tau,T')(\boldsymbol{f}_{O_{u-1}})_{T'}\Big)_T\Big\|_2^2.$$

By Lemma 4.2, it is less than

$$\lesssim \sum_{\tau} \sum_{T\in\bar{\mathbb{T}}_{O_{u+1},\tau}} \Big\| 1_{10T} \sum_{T'\in\mathbb{T}^v_{O_u}:T<T'} \chi(O_{u+1},\tau,T')(\boldsymbol{f}_{O_{u-1}})_{T'}\Big\|_2^2.$$

For fixed τ, $\{10T : \omega(T) = \tau\}$ are parallel tubes that are also finitely overlapping, so the above is less than

$$\lesssim \sum_{\tau} \Big\| 1_{\bigcup_{T\in\bar{\mathbb{T}}_{O_{u+1},\tau}} 10T} \sum_{T'\in\mathbb{T}^v_{O_u}:T<T'} \chi(O_{u+1},\tau,T')(\boldsymbol{f}_{O_{u-1}})_{T'}\Big\|_2^2.$$

Since $\bigcup_{\bar{\mathbb{T}}_{O_{u+1},\tau}} 10T$ is morally contained in $\mathcal{P}_{O_{u+1}}$ which a tube of length r_{u+1} radius $M^{-1}r_{u+1}$, and each $(f_{j,O_{u-1}})_{T'}$ has Fourier support in $N_{R^\beta r_u^{-1}}\Gamma_j$. Since $R^\beta r_u^{-1} r_{u+1} = R^\beta d^{-1}$, by Lemma 2.1 we have the quantity above is less than

$$\lesssim d^{-1}R^\beta \sum_{\tau} \Big\| \sum_{T'\in\mathbb{T}^v_{O_u}:T<T'} \chi(O_{u+1},\tau,T')(\boldsymbol{f}_{O_{u-1}})_{T'}\Big\|_2^2,$$

which further by orthogonality is bounded by the right-hand side of (8.20). Hence we finish the proof of (8.20).

We sum over $O_{u+1} \in \bar{\mathcal{O}}_{u+1}(O_u)$ in (8.20) and plug into (8.19), so that

$$\sum_{O_{u+1}\in\bar{\mathcal{O}}_{u+1}(O_u)} \|\boldsymbol{h}_{O_{u+1}}\|_2^2 \lesssim R^{O(\beta)}d^{-1} \sum_{T'\in\mathbb{T}^v_{O_u}} \sum_{O_{u+1}\in\bar{\mathcal{O}}_{u+1}(O_u)} \sum_{\tau} \chi(O_{u+1},\tau,T')\|(\boldsymbol{f}_{O_{u-1}})_{T'}\|_2^2.$$

By the definition of $\mathbb{T}^v_{O_u}$ and noting that $\sum_\tau \chi(O_{u+1},\tau,T') \le 1$, we have

$$\sum_{O_{u+1}\in\bar{\mathcal{O}}_{u+1}(O_u)} \sum_{\tau} \chi(O_{u+1},\tau,T') \lesssim v,$$

so we obtain

$$\sum_{O_{u+1}\in\bar{\mathcal{O}}_{u+1}(O_u)} \|\boldsymbol{h}_{O_{u+1}}\|_2^2 \lesssim R^{O(\beta)}d^{-1} \sum_{T'\in\mathbb{T}^v_{O_u}} v\|(\boldsymbol{f}_{O_{u-1}})_{T'}\|_2^2.$$

Summing over O_u and noting that $v = v_{O_u}$ depends on O_u, we get

$$\sum_{O_{u+1}\in\bar{\mathcal{O}}_{u+1}} \|\boldsymbol{h}_{O_{u+1}}\|_2^2 \lesssim R^{O(\beta)}d^{-1} \sum_{O_u\in\mathcal{O}_u} \sum_{T'\in\mathbb{T}^{v_{O_u}}_{O_u}} v_{O_u}\|(\boldsymbol{f}_{O_{u-1}})_{T'}\|_2^2.$$

By pigeonholing on v_{O_u}, we can find a number v_{u+1}. If we define $\bar{\mathcal{O}}_u := \{O_u \in \mathcal{O}_u : v_{O_u} \sim v_{u+1}\}$, $\bar{\mathbb{T}}_{O_u} := \mathbb{T}^{v_{u+1}}_{O_u}$ and $\boldsymbol{h}_{O_u} := \sum_{T'\in\bar{\mathbb{T}}_{O_u}} (\boldsymbol{f}_{O_{u-1}})_{T'}$, then we have

$$\sum_{O_{u+1}\in\bar{\mathcal{O}}_{u+1}} \|\boldsymbol{h}_{O_{u+1}}\|_2^2 \lesssim R^{O(\beta)}d^{-1}v_{u+1} \sum_{O_u\in\bar{\mathcal{O}}_u} \|\boldsymbol{h}_{O_u}\|_2^2, \tag{8.21}$$

which verifies (8.3).

Next, we prove (8.4). Actually we will prove the following stronger inequality:

$$\max_{O_u\in\bar{\mathcal{O}}_u} \big|N_{Rr_{u+1}^{-1/2}} L_{a,O_u}\big| \lesssim R^{O(\beta)}d^3 v_{u+1}^{-1} \max_{O_{u+1}\in\bar{\mathcal{O}}_{u+1}} \big|N_{Rr_{u+1}^{-1/2}} L_{a,O_{u+1}}\big|,$$

where the $N_{Rr_u^{-1/2}}$ on the left-hand side of (8.4) is replaced by $N_{Rr_{u+1}^{-1/2}}$.

Fix an $O_u \in \bar{\mathcal{O}}_u$. We choose a maximal collection of $Rr_{u+1}^{-1/2}$-separated points in L_{a,O_u}, denoted by $\{y_l\}_{l=1}^m$. For each y_l, we can pick one tube $T_l \in \bar{\mathbb{T}}_{O_u} = \mathbb{T}_{O_u}^{v_{u+1}}$ so that $y_l \in \mathring{T}_l \cap L_a$. By the definition of $\mathbb{T}_{O_u}^{v_{u+1}}$, for each T_l we can find $\sim v_{u+1}$ sets $\{\bar{\mathbb{T}}_{O_{u+1}} : O_{u+1} < O_u,\ O_{u+1} \in \bar{\mathcal{O}}_{u+1}\}$ and a tube $T_{l,j}$ in each set $\bar{\mathbb{T}}_{O_{u+1}}$, such that $T_{l,j} < T_l$ for $j = 1, \cdots, O(v_{u+1})$. Pick a point $z_{l,j}$ in $L_a \cap \mathring{T}_{l,j}$. Now the total number of points we picked is $\#\{z_{l,j}\} \sim m v_{u+1}$. We will prove

$$\#\{z_{l,j}\} \lesssim \frac{d^3}{(Rr_{u+1}^{-1/2})^2} \max_{O_{u+1}\in\bar{\mathcal{O}}_{u+1}} \left|N_{Rr_{u+1}^{-1/2}} L_{a,O_{u+1}}\right|. \tag{8.22}$$

This immediately implies

$$\max_{O_u\in\bar{\mathcal{O}}_u} \left|N_{10Rr_{u+1}^{-1/2}} L_{a,O_u}\right| \lesssim m(Rr_{u+1}^{-1/2})^2 \lesssim d^3 v_{u+1}^{-1} \max_{O_{u+1}\in\bar{\mathcal{O}}_{u+1}} \left|N_{10Rr_{u+1}^{-1/2}} L_{a,O_{u+1}}\right|$$

as desired.

To prove (8.22), we define for each cell $O_{u+1} \in \bar{\mathcal{O}}_{u+1}(O_u)$ a set $Z_{O_{u+1}}$ consisting of points $z_{l,j}$ whose associated tube $T_{l,j}$ belongs to $\bar{\mathbb{T}}_{O_{u+1}}$:

$$Z_{O_{u+1}} := \{z_{l,j} : T_{l,j} \in \bar{\mathbb{T}}_{O_{u+1}}\}. \tag{8.23}$$

A crucial fact is that the points in a single set $Z_{O_{u+1}}$ have different subscripts l, since they come from different T_l by definition. We fix the point set $Z_{O_{u+1}}$. Via the assumption $T_{l,j} < T_l$ one has $\mathring{T}_{l,j} \cap L_a \subset (Rr_{u+1}^{-1/2})\mathring{T}_l \cap L_a$. Also, since the sets $\{\mathring{T}_l \cap L_a\}$ are $Rr_{u+1}^{-1/2}$ separated, we get that the set $\{\mathring{T}_{l,j} \cap L_a : z_{l,j} \in Z_{O_{u+1}}\}$ are $\sim Rr_{u+1}^{-1/2}$ separated. Hence,

$$\#Z_{O_{u+1}} \sim \frac{1}{(Rr_{u+1}^{-1/2})^2} \left|N_{Rr_{u+1}^{-1/2}}\Big(L_a \bigcap \Big\{\bigcup_{T_{l,j}:z_{l,j}\in Z_{O_{u+1}}} \mathring{T}_{l,j}\Big\}\Big)\right|, \tag{8.24}$$

which is bounded above by

$$\frac{1}{(Rr_{u+1}^{-1/2})^2} \left|N_{Rr_{u+1}^{-1/2}}\Big(L_a \bigcap \Big\{\bigcup_{T\in\bar{\mathbb{T}}_{O_{u+1}}} \mathring{T}\Big\}\Big)\right| \le \frac{1}{(Rr_{u+1}^{-1/2})^2} \left|N_{Rr_{u+1}^{-1/2}} L_{a,O_{u+1}}\right|$$

Finally, note that there are $O(d^3)$ many cells in $\bar{\mathcal{O}}_{u+1}(O_u)$. Since

$$\#\{z_{l,j}\} \le \sum_{O_{u+1}\in\bar{\mathcal{O}}_{u+1}(O_u)} \#Z_{O_{u+1}}, \tag{8.25}$$

we can sum up $\#Z_{O_{u+1}}$ for every cell $O_{u+1} \in \bar{\mathcal{O}}_{u+1}(O_u)$ using the above estimates to prove (8.22). □

Next, we consider the transverse state.

8.3. Transverse state at step $u+1$.

Suppose we are in the situation that STATE$(u+1)$ = trans. So any $O_u \in \mathcal{O}_u$ has $\sim \mu_{u+1}$ children $O_{u+1} \in \mathcal{O}_{u+1}$.

We set $\bar{\mathcal{O}}_u := \mathcal{O}_u$ and $\bar{\mathbb{T}}_{O_u} := \mathbb{T}_{O_u}$, so we have

$$\boldsymbol{h}_{O_u} := \sum_{T\in\bar{\mathbb{T}}_{O_u}} (\boldsymbol{f}_{O_{u-1}})_T, \tag{8.26}$$

which is morally $\boldsymbol{f}_{O_u}$.

Since the wave packets in the summation of $\boldsymbol{h}_{O_{u+1}}$ is a subset of the wave packets in the summation of $\boldsymbol{f}_{O_{u+1}}$, we have

$$\sum_{O_{u+1}\in\bar{\mathcal{O}}_{u+1}} \|\mathrm{Sq}\boldsymbol{h}_{O_{u+1}}\|_2^2 \le \sum_{O_{u+1}\in\bar{\mathcal{O}}_{u+1}} \|\mathrm{Sq}\boldsymbol{f}_{O_{u+1}}\|_2^2. \tag{8.27}$$

Combining it with (6.13), we get

$$\sum_{O_{u+1}\in\bar{\mathcal{O}}_{u+1}} \|\mathrm{Sq}\boldsymbol{h}_{O_{u+1}}\|_2^2 \lesssim \mathrm{Poly}(d)R^{-\delta} \sum_{O_u\in\bar{\mathcal{O}}_u} \|\mathrm{Sq}\boldsymbol{h}_{O_u}\|_2^2. \tag{8.28}$$

We can also trivially obtain

$$\max_{O_u\in\bar{\mathcal{O}}_u} \left|N_{Rr_u^{-1/2}}L_{a,O_u}\right| \lesssim \mu_u \max_{O_{u+1}\in\bar{\mathcal{O}}_{u+1}} \left|N_{Rr_{u+1}^{-1/2}}L_{a,O_{u+1}}\right| \tag{8.29}$$

similarly as we did in the cell case. One just need to note that each O_u have $O(\mu_u)$ many children $O_{u+1}\in\bar{\mathcal{O}}_{u+1}$.

8.4. Conclusion. The backward algorithm stops when $u=1$, from which we obtain the refinements $\bar{\mathcal{O}}_1\subset\mathcal{O}_1$, $\{\bar{\mathbb{T}}_{O_1}\}_{O_1\in\bar{\mathcal{O}}_1}$ and $\{\boldsymbol{h}_{O_1}\}_{O_1\in\bar{\mathcal{O}}_1}$. Set

$$v=\prod_u v_u \lesssim d^{s_c}, \tag{8.30}$$

where the product is taken over all $u\in\{1,\cdots,s\}$ such that STATE(u) =cell.

Note that $(\log R)^{O(s)}R^{O(s_c\beta)}\lesssim R^{O(\delta)}$. Iterate (8.3) and (8.5) and use the base case (8.13) to obtain

$$\sum_{O\in\bar{\mathcal{O}}_s} \|\mathrm{Sq}\boldsymbol{f}_O\|_2^2 \lesssim R^{O(\delta)}d^{-s_c}R^{-s_t\delta}v \sum_{O\in\bar{\mathcal{O}}_1} \|\mathrm{Sq}\boldsymbol{h}_O\|_2^2. \tag{8.31}$$

Since there are less than $R^{O(\delta)}$ cells in $\bar{\mathcal{O}}_1$, by pigeonholing, there exists a cell $O_1\in\bar{\mathcal{O}}_1$ such that

$$\sum_{O\in\bar{\mathcal{O}}_s} \|\mathrm{Sq}(\boldsymbol{f}_O)\|_2^2 \lesssim R^{O(\delta)}d^{-s_c}R^{-s_t\delta}v\|\mathrm{Sq}(\boldsymbol{h}_{O_1})\|_2^2. \tag{8.32}$$

Fix this cell O_1 and define

$$\bar{\mathbb{T}} := \{T\in\mathbb{T}_{O_0} : \exists T'\in\bar{\mathbb{T}}_{O_1}, T'<T\},$$

$$\boldsymbol{h} := \sum_{T\in\bar{\mathbb{T}}} \boldsymbol{f}_T. \tag{8.33}$$

We have

$$\boldsymbol{h}_{O_1} = \sum_{T\in\bar{\mathbb{T}}_{O_1}} \boldsymbol{f}_T = \sum_{T\in\bar{\mathbb{T}}_{O_1}} \boldsymbol{h}_T + \mathrm{RapDec}(R)\|\mathrm{Sq}\boldsymbol{f}\|_2, \tag{8.34}$$

and thus

$$\|\mathrm{Sq}(\boldsymbol{h}_{O_1})\|_2^2 \lesssim \|\mathrm{Sq}\boldsymbol{h}\|_2^2.$$

Plugging back to (8.32), we obtain

$$\sum_{O\in\bar{\mathcal{O}}_s} \|\mathrm{Sq}(\boldsymbol{f}_O)\|_2^2 \lesssim R^{O(\delta)}d^{-s_c}R^{-s_t\delta}v\|\mathrm{Sq}\boldsymbol{h}\|_2^2. \tag{8.35}$$

(8.35) is the L^2 estimate we obtained from the backward algorithm. For the support estimate, define

$$L_{a,\mathcal{P}_R} := L_a \bigcap \Big\{ \bigcup_{T\in\bar{\mathbb{T}}} T \Big\}. \tag{8.36}$$

Iterate (8.4), (8.6) and (8.8) to obtain
(8.37)

$$|N_{R^{1/2}}L_{a,\mathcal{P}_R}| \lesssim R^{O(\delta)}|N_{Rr_1^{-1/2}}L_{a,O_1}| \lesssim R^{O(\delta)}d^{3s_c}\Big(\prod_u \mu_u\Big)v^{-1} \max_{O\in\mathcal{O}_s} \big|N_{Rr_s^{-1/2}}L_{a,O}\big|,$$

where the product is taken over u with STATE(u) =trans.

We need Lemma 8.3 to estimate $|N_{Rr_s^{-1/2}}L_{a,O}|$. Its proof is postponed to the appendix.

LEMMA 8.3. *Fix any cell* $O_s \in \bar{O}_s$. *Recall* $L_{a,O_s} = L_a \bigcap \{\cup_{T\in\mathbb{T}_{O_s}} \mathring{T}\}$, *and set* $r = r_s$. *Then uniformly for every* $a \in [-R, R]$, *one has*

$$\big|N_{Rr_s^{-1/2}}L_{a,O_s}\big| \lesssim M^{-1}R^{O(\delta)}R^2r^{-1/2}. \tag{8.38}$$

Recalling (6.54), we use Lemma 8.3 to bound (8.37) so that

$$|N_{R^{1/2}}L_{a,\mathcal{P}_R}| \lesssim M^{-1}R^{O(\delta)}R^2r^{-1/2}|\mathcal{O}_s|v^{-1}. \tag{8.39}$$

It yields that, if we define

$$X = \bigcup_{T\in\bar{\mathbb{T}}} R^\beta T, \tag{8.40}$$

then we can obtain the support estimate from the backward algorithm

$$|X| \lesssim R^{O(\delta)}R^3 \min(M^{-2}, M^{-1}r^{-1/2}|\mathcal{O}_s|v^{-1}). \tag{8.41}$$

9. Concluding the proof

We conclude the proof of Theorem 3.13 in this section. Let us first estimate the right-hand side of (8.35). For convenience, we define $\bar{\mathbb{T}}_\theta := \{T \in \bar{\mathbb{T}} : \omega(T) = \theta\}$. By (8.33) and the wave packet decomposition, we have

$$\begin{aligned} \|\mathrm{Sq}\boldsymbol{h}\|_2^2 &\lesssim R^{O(\delta)} \sum_{\theta\in\Theta}\sum_{T\in\bar{\mathbb{T}}_\theta} \int |\mathrm{Sq}\boldsymbol{f}_\theta|^2 1_T^* \\ &= R^{O(\delta)} \sum_{\theta\in\Theta}\sum_{j=1}^{R} \int |m_{j,\theta} * f|^2 \Big(\sum_{T\in\bar{\mathbb{T}}_\theta} 1_T^*\Big). \end{aligned} \tag{9.1}$$

Here the kernel $m_{j,\theta}$ was defined in (3.4). Via the kernel estimate (3.14) and Definition 2.2, we see that the kernels $\{R^{-O(\beta)}m_{j,\theta}\}_{j=1}^R$ are adapted to the tube $T_{0,\theta}$, where $T_{0,\theta}$ is the $R^{1/2+\beta}\times R^{1/2+\beta}\times R$ tube centered at the origin with direction c_θ. Now via Lemma 2.3 and recalling (2.29) and (3.21), we can argue similarly as in Lemma 3.9 to obtain

$$\begin{aligned} \|\mathrm{Sq}\boldsymbol{h}\|_2^2 &\lesssim R^{O(\delta)} \sum_{\theta\in\Theta}\sum_{q\in\mathbf{q}(\theta)} \int |\Delta_q f|^2 \widetilde{w}_{T_{0,\theta},N} * \Big(\sum_{T\in\bar{\mathbb{T}}_\theta} 1_T^*\Big) \\ &\lesssim R^{O(\delta)} \sum_{\theta\in\Theta}\sum_{q\in\mathbf{q}(\theta)} \int |\Delta_q f|^2 \Big(\sum_{T\in\bar{\mathbb{T}}_\theta} w_{T,N}\Big). \end{aligned} \tag{9.2}$$

We define X_θ as

$$X_\theta := \bigcup_{T\in\bar{\mathbb{T}}_\theta} R^\beta T, \tag{9.3}$$

so that when $x \in \mathbb{R}^3 \setminus X_\theta$,

$$\sum_{T\in\bar{\mathbb{T}}_\theta} w_{T,N}(x) \le \mathrm{RapDec}(R) w_{\mathcal{P}_R}(x). \tag{9.4}$$

Plugging this back to (9.2), we get

$$\|\mathrm{Sq}\boldsymbol{h}\|_2^2 \lesssim R^{O(\delta)} \sum_{\theta\in\Theta}\sum_{q\in\mathbf{q}(\theta)} \int |\Delta_q f|^2 1_{X_\theta}. \tag{9.5}$$

To bound the right-hand side of (9.5), we trivially bound 1_{X_θ} by 1_X and use Hölder's inequality so that

$$\begin{aligned}\sum_{\theta\in\Theta}\sum_{q\in\mathbf{q}(\theta)} \int |\Delta_q f|^2 1_{X_\theta} &\lesssim \int \Big(\sum_{q\in\mathbf{q}} |\Delta_q f|^2\Big) 1_X \\ &\lesssim \Big(\int \Big(\sum_{q\in\mathbf{q}} |\Delta_q f|^2\Big)^{p/2} w_{\mathcal{P}_R}\Big)^{2/p} |X|^{\frac{p-2}{p}}.\end{aligned} \tag{9.6}$$

Now let us move back to (8.35). Combining (9.5) and (9.6), one can use pigeonhole principle to find a cell $O_s \in \mathcal{O}_s$ such that

$$\begin{aligned}\|\mathrm{Sq}\boldsymbol{f}_{O_s}\|_2^2 \lesssim\ & R^{O(\delta)} d^{-s_c} R^{-s_t\delta} v |\mathcal{O}_s|^{-1} |X|^{\frac{p-2}{p}} \\ &\times \Big(\int \Big(\sum_{q\in\mathbf{q}} |\Delta_q f|^2\Big)^{p/2} w_{\mathcal{P}_R}\Big)^{2/p}.\end{aligned} \tag{9.7}$$

Recall (6.47). To bound $\|1_{O_s}\mathrm{Br}_{A/2^{a(s)}}\mathrm{Sq}\boldsymbol{f}_{O_s}\|_p^p$, we need the following lemma, whose proof is postponed to next section.

Lemma 9.1 (Bilinear estimate). *Let $O \in \mathcal{O}_s$ be any step-s cell (a tangent cell) and let $\boldsymbol{f}_O$ be the associated vector. Then*

$$\int_O \mathrm{Bil}[\mathrm{Sq}\boldsymbol{f}_O]^p \lesssim R^{O(\delta)} r^{-\frac{5(p-2)}{4}} M^{\frac{p-2}{2}} \|\mathrm{Sq}\boldsymbol{f}_O\|_2^p. \tag{9.8}$$

Since we know from (6.48) that $1_{O_s}\mathrm{Br}_{A/2^{a(s)}}\mathrm{Sq}(\boldsymbol{f}_{O_s})$ has morally the same L^p norm. Together with Lemma 3.18, (6.47), (9.7) and (9.8), we obtain

$$\begin{aligned}\int_{\mathcal{P}_R} |\mathrm{Br}_A \mathrm{Sq} f|^p \lesssim\ & R^{O(\delta)} |\mathcal{O}_s|^{\frac{2-p}{2}} M^{\frac{p-2}{2}} r^{\frac{5}{2}-\frac{5p}{4}} (d^{-s_c}R^{-s_t\delta}v)^{\frac{p}{2}} |X|^{\frac{p-2}{2}} \\ &\times \int \Big(\sum_{q\in\mathbf{q}} |\Delta_q f|^2\Big)^{p/2} w_{\mathcal{P}_R}.\end{aligned} \tag{9.9}$$

Since $r = Rd^{-s_c}R^{-s_t\delta}$ and since (8.41), it remains to prove the following lemma so that one can conclude Theorem 3.13 from (9.9).

Lemma 9.2. *Assuming the notation as above, we have*

$$\begin{aligned}&|\mathcal{O}_s|^{\frac{2-p}{2}} (R/r)^{p-3} v^{p/2} \big(\min(M^{-1}r^{1/2}, |\mathcal{O}_s| v^{-1})\big)^{\frac{p-2}{2}} \\ &\lesssim R^{O(\delta)} M^{6-2p} R^{p-3}.\end{aligned} \tag{9.10}$$

Proof. We consider two cases.

9.1. $M^{-1}r^{1/2} \geq |\mathcal{O}_s|v^{-1}$**.** This implies

$$r \geq M^2|\mathcal{O}_s|^2v^{-2}. \tag{9.11}$$

Recall (6.54), we know that $|\mathcal{O}_s| \gtrsim R^{-\beta}d^{3s_c}$, which implies.

$$rR^{O(\delta)} \geq M^2 d^{6s_c}v^{-2} \tag{9.12}$$

Thus, combining the fact $v \lesssim d^{s_c}$ from (8.30), one can estimate (9.10) as

$$\begin{aligned}(9.10) = (R/r)^{p-3}v &\lesssim R^{O(\delta)}M^{6-2p}R^{p-3}(d^{s_c})^{-6(p-3)}v^{2(p-3)+1} \\ &\lesssim R^{O(\delta)}M^{6-2p}R^{p-3}(d^{s_c})^{-6(p-3)+2(p-3)+1} \\ &= R^{O(\delta)}M^{6-2p}R^{p-3}(d^{s_c})^{13-4p}.\end{aligned} \tag{9.13}$$

It implies (9.10) when $p \geq 3.25$.

9.2. $M^{-1}r^{1/2} \leq |\mathcal{O}_s|v^{-1}$**.** Recalling (6.50), we first set

$$\mu = \prod_{\text{STATE}(u)=\text{trans}} \mu_u \tag{9.14}$$

In this case, we have

$$(9.10) = (M|\mathcal{O}_s|)^{\frac{2-p}{2}}R^{p-3}r^{\frac{10-3p}{4}}v^{\frac{p}{2}} \tag{9.15}$$

Note that the condition $M^{-1}r^{1/2} \leq |\mathcal{O}_s|v^{-1}$ implies

$$(M|\mathcal{O}_s|)^{\frac{10-3p}{2}}v^{-\frac{10-3p}{2}}r^{-\frac{10-3p}{2}} \geq 1. \tag{9.16}$$

Multiplying this inequality with (9.15), we get

$$(9.10) \leq R^{p-3}M^{6-2p}|\mathcal{O}_s|^{6-2p}r^{-\frac{10-3p}{4}}v^{2p-5}. \tag{9.17}$$

Since $|\mathcal{O}_s| \gtrsim R^{-\beta}d^{3s_c}$ and $v \lesssim d^{s_c}$, we have

$$|\mathcal{O}_s|^{6-2p}v^{2p-5} \lesssim R^{O(\delta)}(d^{s_c})^{13-4p} \leq R^{O(\delta)},$$

when $p > 3.25$. Also $r^{-\frac{10-3p}{4}} \leq 1$ when $p \leq 10/3$.

We obtain that

$$(9.10) \lesssim R^{O(\delta)}R^{p-3}M^{6-2p} \tag{9.18}$$

when $p \geq 3.25$ and $p \leq 10/3$. □

10. A bilinear estimate

To deal with the tangent case in Lemma 9.1, we need a slightly stronger bilinear estimate than the one in [**Gut16**] Lemma 3.10. Specifically, let Π_1 and Π_2 be two smooth compact curves with positive curvature in $\mathbb{R}^2$, such that any of their normal vectors n_1, n_2 satisfy $\text{Angle}(n_1, n_2) \sim (MK)^{-1}$. For $k = 1, 2$, let $W_k = \{w_k\}$ be a collection of disjoint rectangles of dimensions $\rho^{-1/2} \times M^{-1}\rho^{-1/2}$ contained in $N_{M^{-1}\rho^{-1/2}}(\Pi_j)$ with $\rho > K$ (See Figure 1). Then we have the following geometric lemma.

LEMMA 10.1. *Each $w_1 + w_2$ is contained in a $\rho^{-1/2} \times M^{-1}\rho^{-1/2}$-rectangle. Also, each $w_1 + w_2$ can intersect only $O(K^2)$ many other sets. That is,*

$$\sum_{(w_1,w_2)\in W_1\times W_2} 1_{w_1+w_2} \lesssim K^2. \tag{10.1}$$

In [**Gut16**] Lemma 3.10, Guth proved Lemma 10.1 when Π_1 and Π_2 are both portions of a single smooth curve, and when $M = 1$. The linear version of Lemma 10.1 was first observed in [**Car83**].

PROOF. After rotation and translation, without loss of generality, we can assume that all the normal vectors of Π_1 point to the left-hand side of the vertical axis while all the normal vectors of Π_2 point to the right-hand side. Also, each normal vector of either Π_1 or Π_2 makes an angle $\sim (MK)^{-1}$ with respect to the vertical axis. As a consequence, the tangent vectors of Π_1 make an angle $\sim (MK)^{-1}$ with the horizontal direction, and the tangent vectors of Π_2 also make an angle $\sim (MK)^{-1}$ with the horizontal axis. At this point, each w_j is morally a horizontal rectangle of dimensions $\rho^{-1/2} \times M^{-1}\rho^{-1/2}$.

It suffices to prove (10.1) when its right-hand side is replaced by 1 and assuming the rectangles in W_1, W_2 are K^{-1}-separated. We let $c_{w_1} = ([c_{w_1}]_1, [c_{w_1}]_2)$ and $c_{w_2} = ([c_{w_2}]_1, [c_{w_2}]_2)$ be the center of w_1 and w_2 respectively. Then points in $\{[c_{w_k}]_1\}_{w_k \in W_k}$ are $K\rho^{-1/2}$ separated and points in $\{[c_{w_k}]_2\}_{w_k \in W_k}$ are $M^{-1}\rho^{-1/2}$ separated, for any $k = 1, 2$. Also, for any $w_1, w_1' \in W_1$, $[c_{w_1}]_1 > [c_{w_1'}]_1$ implies $[c_{w_1}]_2 > [c_{w_1'}]_2$. Conversely, for any $w_2, w_2' \in W_2$, $[c_{w_2}]_1 > [c_{w_2'}]_1$ implies $[c_{w_2}]_2 < [c_{w_2'}]_2$.

Clearly $w_1 + w_2$ is contained in a $\rho^{-1/2} \times M^{-1}\rho^{-1/2}$ horizontal rectangle. Note that $(w_1 + w_2) \cap (w_1' + w_2') \neq \varnothing$ only if

$$[c_{w_1}]_1 + [c_{w_2}]_1 = [c_{w_1'}]_1 + [c_{w_2'}]_1 + O(\rho^{-1/2}) \tag{10.2}$$

and

$$[c_{w_1}]_2 + [c_{w_2}]_2 = [c_{w_1'}]_2 + [c_{w_2'}]_2 + O(M^{-1}\rho^{-1/2}). \tag{10.3}$$

By the geometric observation on the sets $\{[c_{w_j}]_1\}$ and $\{[c_{w_j}]_2\}$ above, there are finitely many (w_1', w_2') satisfying $(w_1 + w_2) \cap (w_1' + w_2') \neq \varnothing$. This concludes the proof of the lemma. □

Using Lemma 10.1, we can prove Lemma 9.1. Let us recall Lemma 10.2.

LEMMA 10.2 (Bilinear estimate). *Let $O \in \mathcal{O}_{s+1}$ be any step $s+1$ cell and let $\boldsymbol{f}_O$ be the associated vector. Then*

$$\int_O \mathrm{Bil}[\mathrm{Sq}(\boldsymbol{f}_O)]^p \lesssim R^{O(\delta)} r^{-\frac{5(p-2)}{4}} M^{\frac{p-2}{2}} \|\mathrm{Sq}\boldsymbol{f}_O\|_2^p. \tag{10.4}$$

PROOF. Recall that the radius r always stands for r_s. Until the end of this section, and letters $\omega, \omega_1, \omega_2$ always stand for $r^{-1/2}$-caps in $\mathbb{S}^2$. We remind readers that the definition of the bilinear operator Bil is in (3.39). Since the argument of bilinear estimate has appeared in many references, we sketch the idea of proof. Recall that for any tangent cell $O \in \mathcal{O}_s$, the associated function $\boldsymbol{f}_O$ is concentrated on tangent wave packets. Denote these tangent wave packets by $\mathbb{T}_O$, then by wave packet decomposition we can write

$$\boldsymbol{f}_O = \sum_{T \in \mathbb{T}_O} (\boldsymbol{f}_O)_T + \mathrm{RapDec}(R)\|\mathrm{Sq}\boldsymbol{f}_O\|_2.$$

For any cap $\tau \in \mathbb{S}^2$, we use $\mathbb{T}_{O,\tau}$ to denote the set of tubes T in $\mathbb{T}_O$ whose direction lie in τ.

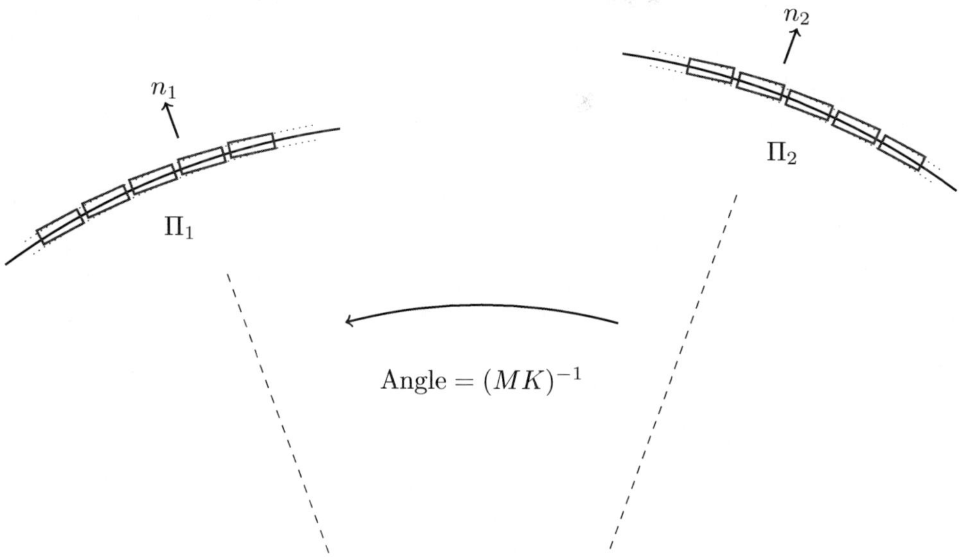

FIGURE 1. Rectangles in $(MK)^{-1}$ bilinear

Since (10.4) is true for $p = 2$, by Hölder's inequality, we only need to prove (10.4) for the endpoint $p = 4$. Since there are $O(K^2)$ many pairs (τ_1, τ_2) in $\mathcal{T}_\sigma \times \mathcal{T}_\sigma$, it suffices to prove for a fixed pair (τ_1, τ_2) that

(10.5)
$$\sum_{1 \le j_1, j_2 \le R} \int_O \Big| \sum_{T_1 \in \mathbb{T}_{O,\tau_1}} (f_{j_1,O})_{T_1} \Big|^2 \Big| \sum_{T_2 \in \mathbb{T}_{O,\tau_2}} (f_{j_2,O})_{T_2} \Big|^2 \lesssim R^{O(\delta)} r^{-\frac{5}{2}} M \|\mathrm{Sq}\boldsymbol{f}_O\|_2^4.$$

Recall from the wave packet decomposition that $(f_{j,O})_T = f_{j,O,\omega} 1_T^*$, where $\omega \subset \mathbb{S}^2$ is the $r^{-1/2}$-cap dual to T. To save notations, we set $\mathbb{T}_1 := \mathbb{T}_{O,\tau_1}$, $\mathbb{T}_2 := \mathbb{T}_{O,\tau_2}$, $f_{j,\omega} := f_{j,O,\omega}$ and $f_{j,T} := (f_{j,O})_T$.

Let $\mathcal{Q} = \{Q\}$ be a collection of finitely overlapping rescaled balls of radius $Mr^{1/2+\beta}$ that forms a cover of O and $Q \cap O \neq \varnothing$. For each $Q \in \mathcal{Q}$, pick a smooth function η_Q such that the Fourier transform of η_Q is supported in the dual slab of Q centered at the origin. We have

$$\sum_{j_1,j_2} \int_O \Big| \sum_{T_1 \in \mathbb{T}_1} f_{j_1,T_1} \Big|^2 \Big| \sum_{T_2 \in \mathbb{T}_2} f_{j_2,T_2} \Big|^2 \lesssim \sum_{j_1,j_2} \sum_{Q \in \mathcal{Q}} \int \eta_Q \Big| \sum_{T_1 \in \mathbb{T}_1} f_{j_1,T_1} \Big|^2 \Big| \sum_{T_2 \in \mathbb{T}_2} f_{j_2,T_2} \Big|^2.$$

Now for each $Q \in \mathcal{Q}$ and each $k = 1, 2$, define $\mathbb{T}_{k,Q}$ as

$$\mathbb{T}_{k,Q} := \{T \in \mathbb{T}_k, T \cap 2Q \neq \varnothing\} \tag{10.6}$$

and define $f_{j_k,Q}$ as the sum of wave packets $f_{j,T}$ that $T \in \mathbb{T}_{k,Q}$. Hence

$$\int \eta_Q \Big| \sum_{T_1 \in \mathbb{T}_1} f_{j_1,T_1} \Big|^2 \Big| \sum_{T_2 \in \mathbb{T}_2} f_{j_2,T_2} \Big|^2 \lesssim \int \eta_Q |f_{j_1,Q}|^2 |f_{j_2,Q}|^2. \tag{10.7}$$

Now for each $r^{-1/2}$-cap ω, we define $\mathbb{T}_{k,Q,\omega} := \{T \in \mathbb{T}_{k,Q}, T \text{ dual to } \omega\}$, so the right-hand side of (10.7) can be rewritten as

$$\int \eta_Q \Big| \sum_{\omega_1 \subset 2\tau_1} \sum_{T \in \mathbb{T}_{1,Q,\omega_1}} f_{j_1,T} \Big|^2 \Big| \sum_{\omega_2 \subset 2\tau_2} \sum_{T \in \mathbb{T}_{2,Q,\omega_2}} f_{j_2,T} \Big|^2. \tag{10.8}$$

By Lemma 5.8, there is a plane V such that tubes in $\mathbb{T}_{k,Q}$ are all contained in $N_{100r^{1/2}}(V)$, which means the tubes in $\mathbb{T}_{k,Q}$ are essentially coplanar. This means the frequency caps ω_1, ω_2 that contribute to the integral (10.8) essentially lie on a one dimensional curve with positive curvature. Now we can use Lemma 10.1 and Cordoba's L^4-argument to conclude

$$(10.8) \lesssim R^{O(\delta)} \int \eta_Q \sum_{\omega_1,\omega_2} \Big| \sum_{T \in \mathbb{T}_{1,Q,\omega_1}} f_{j_1,T} \Big|^2 \Big| \sum_{T \in \mathbb{T}_{2,Q,\omega_2}} f_{j_2,T} \Big|^2, \tag{10.9}$$

Since the tubes in $\mathbb{T}_{1,Q,\omega_1}$ are finitely overlapped, (10.9) is bounded by

$$R^{O(\delta)} \sum_{\omega_1,\omega_2} \sum_{T_1 \in \mathbb{T}_{1,Q,\omega_1}} \sum_{T_2 \in \mathbb{T}_{2,Q,\omega_2}} \int \eta_Q |f_{j_1,T_1}|^2 |f_{j_2,T_2}|^2. \tag{10.10}$$

What follows is a standard application of the essentially constant property of wave packets. For each pair $(f_{j_1,T_1}, f_{j_2,T_2})$, one has

$$\int \eta_Q |f_{j_1,T_1}|^2 |f_{j_2,T_2}|^2 \lesssim R^{O(\delta)} r^{-\frac{5}{2}} M \|f_{j_1,T_1}\|_2^2 \|f_{j_2,T_2}\|_2^2. \tag{10.11}$$

Finally, we can conclude our result. Recall (10.7), (10.8) and (10.9). We first sum up all $T_k \in \mathbb{T}_{k,Q,\omega}$ in the above estimate, then we sum up all $\omega_k \subset \tau_k$ to have

$$\int \eta_Q \Big| \sum_{T_1 \in \mathbb{T}_1} f_{j_1,T_1} \Big|^2 \Big| \sum_{T_2 \in \mathbb{T}_2} f_{j_2,T_2} \Big|^2 \lesssim R^{O(\delta)} r^{-\frac{5}{2}} M \sum_{T_1 \in \mathbb{T}_{1,Q}} \sum_{T_2 \in \mathbb{T}_{2,Q}} \|f_{j_1,T_1}\|_2^2 \|f_{j_2,T_2}\|_2^2$$

Since each pair of tubes (T_1, T_2) belongs to $O(1)$ many cross product $\mathbb{T}_{1,Q} \times \mathbb{T}_{2,Q}$, summing up all $Q \in \mathcal{Q}$ yields

$$\int \Big| \sum_{T_1 \in \mathbb{T}_1} f_{j_1,T_1} \Big|^2 \Big| \sum_{T_2 \in \mathbb{T}_2} f_{j_2,T_2} \Big|^2 \lesssim R^{O(\delta)} r^{-\frac{5}{2}} M \sum_{T_1 \in \mathbb{T}_1} \sum_{T_2 \in \mathbb{T}_2} \|f_{j_1,T_1}\|_2^2 \|f_{j_2,T_2}\|_2^2$$

Finally, we sum up all pairs (j_1, j_2) to conclude (10.5). □

11. Appendix: Intersection of tubes and plane L_a

We prove Lemma 8.3 here. First, recall that we have set $r = r_s$, which is the scale of O_s (i.e. O_s is contained in a rescaled r-ball $\mathcal{P}_{O_s}$). For each cell $O_s \in \mathcal{O}_s$, there is a polynomial P (depending on O_s) of degree $O(d)$, such that the tubes in $\mathbb{T}_{O_s}$ are contained in $N_{R^{O(\delta)}r^{1/2}} Z(P) \cap \mathcal{P}_{O_s}$.

To obtain Lemma 8.3, it suffices to prove the following lemma.

LEMMA 11.1. *Fix an $O_s \in \mathcal{O}_s$, and let $\mathcal{P}_r$ be a rescaled r-ball such that $O_s \subset \mathcal{P}_r$. For each $T \in \mathbb{T}_{O_s}$, let $\mathring{T}$ be the tube with infinite length which is obtained by prolonging T. Let $\mathring{\mathbb{T}}_{O_s} := \{\mathring{T} : T \in \mathbb{T}_{O_s}\}$. Let $\{y_l\}_{l=1}^m$ be an $Rr^{-1/2}$-separated subset of*

$$L_a \bigcap \{\cup_{T \in \mathbb{T}_{O_s}} \mathring{T}\}$$

Then $m \lesssim R^{O(\delta)} r^{1/2} M^{-1}$ uniformly for all $|a| \leq R$.

Although the tubes in $\mathbb{T}_{O_s}$ have dimensions $R^\beta r^{1/2} \times R^\beta r^{1/2} \times r$, but the tiny factor β is harmless in our proof, let us just assume $\beta = 0$ in the rest of this section.

PROOF. For each point y_l, we pick a tube $\mathring{T}_l \in \mathring{\mathbb{T}}_{O_s}$ satisfying $y_l \in \mathring{T}_l$. Let $\mathring{\mathbb{T}}_a := \{\mathring{T}_l\}_{l=1}^m$, and let $\mathbb{T}_a := \{T_l\}_{l=1}^m$. We point out that if two distinct tubes $T_1, T_2 \in \mathbb{T}_a$ intersect, then they make an angle $\gtrsim r^{-1/2}$. Note that for any $T \in \mathbb{T}_{O_s}$, $T \subset N_{R^{O(\delta)}r^{1/2}} Z(P) \cap \mathcal{P}_r$. If we let $\mathcal{Q} = \{Q\}$ to be a collection of finitely overlapping $r^{1/2}$-cube that cover $N_{r^{1/2}R^{O(\delta)}} Z(P) \cap \mathcal{P}_r$. By Wongkew's theorem [**Won93**], $\#\{Q\} \lesssim R^{O(\delta)} r M^{-1}$.

For any $Q \in \mathcal{Q}$, $T_1, T_2 \in \mathbb{T}_a$, we define an incidence function $\chi(Q, T_1, T_2)$ which $= 1$ if $Q \cap T_1 \cap T_2 \neq \varnothing$ and $= 0$ otherwise. We let n_Q be the number of tubes $T \in \mathbb{T}_a$ that intersect Q. By Cauchy-Schwartz,

$$\sum_{Q, T_1, T_2} \chi(Q, T_1, T_2) \geq \sum_Q n_Q^2 \geq \#\{Q\}^{-1} (\sum_Q n_Q)^2 \geq R^{-O(\delta)} M |\mathring{\mathbb{T}}_a|^2.$$

(The last inequality is because each T contains $\sim r^{1/2}$ many Q's.)

If we already have

$$|\mathring{\mathbb{T}}_a| \lesssim R^{O(\delta)} r^{1/2} M^{-1},$$

then we are done. Otherwise we have $|\mathring{\mathbb{T}}_a| > C R^{O(\delta)} r^{1/2} M^{-1}$, which implies a bound on the diagonal term:

$$\sum_{Q,T} \chi(Q, T, T) \leq r^{1/2} |\mathring{\mathbb{T}}_a| \leq \frac{1}{2} R^{-O(\delta)} M |\mathring{\mathbb{T}}_a|^2.$$

The last inequality holds since it is equivalent to

$$|\mathring{\mathbb{T}}_a| \geq R^{O(\delta)} 2 r^{1/2} M^{-1}.$$

So, we have

$$\sum_Q \sum_{T_1 \neq T_2} \chi(Q, T_1, T_2) \gtrsim R^{-O(\delta)} M |\mathring{\mathbb{T}}_a|^2.$$

Notice that when $T_1 \neq T_2$ and $T_1 \cap T_2 \neq \emptyset$, we have $\angle(T_1, T_2) \geq r^{-1/2}$. Thus, there exists a dyadic value $\nu \in [r^{-1/2}, M^{-1}]$ such that

$$\sum_Q \sum_{\angle(T_1, T_2) \sim \nu} \chi(Q, T_1, T_2) \gtrsim R^{-O(\delta)} M |\mathring{\mathbb{T}}_a|^2.$$

By pigeonholing, there exists a $T_1 \in \mathbb{T}_a$ such that for this fixed T_1,

$$\sum_Q \sum_{T_2:\ \angle(T_1, T_2) \sim \nu} \chi(Q, T_1, T_2) \gtrsim R^{-O(\delta)} M |\mathring{\mathbb{T}}_a|.$$

From now on we fix this T_1 and define $\mathbb{T}_\nu := \{T \in \mathbb{T}_a : T \cap T_1 \neq \emptyset, \angle(T_1, T) \sim \nu\}$. First we note that $\sum_Q \chi(Q, T_1, T_2) \lesssim \nu^{-1}$ for $T_2 \in \mathbb{T}_\nu$, so we have

$$|\mathbb{T}_\nu| \gtrsim \nu M R^{-O(\delta)} |\mathring{\mathbb{T}}_a|. \tag{11.1}$$

Next, we define $H := \bigcup_{T \in \mathbb{T}_\nu} T$ which is a hairbrush rooted at T_1. We claim that

$$|\mathbb{T}_\nu| r^2 \lesssim R^{O(\delta)} (\log r) |H|. \tag{11.2}$$

Let us quickly see how these two inequalities combine to give the result. Since H is contained in a fat tube $N_{\nu r}T_1$ of dimensions morally $\nu r \times \nu r \times r$, and since H is contained in the set $N_{R^{O(\delta)}r^{1/2}}Z(P)$. By Wongkew's theorem, we have

$$|H| \lesssim R^{O(\delta)}\nu r^{5/2}.$$

Together with (11.1) and (11.2), we proved

$$|\mathring{\mathbb{T}}_a| \lesssim R^{O(\delta)}r^{1/2}M^{-1}.$$

To prove the claim (11.2), we use the idea of two-ends reduction from Wolff. We will decompose the set $\mathbb{T}_\nu$ into $\sim \nu r^{1/2}$ subsets. To do this, we choose $\sim \nu r^{1/2}$ many planes each of which contains the central line of T_1 and their normal vectors are $\nu^{-1}r^{-1/2}$ separated. Let $\{P_k\}_k$ be the $r^{1/2}$-neighborhood of these planes. We see that all the tubes in $\mathbb{T}_\nu$ lie in $\cup_k P_k$, since all the tubes in $\mathbb{T}_\nu$ lie in a cylinder of dimension $\nu r \times \nu r \times r$ and $\cup_k P_k$ covers this cylinder. For each tube $T \in \mathbb{T}_\nu$, we associate it to a P_k if $T \in P_k$ (if there are many choice of P_k, we just choose one). Denoting by $\mathbb{T}_{\nu,k}$ the tubes that are associated to P_k, we get $\mathbb{T}_\nu = \sqcup_k \mathbb{T}_{\nu,k}$.

For each tube $T \in \mathbb{T}_{\nu,k}$, we let $\tilde{T}$ be the portion $T \setminus N_{\frac{1}{10}r\nu}(T_1)$. Roughly speaking, $\tilde{T}$ still contains a tube of dimensions morally $r^{1/2} \times r^{1/2} \times r$. Let $\tilde{\mathbb{T}}_{\nu,k}$ be the collection of these $\tilde{T}$, and let H_k be the union of tubes in $\tilde{T}_{\nu,k}$. Here comes the key observation: by the separation of P_k, we see $\{H_k\}$ are at most 1000-overlapped. So it suffices to show

$$|\mathbb{T}_{\nu,k}|r^2 \lesssim (\log r)|H_k|. \tag{11.3}$$

To save notations, we let $\mathbb{T} = \mathbb{T}_{\nu,k}$, $H = H_k$ and $P = P_k$ in the rest of the proof. For each $T \in \mathbb{T}$, the intersection of its stretch $\mathring{T}$ with L_a contains a point y_l. By the geometric condition, the set $\{y_l\}$ obtained from $\mathbb{T}$ lie in a $Rr^{-1/2} \times R$ rectangle in L_a. Since y_l's are $Rr^{-1/2}$-separated, we can morally think about these y_l are arranged in a line and two nearby y_l are at least $Rr^{-1/2}$-separated. Now the following argument is quite standard as in the proof of 2-dimensional Kakeya conjecture.

We choose $\mathbb{T}'$ to be a subset of $\mathbb{T}$ such that $\{\mathring{T} \cap L_a : T \in \mathbb{T}'\}$ are $C(\log r)Rr^{-1/2}$-separated and $|\mathbb{T}'| \gtrsim \frac{1}{\log r}|\mathbb{T}|$. For each $T \in \mathbb{T}'$, We have

$$\sum_{T' \in \mathbb{T}',\ T' \neq T} |T \cap T'| \leq r^2 \sum_{j=0}^{r} \frac{1}{Cj(\log r)} \leq \frac{1}{10000}r^2 \leq \frac{1}{2}|\tilde{T}|.$$

This shows that $|\tilde{T} \setminus \cup_{T' \in \mathbb{T}',\ T' \neq T} T'| \geq \frac{1}{2}|\tilde{T}|$ for $T \in \mathbb{T}'$. As a result,

$$|H| \geq |\cup_{T \in \mathbb{T}'} \tilde{T}| \geq \sum_{T \in \mathbb{T}'} |\tilde{T} \setminus \cup_{T' \in \mathbb{T}',\ T' \neq T} T'| \gtrsim r^2|\mathbb{T}'| \gtrsim \frac{r^2}{\log r}|\mathbb{T}|,$$

which finishes the proof of (11.3). □

Acknowledgments

The authors would like to thank Larry Guth, Shaoming Guo, Xiaochun Li, and Andreas Seeger for valuable discussions.

References

[BG11] Jean Bourgain and Larry Guth, *Bounds on oscillatory integral operators based on multilinear estimates*, Geom. Funct. Anal. **21** (2011), no. 6, 1239–1295, DOI 10.1007/s00039-011-0140-9. MR2860188

[Car83] Anthony Carbery, *The boundedness of the maximal Bochner–Riesz operator on $L^4(\mathbf{R}^2)$*, Duke Math. J. **50** (1983), no. 2, 409–416. MR705033

[Cór81] Antonio Córdoba, *Some remarks on the Littlewood–Paley theory*, Proceedings of the Seminar on Harmonic Analysis (Pisa, 1980), Rend. Circ. Mat. Palermo (2) **suppl, suppl. 1** (1981), 75–80. MR639467

[GK15] Larry Guth and Nets Hawk Katz, *On the Erdős distinct distances problem in the plane*, Ann. of Math. (2) **181** (2015), no. 1, 155–190, DOI 10.4007/annals.2015.181.1.2. MR3272924

[GOW+21] Shaoming Guo, Changkeun Oh, Hong Wang, Shukun Wu, and Ruixiang Zhang, *The Bochner-Riesz problem: an old approach revisited*, Preprint, `arXiv:2104.11188`, 2021.

[Gra14] Loukas Grafakos, *Classical Fourier analysis*, 3rd ed., Graduate Texts in Mathematics, vol. 249, Springer, New York, 2014, DOI 10.1007/978-1-4939-1194-3. MR3243734

[GRY20] Shaoming Guo, Joris Roos, and Po-Lam Yung, *Sharp variation-norm estimates for oscillatory integrals related to Carleson's theorem*, Anal. PDE **13** (2020), no. 5, 1457–1500, DOI 10.2140/apde.2020.13.1457. MR4149067

[Gut16] Larry Guth, *A restriction estimate using polynomial partitioning*, J. Amer. Math. Soc. **29** (2016), no. 2, 371–413, DOI 10.1090/jams827. MR3454378

[Gut18] Larry Guth, *Restriction estimates using polynomial partitioning II*, Acta Math. **221** (2018), no. 1, 81–142, DOI 10.4310/ACTA.2018.v221.n1.a3. MR3877019

[Lee04] Sanghyuk Lee, *Improved bounds for Bochner–Riesz and maximal Bochner–Riesz operators*, Duke Math. J. **122** (2004), no. 1, 205–232, DOI 10.1215/S0012-7094-04-12217-1. MR2046812

[LRS12] Sanghyuk Lee, Keith M. Rogers, and Andreas Seeger, *Improved bounds for Stein's square functions*, Proc. Lond. Math. Soc. (3) **104** (2012), no. 6, 1198–1234, DOI 10.1112/plms/pdr067. MR2946085

[LW20] Xiaochun Li and Shukun Wu, *New estimates of the maximal Bochner–Riesz operator in the plane*, Math. Ann. **378** (2020), no. 3-4, 873–890, DOI 10.1007/s00208-020-02038-w. MR4163515

[See88] Andreas Seeger, *Some inequalities for singular convolution operators in L^p-spaces*, Trans. Amer. Math. Soc. **308** (1988), no. 1, 259–272, DOI 10.2307/2000962. MR955772

[Ste58] Elias M. Stein, *Localization and summability of multiple Fourier series*, Acta Math. **100** (1958), 93–147, DOI 10.1007/BF02559603. MR105592

[SW71] Elias M. Stein and Guido Weiss, *Introduction to Fourier analysis on Euclidean spaces*, Princeton Mathematical Series, No. 32, Princeton University Press, Princeton, N.J., 1971. MR0304972

[Wan18] Hong Wang, *A Restriction Estimate in R3*, ProQuest LLC, Ann Arbor, MI, 2019. Thesis (Ph.D.)–Massachusetts Institute of Technology. MR4051415

[Won93] Richard Wongkew, *Volumes of tubular neighbourhoods of real algebraic varieties*, Pacific J. Math. **159** (1993), no. 1, 177–184. MR1211391

[Wu23] Shukun Wu, *On the Bochner–Riesz operator in $\mathbb{R}^3$*, J. Anal. Math. **149** (2023), no. 2, 677–718, DOI 10.1007/s11854-022-0263-y. MR4594404

Deparment of Mathematics, Massachusetts Institute of Technology, Cambridge Massachusetts, 02139

Email address: `shengwen@mit.edu`

Department of Mathematics, University of Illinois at Urbana-Champaign, Urbana Illinois, 61801

Email address: `yifanjing17@gmail.com`

Department of Mathematics, California Institute of Technology, Pasadena California, 91125

Email address: `skwu@caltech.edu`

Contemporary Mathematics
Volume **792**, 2024
https://doi.org/10.1090/conm/792/15894

An introduction to decoupling and harmonic analysis over $\mathbb{Q}_p$

Zane Kun Li

ABSTRACT. The goal of this expository paper is to provide an introduction to decoupling by working in the simpler setting of decoupling for the parabola over $\mathbb{Q}_p$. Over $\mathbb{Q}_p$, commonly used heuristics in decoupling are significantly easier to make rigorous over $\mathbb{Q}_p$ than over $\mathbb{R}$ and such decoupling theorems over $\mathbb{Q}_p$ are still strong enough to derive interesting number theoretic conclusions.

1. Introduction

1.1. Motivation. For an interval $J \subset [0,1]$, let

$$\theta_J := \{(\xi, \eta) \in \mathbb{R}^2 : \xi \in J, |\eta - \xi^2| \leqslant |J|^2\} \tag{1}$$

and let f_J be defined such that $\widehat{f_J} := \widehat{f} 1_{J \times \mathbb{R}}$. For $\delta \in \mathbb{N}^{-1}$, let P_δ be a partition of $[0,1]$ into intervals of length δ. For $2 \leqslant p \leqslant \infty$, let $D_{p,\mathbb{R}}(\delta)$ be the best constant such that

$$\|f\|_{L^p(\mathbb{R}^2)} \leqslant D_{p,\mathbb{R}}(\delta)(\sum_{K \in P_\delta} \|f_K\|^2_{L^p(\mathbb{R}^2)})^{1/2} \tag{2}$$

for all f with Fourier transform supported in $\bigcup_{K \in P_\delta} \theta_K$. A special case of Bourgain and Demeter's paraboloid decoupling theorem [**3**, **4**] gives that

$$D_{p,\mathbb{R}}(\delta) \lesssim_{p,\varepsilon} \delta^{-\varepsilon}(1 + \delta^{-(\frac{1}{2} - \frac{3}{p})}).$$

By interpolation, it suffices to prove the critical estimate that $D_{6,\mathbb{R}}(\delta) \lesssim_\varepsilon \delta^{-\varepsilon}$. In 2020, Guth, Maldague, and Wang in [**13**] improved this estimate to $\lesssim (\log \delta^{-1})^C$ for an unspecified large constant C. One application of their estimate is the following: Let $K(N)$ be the best constant such that

$$\|\sum_{n=1}^{N} a_n e(nx + n^2 t)\|_{L^6_{x,t}(\mathbb{T}^2)} \leqslant K(N)(\sum_{n=1}^{N} |a_n|^2)^{1/2} \tag{3}$$

for all $\{a_n\}_{n=1}^N \subset \mathbb{C}$ and where here $e(z) := e^{2\pi i z}$. They were able to show that $K(N) \lesssim (\log N)^C$ for the same unspecified large constant C. Before this, the best known upper bound was due to Bourgain in [**2**] who showed an upper bound of $\exp(O(\frac{\log N}{\log \log N}))$ using the divisor bound and a lower bound of $(\log N)^{1/6}$ by analyzing major arcs.

2020 *Mathematics Subject Classification.* Primary 43A70.

The author is supported by NSF grants DMS-2037851 and DMS-2311174.

Recently, instead of working over $\mathbb{R}$, decoupling has been studied over $\mathbb{Q}_q$ (we use $\mathbb{Q}_q$ instead of $\mathbb{Q}_p$ as we reserve p for the L^p norm), see for example [**8**] and [**10**]. The prime q throughout will be an odd prime. In [**10**], with Shaoming Guo and Po-Lam Yung, we optimized Guth, Maldague, and Wang's argument and worked over $\mathbb{Q}_q$ instead of $\mathbb{R}$ to prove a parabola decoupling theorem over $\mathbb{Q}_q$ which then was able to reduce C down to 2+, that is, $K(N) \lesssim_\varepsilon (\log N)^{2+\varepsilon}$.

Similar to the advantages of working with the Walsh model in time frequency analysis (see for example [**21**]), there are some clear advantages in working over $\mathbb{Q}_q$ instead of $\mathbb{R}$. The following are some heuristics that are used when thinking about decoupling over $\mathbb{R}$ but need to be made rigorous when writing out the proofs. This often involves working with smooth cutoffs or convolutions to make all the "essentially" statements rigorously true and can be somewhat technical. Over $\mathbb{Q}_q$, however, the statements below are rigorously true with little modification.

(i) The neighborhood of the parabola θ_J is essentially a parallelogram.

(ii) Balls and squares are essentially the same and so the square $[0, R]^2$ is essentially the same as the square $[-R, R]^2$ which in turn is essentially the same as the ball $B(0, R)$. Furthermore, $O(1)$ dilates of geometric objects such as a balls and rectangles are also essentially the same, and so we don't distinguish between $B(0, R)$ and $B(0, 2R)$.

(iii) Consider an $R \times R^2$ rectangle centered at the origin pointing in the direction of (say) $(-1, 1)$. The square $[0, R^2]^2$ can essentially be partitioned by R many translates of these rectangles.

(iv) The Fourier transform of $1_{B(0,1)}$ is essentially $1_{B(0,1)}$.

(v) (Small angle approximation) If $|\xi| \ll 1$, then $e(\xi) \approx 1$.

(vi) (Local constancy property) If f is Fourier supported in θ, then $|f|$ is essentially constant on translates of the polar (or dual) set $\theta^* := \{x \in \mathbb{R}^2 : |x \cdot \xi| \leqslant 1 \text{ for all } \xi \in \theta\}$. Thus if $J = [a, a + \delta]$ and $\theta = \theta_J$, then $|f|$ is essentially constant on translates of the "dual tube" $\{(x, y) \in \mathbb{R}^2 : |x + 2ay| \leqslant \delta^{-1}, |y| \leqslant \delta^{-2}\}$. The a in $x + 2ay$ can be replaced with any point in J.

Additionally, decoupling theorems over $\mathbb{Q}_q$ are still strong enough to imply the applications of decoupling to exponential sums and solution counting such as in Vinogradov's Mean Value Theorem and (3) above as often only even L^p norms are considered. Thus the author believes decoupling over $\mathbb{Q}_q$ provides an ideal setting for someone just starting out in the area.

1.2. Decoupling for the parabola over $\mathbb{Q}_q$. We now redefine our definitions in (1) and (2) to work over $\mathbb{Q}_q$. For $J \subset \mathbb{Z}_q$, let

$$\theta_J := \{(\xi, \eta) \in \mathbb{Q}_q^2 : \xi \in J, |\eta - \xi^2| \leqslant |J|^2\} \tag{4}$$

and let f_J be defined such that $\widehat{f_J} := \widehat{f} 1_{J \times \mathbb{Q}_q}$. For $\delta \in q^{-\mathbb{N}}$, let P_δ be a partition of $\mathbb{Z}_q$ into intervals of length δ. Finally, we let $D_p(\delta) := D_{p,\mathbb{Q}_q}(\delta)$ be the best constant such that

$$\|f\|_{L^p(\mathbb{Q}_q^2)} \leqslant D_{p,\mathbb{Q}_q}(\delta)\Big(\sum_{K \in P_\delta} \|f_K\|_{L^p(\mathbb{Q}_q^2)}^2\Big)^{1/2}$$

for all f with Fourier transform supported in $\bigcup_{K \in P_\delta} \theta_K$. We will prove the following $\mathbb{Q}_q$ analogue of Bourgain and Demeter's decoupling theorem applied to the parabola.

THEOREM 1.1. *For $\delta \in q^{-\mathbb{N}}$, we have*

$$D_{p,\mathbb{Q}_q}(\delta) \lesssim_{p,q,\varepsilon} \delta^{-\varepsilon}(1 + \delta^{-(\frac{1}{2}-\frac{3}{p})}).$$

Once again by interpolation, it suffices to prove that $D_{6,\mathbb{Q}_q}(\delta) \lesssim_{q,\varepsilon} \delta^{-\varepsilon}$. The proof we follow is the efficient congruencing inspired proof of parabola decoupling over $\mathbb{R}$ due to myself in [**16**]. Following the proofs of the Main Conjecture in Vinogradov's Mean Value Theorem using the harmonic analysis method of decoupling by Bourgain, Demeter, and Guth in [**5**] and the number theoretic method of efficient congruencing by Wooley [**26**], it is an interesting question to determine how these two methods are related. This study led to new proofs of decoupling for the parabola [**16**], cubic moment curve [**11**], and the degree k moment curve [**12**], corresponding to [**17**, Section 4.3], [**14**], and [**26**], respectively. See also [**7**] for proof of decoupling for fractal sets on the parabola which was inspired from efficient congruencing for ellipsephic sets [**1**]. Additionally see [**8**] for an interpretation into decoupling of a classical argument (see for example [**22**, Chapter 5] or [**20**, Theorem 13]) that gave partial progress towards the Main Conjecture in Vinogradov's Mean Value Theorem.

This paper is organized as follows: In Sections 2 and 3, we review the needed facts for $\mathbb{Q}_q$ that are relevant to our paper. In particular, in Section 3, we state the version of the locally constant property/wavepacket decomposition that we will use. In Section 4, we show that Theorem 1.1 is sharp up to the ε in the exponent, prove an interpolation theorem that allows use to reduce Theorem 1.1 to just showing the $p = 6$ case, and show that Theorem 1.1 is strong enough to imply an estimate for $K(N)$ in (3). In Section 5, we prove $D_{6,\mathbb{Q}_q}(\delta) \lesssim_{q,\varepsilon} \delta^{-\varepsilon}$ save for one key step (Proposition 5.4). Finally, in Section 6, we give five different proofs of Proposition 5.4.

Notation. We say that f is Fourier supported in Ω if its Fourier transform $\widehat{f}$ is supported in Ω. Given an interval I, we let $P_\delta(I)$ be the partition of I into intervals of length δ. Given two positive expressions X and Y, we write $X \lesssim Y$ if $X \leqslant CY$ for some absolute constant C. If C depends on some additional parameter A, then we write $X \lesssim_A Y$. We write $X \sim Y$ if $X \lesssim Y$ and $Y \lesssim X$. By writing $f(x) = O(g(x))$, we mean $|f(x)| \lesssim g(x)$.

2. Some basic facts about $\mathbb{Q}_q$

We will summarize all the relevant facts about $\mathbb{Q}_q$ in this section and the next. See the Introduction and Section 2 of [**8**], Section 2 of [**10**], Chapters 1 and 2 of [**18**], and Chapter 1, Sections 1 and 4 of [**23**] for additional discussion of analysis on $\mathbb{Q}_q$. The exposition in this section and the next borrows heavily from the presentation in [**8**] and [**10**].

2.1. Definitions and basic geometric observations. The field $\mathbb{Q}_q$ is the completion of $\mathbb{Q}$ under the q-adic norm, defined by $|0| = 0$ and $|q^a b/c| = q^{-a}$ if $a \in \mathbb{Z}$, $b, c \in \mathbb{Z}\backslash\{0\}$ and q is relatively prime to both b and c. Then $\mathbb{Q}_q$ can be identified with the set of all formal series

$$\mathbb{Q}_q = \{\sum_{j=k}^{\infty} a_j q^j : k \in \mathbb{Z}, a_j \in \{0, 1, \ldots, q-1\} \text{ for every } j \geqslant k\}$$

and the q-adic norm on $\mathbb{Q}_q$ satisfies $|\sum_{j=k}^{\infty} a_j q^j| = q^{-k}$ if $a_k \neq 0$. A nice article illustrating one way to visualize $\mathbb{Q}_3$ can be found in [**15**]. Strictly speaking we should be writing $|\cdot|_q$ instead of $|\cdot|$, but we omit this dependence since we will only be dealing with one fixed q. The q-adic norm on $\mathbb{Q}_q$ induces a norm on $\mathbb{Q}_q^2$, which we denote also by $|\cdot|$ by abuse of notation, via $|(x,y)| := \max(|x|,|y|)$. The key property obeyed by the q-adic norm is the ultrametric triangle inequality: $|x+y| \leqslant \max(|x|,|y|)$ with equality if $|x| \neq |y|$.

We adopt all our geometry terminology in analogy from the real setting. An interval $I(a,r)$ in $\mathbb{Q}_q$ is a set of the form $\{\xi \in \mathbb{Q}_q : |\xi - a| \leqslant r\}$ where $a \in \mathbb{Q}_q$ and $r \geqslant 0$; r will then be called the length of the interval. Given an interval I, we will use $|I|$ to denote the length of I. A square in $\mathbb{Q}_q^2$ of side length r is then the product of 2 intervals in $\mathbb{Q}_q$ of length r. Finally, we define the ball $B(a,r) := \{\xi \in \mathbb{Q}_q^2 : |\xi - a| \leqslant r\}$ where $a \in \mathbb{Q}_q^2$ and $r \geqslant 0$ and will say this ball has radius r, in analogy to the real setting.

The ring of integers $\mathbb{Z}_q$ coincides with the unit interval $I(0,1) = \{\xi \in \mathbb{Q}_q : |\xi| \leqslant 1\}$. Elements in $\mathbb{Z}_q$ are exactly the elements in $\mathbb{Q}_q$ of the form $\sum_{j\geqslant 0} a_j q^j$ where $a_j \in \{0,1,\ldots,q-1\}$. As each positive integer has a base q representation, we may embed $\mathbb{N}$ into $\mathbb{Z}_q$. Identifying -1 with the formal power series $\sum_{j\geqslant 0}(q-1)q^j$ in $\mathbb{Z}_q$ allows us to embed $\mathbb{Z}$ into $\mathbb{Z}_q$. Inverting formal power series and that the addition of two formal power series is still a formal power series shows that $\mathbb{Q}$ embeds into $\mathbb{Q}_q$.

Having given the definitions of $\mathbb{Q}_q$ and $\mathbb{Z}_q$, we now give some basic geometric observations. Intervals of length δ partition $[0,1]$ into δ^{-1} many intervals. Not only is the same is true in $\mathbb{Z}_q$, but also two distinct intervals are always separated by a positive distance.

LEMMA 2.1. *Fix $k \in \mathbb{N}$. For $a \in \{0,1,\ldots,q^k-1\}$, the intervals $I(a,q^{-k}) = \{\xi \in \mathbb{Q}_q : |\xi - a| \leqslant q^{-k}\}$ partition $\mathbb{Z}_q$ into q^k many disjoint intervals and each pair of intervals are separated by a distance at least q^{-k+1}.*

PROOF. The $\xi \in \mathbb{Q}_q$ such that $|\xi - a| \leqslant q^{-k}$ are exactly the $\xi \in \mathbb{Z}_q$ such that $\xi \equiv a \pmod{q^k}$ when we write ξ as a formal power series. Since a ranges from 0 to $q^k - 1$, the $I(a,q^{-k})$ partition $\mathbb{Z}_q$ into q^k many disjoint intervals of length q^{-k}.

Next, suppose $|\xi_1 - a| \leqslant q^{-k}$ and $|\xi_2 - b| \leqslant q^{-k}$ for some $a \neq b$. Since $|a-b| \geqslant q^{-k+1}$ and $|(\xi_1 - \xi_2) - (a-b)| \leqslant q^{-k}$, it follows from the equality case of the ultrametric triangle inequality that $|\xi_1 - \xi_2| = |a-b| \geqslant q^{-k+1}$. □

Next, from how $|\cdot|$ is defined on $\mathbb{Q}_q^2$, squares of side length r and balls of radius r are the same. More precisely:

LEMMA 2.2. *Fix an arbitrary $(A,B) \in \mathbb{Q}_q^2$. The square $\{(x,y) \in \mathbb{Q}_q^2 : |x-A| \leqslant r, |y-B| \leqslant r\}$ of side length r is the same as the ball $\{(x,y) \in \mathbb{Q}_q^2 : |(x,y)-(A,B)| \leqslant r\}$ of radius r.*

By the ultrametric triangle inequality, intervals can be centered at any point in the interval.

LEMMA 2.3. *Fix an interval $I(x_0,r)$. For any $b \in I(x_0,r)$, we have $I(x_0,r) = I(b,r)$.*

PROOF. Fix arbitrary $b \in I(x_0,r)$, then $|b - x_0| \leqslant r$. We only show that $\{\xi \in \mathbb{Q}_q : |\xi - b| \leqslant r\} \subset I(x_0,r)$. But this follows from the ultrametric triangle inequality: $|\xi - x_0| = |\xi - b + b - x_0| \leqslant \max(|\xi - b|, |b - x_0|) \leqslant r$. The reverse containment is true by interchanging the roles of x_0 and b. $\square$

Lemma 2.3 then is able to show that two intervals of the same length are either exactly the same or completely disjoint.

COROLLARY 2.4. *If two intervals (potentially of different lengths) intersect, then one is contained inside the other.*

PROOF. Let I_1 and I_2 be two intervals with $|I_1| = r \leqslant s = |I_2|$. By hypothesis, we can find an $a \in I_1 \cap I_2$. By Lemma 2.3, $I_1 = I(a, r)$ and $I_2 = I(a, s)$ and hence $I_1 \subset I_2$. $\square$

Corollary 2.4 shows that intervals in $\mathbb{Q}_q$ act like dyadic intervals in $\mathbb{R}$. Furthermore, Lemma 2.1 shows that two distinct intervals of two (potentially different) lengths are going to be at least separated by q times the length of the smaller interval. We can also extend Corollary 2.4 to products of intervals.

COROLLARY 2.5. *If two squares (potentially of different side lengths) intersect, then one is contained inside the other.*

PROOF. Let $B = I \times J$ and $B' = I' \times J'$ with $|I| = |J| = r$ and $|I'| = |J'| = s$. Without loss of generality we may assume that $r \leqslant s$. We can find a $(x_0, y_0) \in B \cap B'$. Therefore $x_0 \in I \cap I'$ and $y_0 \in J \cap J'$ and by Corollary 2.4, $I \subset I'$ and $J \subset J'$. Therefore $B \subset B'$. $\square$

We end with the observation that the sum of two balls of radius r is still a ball of radius r and that the shear of a ball is still a ball. This should be compared to the real setting where the sum of two balls of radius r is a ball of radius $2r$ and the shear of a ball or a square is an ellipse or parallelogram, respectively.

LEMMA 2.6. *We have the following properties:*

(i) *For $a, b \in \mathbb{Q}_q$, if $I = I(a, r)$ and $I' = I(b, r)$, then $I + I' = I(a + b, r)$. Similarly, for $c, d \in \mathbb{Q}_q^2$, $B(c, r) + B(d, r) = B(c + d, r)$.*

(ii) *If $S_a = \begin{pmatrix} 1 & a \\ 0 & 1 \end{pmatrix}$ with $|a| \leqslant 1$, then $S_a B(0, 1) = B(0, 1)$.*

PROOF. To prove the first part of (i), if α and β were such that $|\alpha - a| \leqslant r$ and $|\beta - b| \leqslant r$, then $|(\alpha + \beta) - (a + b)| \leqslant \max(|\alpha - a|, |\beta - b|) \leqslant r$ and hence $\alpha + \beta \in I(a+b, r)$. On the other hand, if ξ is such that $|\xi - (a+b)| \leqslant r$, then write $\xi = (\xi - b) + b$. We have $\xi - b \in I$ and $b \in I'$. The second part of (i) follows from the first part of (i) and the observation that balls of radius r are just products of two intervals of length r.

Next to prove (ii), since $S_a^{-1} = \begin{pmatrix} 1 & -a \\ 0 & 1 \end{pmatrix}$, it just suffices to show that $S_a B(0, 1) \subset B(0, 1)$. But this follows from that $|x + 2ay| \leqslant \max(|x|, |2ay|) \leqslant 1$ for $|x|, |y| \leqslant 1$ where here we have also used that q is an odd prime and so $|2| = 1$. $\square$

2.2. The Fourier transform on $\mathbb{Q}_q$. We now summarize the relevant facts about the Fourier transform over $\mathbb{Q}_q$. Let χ be the additive character of $\mathbb{Q}_q$ that is equal to 1 on $\mathbb{Z}_q$ and nontrivial on $q^{-1}\mathbb{Z}_q$. This is the analogue of $e^{2\pi i x}$. Up to isomorphism, the character is essentially unique, given by $\chi(x) := e(\sum_{j=k}^{-1} a_j q^j)$ if $x = \sum_{j \geqslant k} a_j q^j$. Therefore the small angle approximation in $\mathbb{Q}_q$: "If $|\xi| \leqslant 1$, then $\chi(\xi) = 1$." is rigorously true.

In $\mathbb{Q}_q^d$, the Schwartz functions are finite linear combinations of characteristic functions of cubes in $\mathbb{Q}_q^d$. The Fourier transform of such a function f is defined by

$$\widehat{f}(\xi) := \int_{\mathbb{Q}_q^d} f(x)\chi(-x\cdot\xi)\,dx$$

where here dx is the Haar measure on $\mathbb{Q}_q^d$ normalized so that $\mathbb{Z}_q^d$ has volume 1. The theory of integration and the Fourier transform in $\mathbb{Q}_q$ is essentially the same as in $\mathbb{R}$. We refer the interested reader to [**23**, Section 4] for more details.

The key property obeyed by the Fourier transform is that the Fourier transform of the unit interval is the unit interval and so the indicator function of the unit intervals plays the same role as Gaussians in $\mathbb{R}$. This should be compared to the real setting where the Fourier transform of the unit interval is a sinc function. More precisely:

PROPOSITION 2.7. *We have that* $\widehat{1}_{\mathbb{Z}_q} = 1_{\mathbb{Z}_q}$.

PROOF. We need to show that

$$\int_{\mathbb{Z}_q} \chi(-x\cdot\xi)\,dx = 1_{\mathbb{Z}_q}(\xi). \tag{5}$$

If $\xi \in \mathbb{Z}_q$, then $|-x\xi| \leqslant 1$ and hence $-x\xi \in \mathbb{Z}_q$. Since χ is trivial on $\mathbb{Z}_q$ and $\mathbb{Z}_q$ has volume 1, it follows that the left hand side of (5) is equal to 1.

On the other hand if $\xi \notin \mathbb{Z}_q$, choose an x_0 such that $|x_0| = 1$. Since if $|\xi| > 1$, then $|\xi| \geqslant q$, this implies that $|-x_0\xi| \geqslant q$ and so $\chi(-x_0\xi) \neq 1$. Additionally, the ultrametric triangle inequality gives that $\mathbb{Z}_q = x_0 + \mathbb{Z}_q$. We then have

$$\begin{aligned}\int_{\mathbb{Z}_q} \chi(-x\xi)\,dx &= \chi(-x_0\xi)\int_{\mathbb{Z}_q}\chi(-(x-x_0)\xi)\,dx\\ &= \chi(-x_0\xi)\int_{x_0+\mathbb{Z}_q}\chi(-y\xi)\,dy = \chi(-x_0\xi)\int_{\mathbb{Z}_q}\chi(-y\xi)\,dy.\end{aligned}$$

But since $\chi(-x_0\xi) \neq 1$, we must have $\int_{\mathbb{Z}_q}\chi(-x\xi)\,dx = 0$. □

Since $B(0,1) = \mathbb{Z}_q^2$, Proposition 2.7 immediately shows that the Fourier transform of $1_{B(0,1)}$ is also $1_{B(0,1)}$. Additionally, by a change of variables, this shows for example that the Fourier transform of $1_{B(0,R)}$ is $R^2 1_{B(0,1/R)}$.

3. Geometry for parabola decoupling and wavepacket decomposition

Having set up the basic geometric and analytic facts over $\mathbb{Q}_q$, we now move to observations that are most relevant to parabola decoupling over $\mathbb{Q}_q$.

3.1. The neighborhood of the parabola. We first finish some easy consequences about θ_J. In addition to (4), for $J \subset \mathbb{Z}_q$, let $\tau_J := B((a,a^2),|J|)$ for any $a \in J$. The definition of τ_J is independent of the choice of $a \in J$ by the ultrametric triangle inequality. Additionally, $\theta_J \subset \tau_J$ since if $(\xi,\eta) \in \theta_J$, then $|\xi - a| \leqslant |J|$ since $\xi \in J$ and

$$|\eta - a^2| = |\eta - \xi^2 + \xi^2 - a^2| \leqslant \max(|\eta-\xi^2|, |\xi-a||\xi+a|) \leqslant |J|$$

where in the last inequality we have also used that $\xi, a \in \mathbb{Z}_q$.

While θ_J is the $|J|^2$ neighborhood of the piece of parabola above J, it also turns out at this scale, the curvature of θ_J is no longer visible.

LEMMA 3.1. *For any $J \subset \mathbb{Z}_q$, θ_J coincides with the parallelogram*

$$\{(\xi, \eta) \in \mathbb{Q}_q^2 : |\xi - a| \leqslant |J|, |\eta - 2a\xi + a^2| \leqslant |J|^2\}$$

where a is any point in J.

PROOF. Fix arbitrary $a \in J$. By Lemma 2.3, $J = \{\xi \in \mathbb{Q}_q : |\xi - a| \leqslant |J|\}$. Next, $|\eta - \xi^2| = |(\eta - 2a\xi + a^2) - (\xi - a)^2|$ and hence if $|\xi - a| \leqslant |J|$, then $|\eta - \xi^2| \leqslant |J|^2$ if and only if $|\eta - 2a\xi + a^2| \leqslant |J|^2$. □

Finally, the following lemma will be useful later in computing the support of a convolution.

LEMMA 3.2. *For any $J \subset \mathbb{Z}_q$, $\theta_J + B(0, |J|^2) = \theta_J$.*

PROOF. The $\supset$ direction is immediate. Let $(\xi_0, \eta_0) \in \theta_J$ and $(r, s) \in B(0, |J|^2)$. Write $J = \{\xi \in \mathbb{Q}_q : |\xi - a| \leqslant |J|\}$ for some $a \in J$. Then $|(\xi_0 + r) - a| \leqslant \max(|\xi_0 - a|, |r|) \leqslant |J|$ and hence $\xi_0 + r \in J$. Next,

$$\begin{aligned}|(\eta_0 + s) - (\xi_0 + r)^2| &= |(\eta_0 - \xi_0^2) + (s - 2r\xi_0 - r^2)| \\ &\leqslant \max(|\eta_0 - \xi_0^2|, |s|, |2r\xi_0|, |r^2|) \leqslant |J|^2\end{aligned}$$

where here we have also used that q is an odd prime and so $|2| = 1$. □

3.2. Geometry of tubes and wavepacket decomposition. Fix a $\delta \in q^{-\mathbb{N}}$. We define the shorthand

$$\theta_\delta := \delta\mathbb{Z}_q \times \delta^2\mathbb{Z}_q \quad \text{and} \quad T_\delta := \delta^{-1}\mathbb{Z}_q \times \delta^{-2}\mathbb{Z}_q.$$

They are dual to each other in the sense that $T_\delta = \{x \in \mathbb{Q}_q^2 : |x \cdot \xi| \leqslant 1 \text{ for all } \xi \in \theta_\delta\}$. Since $\delta\mathbb{Z}_q$ can be partitioned into δ^{-1} many intervals of length δ^2, it follows that θ_δ is the union of δ^{-1} many squares of side length δ^2. Similarly, T_δ is the union of δ^{-1} many squares of side length δ^{-1}.

For $a \in \mathbb{Z}_q$, let $M_a = \left(\begin{smallmatrix} 1 & 0 \\ 2a & 1 \end{smallmatrix}\right)$. Then by Lemma 3.1, for each $K \in P_\delta$, we have

$$\theta_K = \begin{pmatrix} a \\ a^2 \end{pmatrix} + M_a\theta_\delta \tag{6}$$

for any $a \in K$. The right hand side is independent of $a \in K$ since if $b \in K$, then

$$\begin{pmatrix} b \\ b^2 \end{pmatrix} = \begin{pmatrix} a \\ a^2 \end{pmatrix} + \begin{pmatrix} 1 \\ 2a \end{pmatrix}(b - a) + \begin{pmatrix} 0 \\ 1 \end{pmatrix}(b - a)^2 \in \begin{pmatrix} a \\ a^2 \end{pmatrix} + M_a\theta_\delta$$

and $M_a = M_b\left(\begin{smallmatrix} 1 & 0 \\ 2(a-b) & 1 \end{smallmatrix}\right)$ and this latter matrix preserves θ_δ by the ultrametric triangle inequality.

For $K \in P_\delta$ and $a \in K$, let $T_{0,K}$ be the dual parallelogram to θ_K centered at the origin given by

$$T_{0,K} = \{x \in \mathbb{Q}_q^2 : |x \cdot (\xi - \begin{pmatrix} a \\ a^2 \end{pmatrix})| \leqslant 1 \text{ for all } \xi \in \theta_K\}.$$

Using (6),

$$T_{0,K} = \{x \in \mathbb{Q}_q^2 : |x_1 + 2ax_2| \leqslant \delta^{-1}, |x_2| \leqslant \delta^{-2}\} = \{x \in \mathbb{Q}_q^2 : M_a^T x \in T_\delta\} = M_a^{-T}T_\delta$$

for any $a \in K$. This parallelogram depends only on K but not on the choice of $a \in K$ since for any other $b \in K$, $M_a^{-T} = M_b^{-T}\left(\begin{smallmatrix} 1 & 2(b-a) \\ 0 & 1 \end{smallmatrix}\right)$ and $|2(b - a)| \leqslant \delta$ and so this second matrix preserves T_δ by the ultrametric triangle inequality. In analogy to the real setting, we will call $(-2a, 1)$ the direction of $T_{0,K}$.

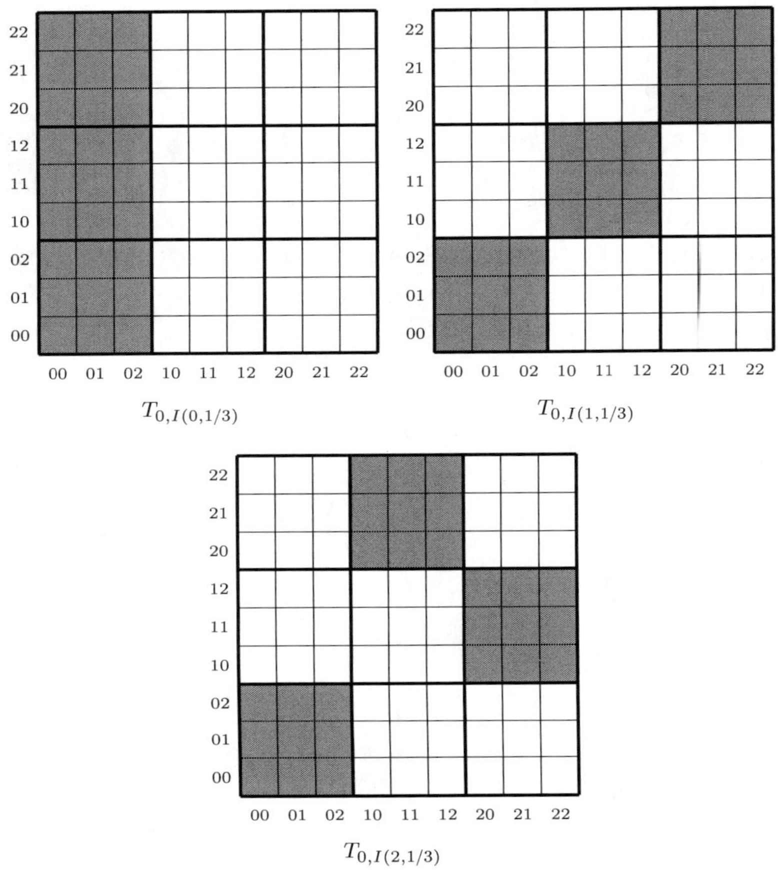

FIGURE 1. In $\mathbb{Q}_3$, the three possible 3×9 parallelograms containing the origin in $B(0,9)$ are: $T_{0,I(0,1/3)} := \{x \in \mathbb{Q}_3^2 : |x_1|_3 \leqslant 3, |x_2|_3 \leqslant 9\}$, $T_{0,I(1,1/3)} := \{x \in \mathbb{Q}_3^2 : |x_1+2x_2|_3 \leqslant 3, |x_2|_3 \leqslant 9\}$, and $T_{0,I(2,1/3)} := \{x \in \mathbb{Q}_3^2 : |x_1 + 4x_2|_3 \leqslant 3, |x_2|_3 \leqslant 9\}$. Here for example, "12" denotes the elements in $\mathbb{Q}_3$ given by $1 \cdot 3^{-2} + 2 \cdot 3^{-1} + \cdots$.

Since T_δ is the union of δ^{-1} many squares of side length δ^{-1}, by Lemma 2.6(ii), it follows that $T_{0,K}$ is also the union of δ^{-1} many squares of side length δ^{-1}. Heuristically, it is useful to think of $T_{0,K}$ as a $\delta^{-1} \times \delta^{-2}$ "tube" pointing in the direction $(-2a, 1)$.

The translates of $T_{0,K}$ pointing in any direction completely partition any square of side length δ^{-2}.

LEMMA 3.3. *Let $\delta \in q^{-\mathbb{N}}$ and fix $K \in P_\delta$. Then a $\delta^{-2} \times \delta^{-2}$ square in $\mathbb{Q}_q^2$ can be partitioned into δ^{-1} many translates $T_{0,K}$.*

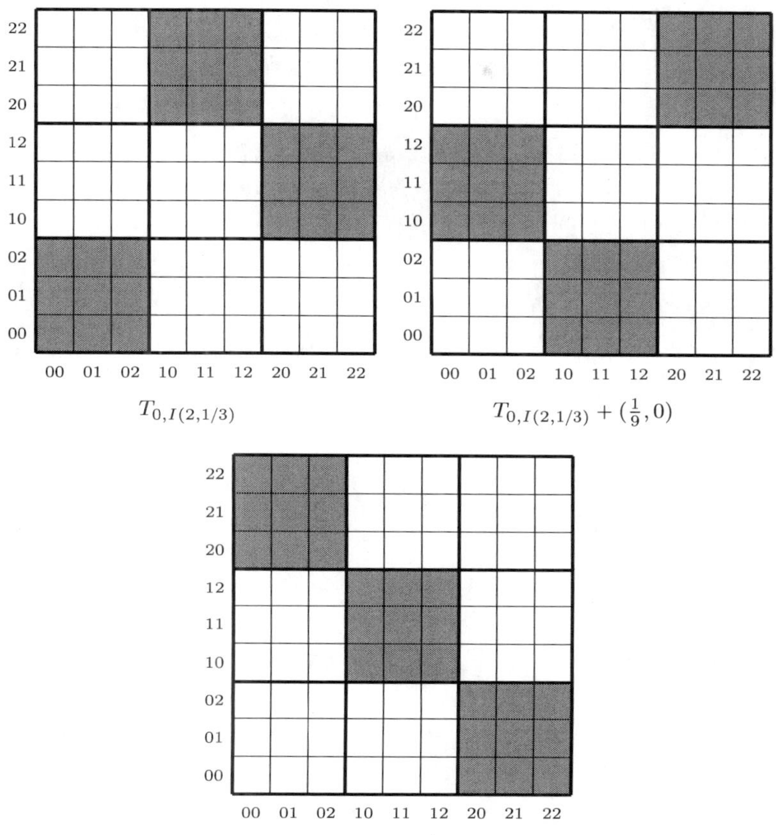

FIGURE 2. The three translates of $T_{0,I(2,1/3)}$ that partition $B(0,9)$. Informally, we are just shifting $T_{0,I(2,1/3)} + (\frac{2}{9}, 0)$ and having the boundary of $B(0,9)$ wrap around.

PROOF. Let B be the square of side length δ^{-2}. Without loss of generality, we may assume that $B = B(0, \delta^{-2})$. Since

$$\{x \in \mathbb{Q}_q : |x| \leqslant \delta^{-2}\} = \bigcup_{t \in \mathcal{S}} \{x \in \mathbb{Q}_q : |x - t| \leqslant \delta^{-1}\}$$

where $\mathcal{S} = \{\sum_{-2N \leqslant j < -N} a_j q^j : a_j \in \{0, 1, \ldots, q-1\}\}$ and $\delta = q^{-N}$, we can write $B = \bigcup_t T_\delta + (t, 0)$. Applying M_a^{-T} to both sides, we obtain that

$$B = M_a^{-T} B = \bigcup_t M_a^{-T} T_\delta + (t, 0) = \bigcup_t T_{0,K} + (t, 0)$$

where in the first equality we have used the ultrametric triangle inequality. $\square$

Lemma 3.3 implies that we can tile $\mathbb{Q}_q^2$ by translates of $T_{0,K}$ for each K. Let $\mathbb{T}(K)$ denote this tiling.

LEMMA 3.4 (Wavepacket decomposition). *Let $\delta \in q^{-\mathbb{N}}$ and fix $K \in P_\delta$. Let f be a Schwartz function with Fourier transform supported in θ_K. Then $|f|$ is constant* *on every $T \in \mathbb{T}(K)$ and $\widehat{f1_T}$ is supported on θ_K for every $T \in \mathbb{T}(K)$. Hence it is natural to write*

$$f = \sum_{T \in \mathbb{T}(K)} f1_T \tag{7}$$

where each term $f1_T$ (which we will call a "wavepacket") is Fourier supported on θ_K and has constant modulus on every $T \in \mathbb{T}(K)$. It also follows that if $\mathcal{T}$ is any subset of $\mathbb{T}(K)$, then $\sum_{T\in\mathcal{T}} f1_T$ is Fourier supported in θ_K.

REMARK 1. In Lemma 3.4, that $|f|$ is constant on every $T \in \mathcal{T}(K)$ is sometimes called the "locally constant property."

PROOF. We will show that $|f(x)|$ is constant for all $x \in A + T_{0,K}$ for any $A \in \mathbb{Q}_q^2$. By Fourier inversion, we have that

$$\begin{aligned} |f(x)| &= \Big|\int_{\theta_K} \widehat{f}(\xi)\chi(\xi \cdot x)\, d\xi\Big| \\ &= \Big|\int_{|s|\leqslant\delta, |t|\leqslant\delta^2} \widehat{f}(s+a, (s+a)^2+t)\chi((s+a)x_1 + (s+a)^2 x_2)\chi(tx_2)\, ds\, dt\Big| \\ &= \Big|\int_{|s|\leqslant\delta, |t|\leqslant\delta^2} \widehat{f}(s+a, (s+a)^2+t)\chi(s(x_1+2ax_2) + s^2x_2)\chi(tx_2)\, ds\, dt\Big|. \end{aligned} \tag{8}$$

For $x \in A + T_{0,K}$, write $x = A + M_a^{-T}y$ for $y \in T_\delta$. Then $M_a^T x = M_a^T A + y$ and hence

$$\chi(s(x_1 + 2ax_2) + s^2x_2) = \chi((s, s^2)\cdot M_a^T A)\chi((s,s^2)\cdot y) = \chi((s,s^2)\cdot M_a^T A)$$

since $sy_1, s^2y_2 \in \mathbb{Z}_q$. Next $\chi(tx_2) = \chi(tA_2)\chi(ty_2) = \chi(tA_2)$ since $ty_2 \in \mathbb{Z}_q$. Therefore (8) is equal to

$$\Big|\int_{|s|\leqslant\delta, |t|\leqslant\delta^2} \widehat{f}(s+a, (s+a)^2+t)\chi((s,s^2)\cdot M_a^T A)\chi(tA_2)\, ds\, dt\Big|$$

which is a constant independent of x and only depends on A, K, and f.

Next to prove $\widehat{f1_T}$ is supported on θ_K, it suffices to observe that $\widehat{f1_T} = \widehat{f} * \widehat{1_T}$. Observe that $\widehat{1_T}$ is a modulation of $\widehat{1_{T_{0,K}}}$ and

$$\widehat{1_{T_{0,K}}} = \int_{M_a^{-T}T_\delta} \chi(-x\cdot\xi)\, dx = \int_{T_\delta} \chi(-M_a^{-T}y\cdot\xi)\, dy = \delta^3 1_{\theta_\delta}(M_a^{-1}\xi).$$

Thus $\widehat{1_T}$ is supported on $M_a\theta_\delta = \theta_K - \theta_K$.

Finally, the decomposition (7) follows since translates of $T_{0,K}$ tiles $\mathbb{Q}_q^2$. This completes the proof. □

COROLLARY 3.5. *If $K \in P_\delta$ and f is Fourier supported on θ_K, then $|f|$ is constant on any ball of radius δ.*

PROOF. Observe that $B(a,\delta) \subset a + T_{0,K}$ and $|f|$ is constant on $a + T_{0,K}$ and hence also on $B(a,\delta)$. □

We now end with a lemma about the intersection of two nonparallel tubes (of course we only use this bound when they intersect).

LEMMA 3.6. *Let $\delta \in q^{-\mathbb{N}}$ and $K, K' \in P_\delta$ be two distinct intervals. Consider a translate T of $T_{0,K}$ and a translate T' of $T_{0,K'}$. Then $|T \cap T'| \leqslant \delta^{-2}/d(K,K')$.*

PROOF. Write $T = T_{0,K} + (r,s)$ and $T' = T_{0,K'} + (r',s')$. Without loss of generality, we may assume that $(r,s) = (0,0)$. Applying M_a^T we may assume that $T = T_\delta$. Then for some $b \in K'$,

$$T \cap T' \subset \{x \in \mathbb{Q}_q^2 : |x_1| \leqslant \delta^{-1}, |x_1 + 2bx_2 - r'| \leqslant \delta^{-1}\}$$
$$\subset \{x \in \mathbb{Q}_q^2 : |x_1| \leqslant \delta^{-1}, |x_2 - \frac{r'}{2b}| \leqslant \frac{\delta^{-1}}{|b|}\}$$

which has area $\leqslant \delta^{-2}/|b| = \delta^{-2}/d(K,K')$. □

4. Basics of decoupling

For ease of notation, we will now let $D_p(\delta) := D_{p,\mathbb{Q}_q}(\delta)$.

4.1. Sharpness of Theorem 1.1. Before we start our proof of Theorem 1.1, we first prove that up to a $\delta^{-\varepsilon}$ factor, the estimate is sharp. We first start with a computation.

LEMMA 4.1. *We have*

$$\check{1}_{\theta_K}(x) = \chi(ax_1 + a^2x_2)\delta^3 1_{T_{0,K}}$$

for any $a \in K$.

REMARK 2. The right hand side is well-defined since for $x \in T_{0,K}$,

$$\chi((a-b)x_1 + (a^2-b^2)x_2) = \chi((a-b)(x_1+2ax_2) - (a-b)^2x_2) = 1$$

if $a, b \in K$ (and so $|a-b| \leqslant \delta$).

PROOF. We compute

$$\check{1}_{\theta_K} = \int_{\theta_K} \chi(x\cdot\xi)\,d\xi = \int_{|s|\leqslant\delta, |t|\leqslant\delta^2} e((s+a)x_1 + (s+a)^2x_2)e(tx_2)\,ds\,dt$$
$$= \chi(ax_1 + a^2x_2)\int_{|t|\leqslant\delta^2}\chi(tx_2)\,dt\int_{|s|\leqslant\delta}\chi(s(x_1+2ax_2)+s^2x_2)\,ds.$$

By Proposition 2.7, the integral in t is equal to $\delta^2 1_{|x_2|\leqslant\delta^{-2}}$ and the integral in s is equal to $\delta 1_{|x_1+2ax_2|\leqslant\delta^{-1}}$. This completes the proof of the lemma. □

This then allows us to show the following result.

PROPOSITION 4.2. *We have $D_p(\delta) \gtrsim 1 + \delta^{-(\frac{1}{2}-\frac{3}{p})}$.*

PROOF. To see that $D_p(\delta) \geqslant 1$, take $f = \check{1}_{\theta_\delta}$. Then

$$\|f\|_{L^p(\mathbb{Q}_q^2)} = \|\check{1}_{\theta_\delta}\|_{L^p(\mathbb{Q}_q^2)} = (\sum_{K\in P_\delta}\|f_K\|^2_{L^p(\mathbb{Q}_q^2)})^{1/2}$$

and hence

$$D_p(\delta) \geqslant \frac{\|f\|_{L^p(\mathbb{Q}_q^2)}}{(\sum_{K\in P_\delta}\|f_K\|^2_{L^p(\mathbb{Q}_q^2)})^{1/2}} = 1.$$

Next, to see that $D_p(\delta) \geqslant \delta^{-(\frac{1}{2}-\frac{3}{p})}$, take $f = \sum_{K\in P_\delta} f_K$ where f_K is such that $\widehat{f_K} = \delta^{-3}1_{\theta_K}$. Therefore $f_K = \chi(a_Kx_1 + a_K^2x_2)1_{T_{0,K}}$ for any $a_K \in K$. This implies that $\|f_K\|_{L^p(\mathbb{Q}_q^2)} = \delta^{-3/p}$ and so

$$(\sum_{K\in P_\delta}\|f_K\|^2_{L^p(\mathbb{Q}_q^2)})^{1/2} = \delta^{-1/2-3/p}.$$

We also have

$$f=\sum_{K\in P_\delta} f_K=\sum_{K\in P_\delta}\chi(a_Kx_1+a_K^2x_2)1_{T_{0,K}}.$$

For $x\in\mathbb{Z}_q^2$, since $a_K\in K\subset\mathbb{Z}_q$, $a_Kx_1+a_K^2x_2\in\mathbb{Z}_q$ and hence $\chi(a_Kx_1+a_K^2x_2)=1$ for $x\in\mathbb{Z}_q^2$. Furthermore, $\mathbb{Z}_q^2\subset T_{0,K}$ for every $K\in P_\delta$. Thus

$$\|f\|_{L^p(\mathbb{Q}_q^2)}\geqslant\|f\|_{L^p(\mathbb{Z}_q^2)}\geqslant\delta^{-1}$$

and hence

$$D_p(\delta)\geqslant\frac{\|f\|_{L^p(\mathbb{Q}_q^2)}}{(\sum_{K\in P_\delta}\|f_K\|^2_{L^p(\mathbb{Q}_q^2)})^{1/2}}\geqslant\delta^{-1-(-\frac{1}{2}-\frac{3}{p})}=\delta^{-(\frac{1}{2}-\frac{3}{p})}.$$

□

4.2. Interpolation of decoupling estimates. Now we reduce Theorem 1.1 to just showing the case when $p=6$. To do this, we need to be able to interpolate decoupling estimates. First we need that Fourier restriction to θ_K is a bounded operator on L^p.

LEMMA 4.3. *Let $K\in P_\delta$ and f be Fourier supported in $\bigcup_{K\in P_\delta}\theta_K$. Then*

$$\|f_K\|_{L^p(\mathbb{Q}_q^2)}\leqslant\|f\|_{L^p(\mathbb{Q}_q^2)}$$

for all $1\leqslant p\leqslant\infty$.

PROOF. By definition, $f_K=f*\check{1}_{\theta_K}$ and hence Young's inequality gives that

$$\|f_K\|_{L^p(\mathbb{Q}_q^2)}=\|f*\check{1}_{\theta_K}\|_{L^p(\mathbb{Q}_q^2)}\leqslant\|f\|_{L^p(\mathbb{Q}_q^2)}\|\check{1}_{\theta_K}\|_{L^1(\mathbb{Q}_q^2)}.$$

Applying Lemma 4.1 shows that $\|\check{1}_{\theta_K}\|_{L^1(\mathbb{Q}_q^2)}=1$ which completes the proof of the lemma. □

Now we are ready to show that one can interpolate decoupling estimates.

PROPOSITION 4.4. *If $\frac{1}{p}=\frac{\theta}{p_0}+\frac{1-\theta}{p_1}$, then*

$$D_p(\delta)\leqslant D_{p_0}(\delta)^\theta D_{p_1}(\delta)^{1-\theta}.$$

PROOF. Write $P_\delta:=\{I_1,\dots,I_N\}$ for some $N=\delta^{-1}\in q^{\mathbb{N}}$. Define the operator

$$T[(f_j)_{j=1}^N]:=\sum_{j=1}^N P_{\theta_{I_j}}f_j$$

where $P_{\theta_{I_j}}f:=f*\check{1}_{\theta_{I_j}}$ is the Fourier projection of f to θ_{I_j}. Then for $r=p_0,p_1$, we have

$$\begin{aligned}\|T[(f_j)]\|_{L^r(\mathbb{Q}_q^2)}&\leqslant D_r(\delta)(\sum_{k=1}^N\|(\sum_{j=1}^N P_{\theta_{I_j}}f_j)_{I_k}\|^2_{L^r(\mathbb{Q}_q^2)})^{1/2}\\&=D_r(\delta)(\sum_{k=1}^N\|P_{\theta_{I_k}}f_k\|^2_{L^r(\mathbb{Q}_q^2)})^{1/2}\\&\leqslant D_r(\delta)(\sum_{k=1}^N\|f_k\|^2_{L^r(\mathbb{Q}_q^2)})^{1/2}=D_r(\delta)\|(f_j)\|_{\ell^2L^r(\mathbb{Q}_q^2)}\end{aligned}$$

where in the second inequality we have used Lemma 4.3. Therefore the operator T sends $\ell^2(L^{p_0}(\mathbb{Q}_q^2))$ into $L^{p_0}(\mathbb{Q}_q^2)$ and also $\ell^2(L^{p_1}(\mathbb{Q}_q^2))$ into $L^{p_1}(\mathbb{Q}_q^2)$. Thus by Riesz-Thorin interpolation, $T : \ell^2(L^p(\mathbb{Q}_q^2)) \to L^p(\mathbb{Q}_q^2)$ for all $p \in [p_0, p_1]$ and the operator norm of T, that is, $D_p(\delta)$, is $\leqslant D_{p_0}(\delta)^\theta D_{p_1}(\delta)^{1-\theta}$. $\square$

Therefore if we could show that $D_6(\delta) \lesssim_\varepsilon \delta^{-\varepsilon}$, then interpolating this with the estimate that $D_2(\delta) = 1$, which follows from Plancherel, and $D_\infty(\delta) \leqslant \delta^{-1/2}$ which follows from Cauchy-Schwarz proves Theorem 1.1. Thus it remains to just show the l^2L^6 decoupling estimate.

4.3. Parabola decoupling over $\mathbb{Q}_q$ implies quadratic VMVT. Before we begin our proof of l^2L^6 parabola decoupling, we show that such a decoupling theorem over $\mathbb{Q}_q$ is strong enough to imply exponential sum estimates.

LEMMA 4.5. *Suppose for some odd prime q, we knew that*

$$\|f\|_{L^6(\mathbb{Q}_q^2)} \lesssim_{q,\varepsilon} \delta^{-\varepsilon}\Big(\sum_{J\in P_\delta} \|f_J\|^2_{L^6(\mathbb{Q}_q^2)}\Big)^{1/2} \tag{9}$$

for all f Fourier supported in $\bigcup_{K\in P_\delta}\theta_K$. Then $K(N) \lesssim_{q,\varepsilon} N^\varepsilon$.

PROOF. By zeroing out entries, $K(N)$ is increasing in N. Thus it suffices to only show the desired estimate in the case when $N = q^t$ for some $t \in \mathbb{N}$. Indeed, suppose we knew this for all $N \in q^{\mathbb{N}}$. Then for $N \notin q^{\mathbb{N}}$, we can find an s such that $q^s < N < q^{s+1}$ and hence for some C_ε, $K(N) \leqslant K(q^{s+1}) \leqslant C_{\varepsilon,q}(q^{s+1})^\varepsilon = C_{\varepsilon,q}q^\varepsilon q^{s\varepsilon} \leqslant C_{\varepsilon,q}q^\varepsilon N^\varepsilon$.

Choose $\delta = q^{-t}$ and F such that

$$\widehat{F}(\xi,\eta) = \sum_{n=1}^{q^t} q^{20t} a_n 1_{(n,n^2)+B(0,q^{-10t})}(\xi,\eta).$$

We first check that $\widehat{F}$ is supported in $\bigcup_{K\in P_{q^{-t}}}\theta_K$. It suffices to show that for each $K = I(n,q^{-t})$, we have $(n,n^2) + B(0,q^{-10t}) \subset \theta_K$. Indeed if $|\xi - n| \leqslant q^{-10t}$ and $|\eta - n^2| \leqslant q^{-10t}$, then $\xi \in K$ and

$$|\xi^2 - \eta| = |(\xi-n)^2 + 2n(\xi-n) + n^2 - \eta| \leqslant \max(|\xi-n|^2, |2n||\xi-n|, |n^2-\eta|).$$

Since $q \geqslant 3$ is an odd prime, $|2n| \leqslant 1$ and so the above is $\leqslant q^{-10t} \leqslant q^{-2t}$.

Inverting the Fourier transform gives that

$$F(x) = \Big(\sum_{n=1}^{q^t} a_n\chi(x_1n + x_2n^2)\Big)1_{B(0,q^{10t})}(x)$$

and $F_K = a_n\chi(x_1n + x_2n^2)1_{B(0,q^{10t})}(x)$ with $K = I(n,q^{-t})$. This gives that $\|F_K\|_{L^6(\mathbb{Q}_q^2)} = q^{20t/6}|a_n|$ and so

$$\Big(\sum_{K\in P_\delta} \|F_K\|^2_{L^6(\mathbb{Q}_q^2)}\Big)^3 = q^{20t}\Big(\sum_{n=1}^{q^t}|a_n|^2\Big)^3. \tag{10}$$

On the other hand, expanding yields

$$\|F\|_{L^6(\mathbb{Q}_q^2)}^6 = \sum_{n_1,\dots,n_6=1}^{q^t} a_{n_1}a_{n_2}a_{n_3}\overline{a_{n_4}a_{n_5}a_{n_6}}\times$$

$$\int_{B(0,q^{10t})} \chi((n_1+n_2+n_3-n_4-n_5-n_6)x_1+(n_1^2+n_2^2+n_3^2-n_4^2-n_5^2-n_6^2)x_2)\,dx$$

and applying Proposition 2.7 gives that the above is equal to

$$q^{20t}\sum_{n_1,\dots,n_6=1}^{q^t} a_{n_1}a_{n_2}a_{n_3}\overline{a_{n_4}a_{n_5}a_{n_6}}\times 1_{|n_1+n_2+n_3-n_4-n_5-n_6|\leqslant q^{-10t},|n_1^2+n_2^2+n_3^2-n_4^2-n_5^2-n_6^2|\leqslant q^{-10t}}. \tag{11}$$

Note that $(n_1,\dots,n_6)\in([1,q^t]\cap\mathbb{N})^6$ satisfies

$$\begin{aligned} n_1+n_2+n_3-n_4-n_5-n_6 &\equiv 0 \pmod{q^{10t}}\\ n_1^2+n_2^2+n_3^2-n_4^2-n_5^2-n_6^2 &\equiv 0 \pmod{q^{10t}}\end{aligned} \tag{12}$$

if and only if it satisfies

$$\begin{aligned} n_1+n_2+n_3-n_4-n_5-n_6 &= 0\\ n_1^2+n_2^2+n_3^2-n_4^2-n_5^2-n_6^2 &= 0.\end{aligned} \tag{13}$$

Indeed, since $n_i\in[1,q^t]\cap\mathbb{N}$, we have $n_1+n_2+n_3-n_4-n_5-n_6\in[-3q^t,3q^t]\cap\mathbb{N}$ and $n_1^2+n_2^2+n_3^2-n_4^2-n_5^2-n_6^2\in[-3q^{2t},3q^{2t}]\cap\mathbb{N}$. Since $3q^t\leqslant 3q^{2t}\leqslant q^{10t}$, the only integer $\equiv 0 \pmod{q^{10t}}$ in $[-3q^t,3q^t]$ and $[-3q^{2t},3q^{2t}]$ is 0 itself. Thus (11) is equal to

$$q^{20t}\|\sum_{n=1}^{q^t} a_n e(nx_1+n^2x_2)\|_{L^6(\mathbb{T}^2)}^6.$$

Combining this with (10) and (9), then shows that $K(N)\lesssim_{q,\varepsilon} N^{\varepsilon}$ as desired. □

REMARK 3. The equivalence of (12) with (13) for $n_i\in[1,q^t]\cap\mathbb{N}$ is the same equivalence used in efficient congruencing, see for example [**26**, Eq. (2.9)].

5. A proof of l^2L^6 parabola decoupling

Our main goal now is to prove that $D_6(\delta)\lesssim_\varepsilon \delta^{-\varepsilon}$. For ease of notation, we will let $D(\delta):=D_6(\delta)=D_{6,\mathbb{Q}_q}(\delta)$.

5.1. Parabolic rescaling.

PROPOSITION 5.1. *Let f be a Schwartz function on $\mathbb{Q}_q^2$ which is Fourier supported in $\bigcup_{K\in P_\delta}\theta_K$. Then for any interval $I\subset\mathbb{Z}_q$ of length $\sigma\geqslant\delta$, we have*

$$\|f_I\|_{L^p(\mathbb{Q}_q^2)}\leqslant D_p(\frac{\delta}{\sigma})(\sum_{K\in P_\delta(I)}\|f_K\|_{L^p(\mathbb{Q}_q^2)}^2)^{1/2}.$$

PROOF. Write $I=I(a,\sigma)$ for some $a\in I$. We have

$$|f_I(x)|=|\int_{|t|\leqslant\delta^2}\int_{I(a,\sigma)}\widehat{f}(s,s^2+t)\chi(sx_1+s^2x_2)e(tx_2)\,ds\,dt|.$$

Let $s' := \sigma^{-1}(s-a)$ and $t' := \sigma^{-2}t$. Then the above is equal to

$$|\sigma^3 \int_{|t'|\leqslant(\delta/\sigma)^2}\int_{\mathbb{Z}_q} \widehat{f}(\sigma s'+a,(\sigma s'+a)^2+\sigma^2 t')$$
$$\chi(\sigma s'(x_1+2ax_2)+\sigma^2 s'^2 x_2)\chi(\sigma^2 t' x_2)\,ds'\,dt'|.$$

Write $T_{a,\sigma} = \begin{pmatrix} \sigma & 2a\sigma \\ 0 & \sigma^2 \end{pmatrix}$, $f_a(x) := f(x)\chi(ax_1+a^2x_2)$, and $g_a(y) = \sigma^{-3}f_a(T_{a,\sigma}^{-1}y)$, then $\widehat{g_a}(s', s'^2+t') = \widehat{f}(\sigma s'+a,(\sigma s'+a)^2+\sigma^2 t')$ and g_a is Fourier supported in $\bigcup_{K'\in P_{\delta/\sigma}}\theta_{K'}$. Therefore $\|f_I\|_{L^p(\mathbb{Q}_q^2)} = \sigma^{3-3/p}\|g_a\|_{L^p(\mathbb{Q}_q^2)}$. Applying the definition of $D_p(\delta/\sigma)$ and undoing the above change of variables finishes the proof. □

5.2. Bilinear reduction. Let $\nu\in q^{-\mathbb{N}}$ be an intermediate scale chosen later such that $\delta \ll \nu \ll 1$. On a first read, it may be useful to set $\nu = 1/q$ though we will later choose it to be something larger, see the proof of Proposition 5.11. Suppose a and b were integers such that $\delta \leqslant \min(\nu^a,\nu^b)$. Let $M_{a,b}(\delta,\nu)$ be the best constant such that

$$\int_{\mathbb{Q}_q^2}|f_I|^2|f_{I'}|^4 \leqslant M_{a,b}(\delta,\nu)^6\Big(\sum_{K\in P_\delta(I)}\|f_K\|^2_{L^6(\mathbb{Q}_q^2)}\Big)\Big(\sum_{K'\in P_\delta(I')}\|f_{K'}\|^2_{L^6(\mathbb{Q}_q^2)}\Big)^2 \tag{14}$$

for all f Fourier supported in $\bigcup_{K\in P_\delta}\theta_K$ and all intervals $I, I'\subset\mathbb{Z}_q$ with $|I| = \nu^a$, $|I'| = \nu^b$ and $d(I,I') > \nu$.

If we think of $\nu \approx \delta^\varepsilon$ and a and b are small, then we essentially have $M_{a,b}(\delta,\nu)\approx D(\delta)$ as I and I' are rather large intervals and we have not decoupled much. However, if a and b were so large such that ν^a and ν^b are basically the same size as δ, then $M_{a,b}(\delta,\nu)\approx 1$ after applying the triangle inequality. Thus our goal will be to pass from small (a,b) to large (a,b) efficiently.

The following trivial bound will be useful when we reach large (a,b).

LEMMA 5.2. *If $\delta\leqslant\min(\nu^a,\nu^b)$, then*

$$M_{a,b}(\delta,\nu)\leqslant D\Big(\frac{\delta}{\nu^a}\Big)^{1/3}D\Big(\frac{\delta}{\nu^b}\Big)^{2/3}.$$

PROOF. This follows immediately from $\int f^2g^4 \leqslant (\int f^6)^{1/3}(\int g^6)^{2/3}$, parabolic rescaling, and applying the definitions of $D(\delta)$ and $M_{a,b}(\delta,\nu)$. □

The previous lemma also implies that if we knew a good bound for $D(\delta)$, then we would know a good bound for $M_{a,b}(\delta,\nu)$. However, what we want is the opposite. This is accomplished in the next proposition and essentially reduces the linear decoupling problem to a bilinear decoupling problem.

PROPOSITION 5.3. *If $\delta\leqslant\nu$, then*

$$D(\delta)\lesssim D\Big(\frac{\delta}{\nu}\Big)+\nu^{-O(1)}M_{1,1}(\delta,\nu).$$

PROOF. Fix arbitrary f Fourier supported in $\bigcup_{K\in P_\delta}\theta_K$. We have

$$\int_{\mathbb{Q}_q^2}|f|^6 = \int_{\mathbb{Q}_q^2}\Big|\sum_{J,J'\in P_\nu} f_Jf_{J'}\Big|^3 \lesssim \int_{\mathbb{Q}_q^2}\Big|\sum_{J\in P_\nu} f_J^2\Big|^3 + \int_{\mathbb{Q}_q^2}\Big|\sum_{\substack{J,J'\in P_\nu\\ d(J,J')>\nu}} f_Jf_{J'}\Big|^3.$$

By Minkowski's inequality and parabolic rescaling, the first term above is

$$\leqslant \Big(\sum_{J\in P_\nu} \|f_J\|_{L^6(\mathbb{Q}_q^2)}^2\Big)^{\frac{1}{2}\cdot 6} \leqslant D\Big(\frac{\delta}{\nu}\Big)^6\Big(\sum_{K\in P_\delta} \|f_K\|_{L^6(\mathbb{Q}_q^2)}^2\Big)^{\frac{1}{2}\cdot 6}. \tag{15}$$

On the other hand, the second term is

$$\leqslant \nu^{-O(1)} \max_{\substack{J,J'\in P_\nu\\ d(J,J')>\nu}} \int_{\mathbb{Q}_q^2} |f_J|^3|f_{J'}|^3. \tag{16}$$

Observing that $\int f^3g^3 \leqslant (\int f^2g^4)^{1/2}(\int f^4g^2)^{1/2}$ and then applying the definition of $M_{1,1}(\delta,\nu)$ to each term then shows that (16) is

$$\leqslant \nu^{-O(1)}M_{1,1}(\delta,\nu)^6\Big(\sum_{K\in P_\delta} \|f_K\|_{L^6(\mathbb{Q}_q^2)}^2\Big)^3.$$

Combining this with (15) and applying the definition of $D(\delta)$ completes the proof. □

As our argument is bilinear in nature, we could use Hölder's inequality to reduce to the bilinear decoupling problem. Before we continue, we give a second proof of Proposition 5.3 that relies on the broad/narrow decomposition first introduced by Bourgain and Guth in [**6**]. The broad/narrow decomposition is more commonly used to reduce linear decoupling problems to multilinear decoupling problems. Let f be Fourier supported in $\bigcup_{K\in P_\delta}\theta_K$. For each $x_0 \in \mathbb{Q}_q^2$, let $\mathcal{I}_{x_0}$ be the set of all intervals $J' \in P_\nu$ such that $|f_{J'}(x_0)| \geqslant \nu\max_{J\in P_\nu}|f_J(x_0)|$. We have two cases: $\#\mathcal{I}_{x_0} \geqslant 2$ or $\#\mathcal{I}_{x_0} < 2$.

If $\#\mathcal{I}_{x_0} \geqslant 2$, let J_1 and J_2 be two distinct intervals in $\mathcal{I}_{x_0}$. Note that the choice of J_1 and J_2 depends on x_0. We then have

$$\begin{aligned}|f(x_0)| &\leqslant \sum_{J\in P_\nu}|f_J(x_0)| \leqslant \nu^{-1}\max_{J\in P_\nu}|f_J(x_0)|\\ &\leqslant \nu^{-2}|f_{J_1}(x_0)|^{1/2}|f_{J_2}(x_0)|^{1/2} \leqslant \nu^{-2}\max_{\substack{J,J'\in P_\nu\\ d(J,J')>\nu}}|f_J(x_0)f_{J'}(x_0)|^{1/2}\end{aligned}$$

where in the third inequality we have used that $J_1, J_2 \in \mathcal{I}_{x_0}$ and in the last inequality we have used that J_1 and J_2 are distinct and so $d(J_1,J_2) \geqslant q\nu$ by Lemma 2.1. On the other hand if $\#\mathcal{I}_{x_0} < 2$, then we have

$$\begin{aligned}|f(x_0)| &\leqslant \Big|\sum_{J\in\mathcal{I}_{x_0}} f_J(x_0)\Big| + \Big|\sum_{J\notin\mathcal{I}_{x_0}} f_J(x_0)\Big|\\ &< 2\max_{J\in P_\nu}|f_J(x_0)| + \sum_{J\notin\mathcal{I}_{x_0}}|f_J(x_0)| \leqslant 3\max_{J\in P_\nu}|f_J(x_0)|.\end{aligned}$$

Therefore we have shown that for each $x \in \mathbb{Q}_q^2$, we have

$$|f(x)| \leqslant 3\max_{J\in P_\nu}|f_J(x)| + \nu^{-2}\max_{\substack{J,J'\in P_\nu\\ d(J,J')>\nu}}|f_J(x)f_{J'}(x)|^{1/2}.$$

Raising both sides to the 6th power and integrating gives

$$\begin{aligned}\int_{\mathbb{Q}_q^2} |f|^6 &\lesssim \int_{\mathbb{Q}_q^2} \max_{J \in P_\nu} |f_J(x)|^6\, dx + \nu^{-O(1)} \int_{\mathbb{Q}_q^2} \max_{\substack{J,J' \in P_\nu \\ d(J,J') > \nu}} |f_J(x) f_{J'}(x)|^3 \\ &\lesssim \int_{\mathbb{Q}_q^2} \Big(\sum_{J \in P_\nu} |f_J(x)|^2\Big)^3 + \nu^{-O(1)} \max_{\substack{J,J' \in P_\nu \\ d(J,J') > \nu}} \int_{\mathbb{Q}_q^2} |f_J(x) f_{J'}(x)|^3.\end{aligned}$$

Now proceeding as in (15) and (16) then finishes the proof.

5.3. The key step. The following lemma turns out to be the efficient way to climb up from small (a, b) to large (a, b).

PROPOSITION 5.4. *If $1 \leqslant a \leqslant 2b$ and $\delta \leqslant \nu^{2b}$, then*

$$M_{a,b}(\delta, \nu) \lesssim \nu^{-O(1)} M_{2b,b}(\delta, \nu).$$

To prove Proposition 5.4, it suffices to show the following lemma.

LEMMA 5.5. *If $1 \leqslant a \leqslant 2b$, $|I| = \nu^a$, $|I'| = \nu^b$, $d(I, I') \geqslant q\nu$, and $\delta \leqslant \nu^{2b}$ we have*

$$\int_{\mathbb{Q}_q^2} |f_I|^2 |f_{I'}|^4 \lesssim \nu^{-O(1)} \sum_{J \in P_{\nu^{2b}}(I)} \int_{\mathbb{Q}_q^2} |f_J|^2 |f_{I'}|^4$$

for all f Fourier supported in $\bigcup_{K \in P_\delta} \theta_K$.

Indeed, Proposition 5.4 follows from Lemma 5.5 by applying the definition of $M_{2b,b}(\delta, \nu)$ to $\int_{\mathbb{Q}_q^2} |f_J|^2 |f_{I'}|^4$ and then applying the definition of $M_{a,b}(\delta, \nu)$. Next, by the proof of parabolic rescaling and shifting I' to the origin, to prove Lemma 5.5 it suffices to instead prove the following result.

LEMMA 5.6. *If $1 \leqslant a \leqslant 2b$, $|I| = \nu^a$, $d(I, 0) \geqslant q\nu$, and $\delta \leqslant \nu^{2b}$ we have*

$$\int_{\mathbb{Q}_q^2} |f_I|^2 |f_{I(0,\nu^b)}|^4 \lesssim \nu^{-O(1)} \sum_{J \in P_{\nu^{2b}}(I)} \int_{\mathbb{Q}_q^2} |f_J|^2 |f_{I(0,\nu^b)}|^4 \tag{17}$$

for all f Fourier supported in $\bigcup_{K \in P_\delta} \theta_K$.

We will give five somewhat different proofs of Lemma 5.6 later in Section 6. In some of the proofs we will prove instead the following reduction.

LEMMA 5.7. *If $1 \leqslant a \leqslant 2b$, $|I| = \nu^a$, $d(I, 0) \geqslant q\nu$, and $\delta \leqslant \nu^{2b}$, we have*

$$\int_{B(0,\nu^{-2b})} |f_I|^2 |f_{I(0,\nu^b)}|^4 \lesssim \nu^{-O(1)} \sum_{J \in P_{\nu^{2b}}(I)} \int_{B(0,\nu^{-2b})} |f_J|^2 |f_{I(0,\nu^b)}|^4 \tag{18}$$

for all f Fourier supported in $\bigcup_{K \in P_\delta} \theta_K$.

To see that Lemma 5.7 implies Lemma 5.6 (and hence Proposition 5.4), since (18) is true for all f Fourier supported in $\bigcup_{K \in P_\delta} \theta_K$, by translation we have that

$$\int_B |f_I|^2 |f_{I(0,\nu^b)}|^4 \lesssim \nu^{-O(1)} \sum_{J \in P_{\nu^{2b}}(I)} \int_B |f_J|^2 |f_{I(0,\nu^b)}|^4 \tag{19}$$

for all squares B of side length ν^{-2b} and all f Fourier supported in $\bigcup_{K \in P_\delta} \theta_K$. Since squares of side length ν^{-2b} tile $\mathbb{Q}_q^2$, summing (19) over B then gives Lemma 5.6.

5.4. Iteration. Before we give its proof, we now show that if we had Lemma 5.6 (and hence Proposition 5.4), then we will have shown $D(\delta) \lesssim_{q,\varepsilon} \delta^{-\varepsilon}$ which by the interpolation from Proposition 4.4 proves Theorem 1.1. Proposition 5.4 allows us to climb in the first index in $M_{a,b}(\delta,\nu)$. To move this increase to the second index, we need the following.

PROPOSITION 5.8. *If $\delta \leqslant \min(\nu^a,\nu^b)$, then*

$$M_{a,b}(\delta,\nu) \leqslant M_{b,a}(\delta,\nu)^{1/2} D(\frac{\delta}{\nu^b})^{1/2}.$$

PROOF. This follows from $\int f^2g^4 \leqslant (\int f^4g^2)^{1/2}(\int g^6)^{1/2}$ and applying the definition of $M_{b,a}(\delta,\nu)$ to the first term on the right hand side and parabolic rescaling to the second term on the right hand side. □

LEMMA 5.9. *If $N \in \mathbb{N}$ and $\delta, \nu \in q^{-\mathbb{N}}$ are such that $\delta \leqslant \nu^{2^N}$, then*

$$M_{1,1}(\delta,\nu) \leqslant C_1^2\nu^{-O(1)} D(\frac{\delta}{\nu^{2^{N-1}}})^{\frac{1}{3\cdot 2^N}} D(\frac{\delta}{\nu^{2^N}})^{\frac{2}{3\cdot 2^N}} \prod_{j=0}^{N-1} D(\frac{\delta}{\nu^{2^j}})^{\frac{1}{2^{j+1}}}$$

where C_1 is the implied absolute constant from Proposition 5.4.

PROOF. Repeatedly applying Proposition 5.4 followed by Proposition 5.8 a total of N times gives that

$$M_{1,1}(\delta,\nu) \leqslant (C_1\nu^{-O(1)})^2 M_{2^{N-1},2^N}(\delta,\nu)^{1/2^N} \prod_{j=0}^{N-1} D(\frac{\delta}{\nu^{2^j}})^{1/2^{j+1}}.$$

Applying Lemma 5.2, then completes the proof. □

Combining this with Proposition 5.3 gives the following.

LEMMA 5.10. *If $N \in \mathbb{N}$ and $\delta, \nu \in q^{-\mathbb{N}}$ are such that $\delta \leqslant \nu^{2^N}$, then*

$$D(\delta) \leqslant C_2 D(\frac{\delta}{\nu}) + C_2C_1^2\nu^{-O(1)} D(\frac{\delta}{\nu^{2^{N-1}}})^{\frac{1}{3\cdot 2^N}} D(\frac{\delta}{\nu^{2^N}})^{\frac{2}{3\cdot 2^N}} \prod_{j=0}^{N-1} D(\frac{\delta}{\nu^{2^j}})^{\frac{1}{2^{j+1}}}$$

where C_1 is the implied absolute constant from Proposition 5.4 *and C_2 is the implied absolute constant from Proposition* 5.3.

We now just need to iterate Lemma 5.10. To avoid having to deal with the constraints that $\delta, \nu \in q^{-\mathbb{N}}$, it will be more convenient to bound

$$\mathfrak{D}(\delta) := \sup_{\delta_0 \in q^{-\mathbb{N}} \cap [\delta,1]} D(\delta_0) \tag{20}$$

instead of $D(\delta)$ as $\mathfrak{D}(\delta)$ is defined for all real $\delta \in (0,1]$ (rather than just for $\delta \in q^{-\mathbb{N}}$) and is monotonic, that is, $\mathfrak{D}(\delta_L) \leqslant \mathfrak{D}(\delta_S)$ if $\delta_L \geqslant \delta_S$. (A warning: The notation in (20) is reverse from what it is in [**8**, Section 5].) Trivially we also have that if $\delta \in q^{-\mathbb{N}}$, then $D(\delta) \leqslant \mathfrak{D}(\delta)$. Thus to show that $D(\delta) \lesssim_{q,\varepsilon} \delta^{-\varepsilon}$ for all $\delta \in q^{-\mathbb{N}}$, it suffices to show that $\mathfrak{D}(\delta) \lesssim_{q,\varepsilon} \delta^{-\varepsilon}$ for all $\delta \in (0,1]$.

PROPOSITION 5.11. *If $N \in \mathbb{N}$ and $\delta \in (0,1]$, then*

$$\begin{aligned}\mathfrak{D}(\delta) \leqslant &(q^{2^{N-1}} + C_2)\mathfrak{D}(q^{-1}\delta^{1-\frac{1}{2^N}})\\ &+ (q^{2^{N-1}} + C_2C_1^2q^{1/3})\delta^{-\frac{C_3}{2^N}} \mathfrak{D}(q^{-2^{N-1}}\delta^{1/2})^{\frac{1}{3\cdot 2^N}} \prod_{j=0}^{N-1} \mathfrak{D}(q^{-2^j}\delta^{1-\frac{1}{2^{N-j}}})^{\frac{1}{2^{j+1}}}\end{aligned}$$

where C_1 is the implied absolute constant from Proposition 5.4*, C_2 is the implied absolute constant from Proposition* 5.3*, and C_3 is the implied absolute constant from the $\nu^{-O(1)}$ in Lemma* 5.10.

PROOF. Fix $N \in \mathbb{N}$ and $\delta_0 \in q^{-\mathbb{N}} \cap [\delta, 1]$. Suppose first that $\delta_0 \leqslant q^{-2^N}$. Let $\nu := q^{\lceil \log_q \delta_0^{1/2^N} \rceil} \geqslant \delta_0^{1/2^N} \geqslant \delta^{1/2^N}$. Since $\delta_0 \leqslant q^{-2^N}$, $\nu \in q^{-\mathbb{N}}$. Note that for $j = 0, 1, 2, \ldots, N$,

$$\frac{\delta_0}{\nu^{2^j}} = q^{\log_q \delta_0 - \lceil \log_q \delta_0^{1/2^N} \rceil 2^j} \geqslant \delta_0^{1-\frac{1}{2^{N-j}}} q^{-2^j} \geqslant \delta^{1-\frac{1}{2^{N-j}}} q^{-2^j}$$

and hence

$$D(\frac{\delta_0}{\nu^{2^j}}) \leqslant \mathfrak{D}(q^{-2^j}\delta^{1-\frac{1}{2^{N-j}}}).$$

Thus applying Lemma 5.10 gives that

$$\begin{aligned} D(\delta_0) &\leqslant C_2 D(\frac{\delta_0}{\nu}) + C_2 C_1^2 \nu^{-C_3} D(\frac{\delta_0}{\nu^{2^{N-1}}})^{\frac{1}{3\cdot 2^N}} D(\frac{\delta_0}{\nu^{2^N}})^{\frac{2}{3\cdot 2^N}} \prod_{j=0}^{N-1} D(\frac{\delta_0}{\nu^{2^j}})^{\frac{1}{2^{j+1}}} \\ &\leqslant C_2 \mathfrak{D}(q^{-1}\delta^{1-\frac{1}{2^N}}) \\ &\quad + C_2 C_1^2 \delta^{-\frac{C_3}{2^N}} \mathfrak{D}(q^{-2^{N-1}}\delta^{1/2})^{\frac{1}{3\cdot 2^N}} \mathfrak{D}(q^{-2^N})^{\frac{2}{3\cdot 2^N}} \prod_{j=0}^{N-1} \mathfrak{D}(q^{-2^j}\delta^{1-\frac{1}{2^{N-j}}})^{\frac{1}{2^{j+1}}}. \end{aligned}$$

Applying the trivial bound that $\mathfrak{D}(q^{-2^N}) \leqslant q^{2^N/2}$ then shows that

$$\begin{aligned} D(\delta_0) &\leqslant C_2 \mathfrak{D}(q^{-1}\delta^{1-\frac{1}{2^N}}) \\ &\quad + C_2 C_1^2 q^{1/3} \delta^{-\frac{C_3}{2^N}} \mathfrak{D}(q^{-2^{N-1}}\delta^{1/2})^{\frac{1}{3\cdot 2^N}} \prod_{j=0}^{N-1} \mathfrak{D}(q^{-2^j}\delta^{1-\frac{1}{2^{N-j}}})^{\frac{1}{2^{j+1}}} \end{aligned}$$

as long as $\delta_0 \leqslant q^{-2^N}$. On the other hand if $\delta_0 > q^{-2^N}$, we apply the trivial bound and obtain that $D(\delta_0) \leqslant \delta_0^{-1/2} \leqslant q^{2^N/2}$. Since $\mathfrak{D}(\delta) \geqslant 1$ by Proposition 4.2, combining the bounds obtained in the two regimes above and then taking the supremum over all $\delta_0 \in q^{-\mathbb{N}} \cap [\delta, 1]$, the completes the proof of the lemma. □

We now are ready to show that $\mathfrak{D}(\delta) \lesssim_{q,\varepsilon} \delta^{-\varepsilon}$. Let $\lambda \geqslant 0$ be the smallest constant such that $\mathfrak{D}(\delta) \lesssim_{q,\varepsilon} \delta^{-\lambda-\varepsilon}$ for all $\delta \in (0, 1]$. From the trivial bound on $D(\delta)$, $\lambda \leqslant 1/2$. If $\lambda = 0$, then we are done and so we assume that $\lambda > 0$. Choose N such that

$$\frac{5}{6} + \frac{N}{2} - \frac{C_3}{\lambda} \geqslant 1.$$

Then Proposition 5.11 implies that for all $\delta \in (0, 1]$, we have

$$\begin{aligned} \mathfrak{D}(\delta) &\lesssim_{q,N,\varepsilon} \delta^{-\lambda(1-\frac{1}{2^N})-\varepsilon} + \delta^{-\frac{C_3}{2^N} - \frac{\lambda}{6\cdot 2^N} - \sum_{j=0}^{N-1}(1-\frac{1}{2^{N-j}})\frac{1}{2^{j+1}}\lambda - \varepsilon} \\ &\leqslant \delta^{-\lambda(1-\frac{1}{2^N})-\varepsilon} + \delta^{-\lambda(1-\frac{1}{2^N}(\frac{5}{6}+\frac{N}{2}-\frac{C_3}{\lambda}))-\varepsilon} \leqslant \delta^{-\lambda(1-\frac{1}{2^N})-\varepsilon} \end{aligned}$$

where the last inequality is by our choice of N. But this contradicts the minimality of λ. Therefore we must have $\lambda = 0$ which finishes the proof.

6. Five proofs of the key step

In this section we give five different proofs of Lemma 5.6 which in turn implies Proposition 5.4. Before we begin our proofs, we first discuss how the efficient congruencing proof in the quadratic Vinogradov case found in [**17**, Section 4.3] motivated the definition of $M_{a,b}(\delta, \nu)$ and Proposition 5.4.

6.1. Motivation. The inspiration for the definition of $M_{a,b}(\delta, \nu)$ and Proposition 5.4 came from [**17**, Section 4.3] which is the quadratic Vinogradov version of Heath-Brown's simplification [**14**] of Wooley's proof [**25**] of cubic Vinogradov. See [**17**] and [**24**] for more details on Vinogradov's Mean Value Theorem.

The main bilinear object considered in [**17**, Section 4.3] is

$$I_1(X; a, b)$$
$$:= \max_{\xi \not\equiv \eta \pmod{p}} \int_{[0,1]^2} |\sum_{\substack{n \leqslant X \\ n \equiv \xi \pmod{p^a}}} e(\alpha n + \beta n^2)|^2 | \sum_{\substack{n \leqslant X \\ n \equiv \eta \pmod{p^b}}} e(\alpha n + \beta n^2)|^4 \, d\alpha \, d\beta.$$

For fixed residue classes $\xi \pmod{p^a}$ and $\eta \pmod{p^b}$ with $\xi \not\equiv \eta \pmod{p}$, the integral counts the number of solutions to

$$\begin{aligned} x_1 + x_2 + x_3 &= x_4 + x_5 + x_6 \\ x_1^2 + x_2^2 + x_3^2 &= x_4^2 + x_5^2 + x_6^2 \end{aligned} \tag{21}$$

with $1 \leqslant x_i \leqslant X$, $x_1 \equiv x_4 \equiv \xi \pmod{p^a}$, the other $x_i \equiv \eta \pmod{p^b}$, and $\xi \not\equiv \eta \pmod{p}$. The key estimate in [**17**, Section 4.3] with regards to $I_1(X; a, b)$ is that if $1 \leqslant a \leqslant 2b$ and $p^{2b} \leqslant X$, then

$$I_1(X; a, b) \leqslant p^{2b-a} I_1(X; 2b, b). \tag{22}$$

This should be immediately compared to Proposition 5.4. This loss of p^{2b-a} is sharp. To see this, heuristically we expect $I_1(X; a, b) \approx X^3/p^{a+2b}$ (here we use $\approx$ somewhat informally). Thus if we define $I_1'(X; a, b) := (p^{a+2b}/X^3) I_1(X; a, b)$, then (22) now reads $I_1'(X; a, b) \leqslant I_1'(X; 2b, b)$ and we expect both sides to be ≈ 1. Since we will want to prove $D(\delta) \approx 1$ (and hence $M_{a,b}(\delta, \nu) \approx 1$), we should expect $I_1'(X; a, b)$ to be similar to $M_{a,b}(\delta, \nu)$.

We now turn to the proof of (22) from [**17**, Lemma 4.4] as this will motivate our proofs of Lemma 5.6. Since solutions to (21) are translation invariant, by subtracting η, we may assume that $\eta = 0$ and so $\xi \not\equiv 0 \pmod{p}$. Write $x_i = \xi + p^a y_i$ for $i = 1, 4$ and $x_i = p^b y_i$ for $i = 2, 3, 5, 6$. Since $x_1^2 + x_2^2 + x_3^2 = x_4^2 + x_5^2 + x_6^2$ we must have

$$(\xi + p^a y_1)^2 \equiv (\xi + p^a y_4)^2 \pmod{p^{2b}}. \tag{23}$$

Then

$$p^a (y_1 - y_4)(2\xi + p^a(y_1 + y_4)) \equiv 0 \pmod{p^{2b}}.$$

Since $\xi \not\equiv 0 \pmod{p}$, we have $2\xi + p^a(y_1 + y_2) \not\equiv 0 \pmod{p}$ and hence we obtain an extra congruence $y_1 \equiv y_4 \pmod{p^{2b-a}}$. This will allow us to count the solutions efficiently. Fix $y_4 \pmod{p^{2b-a}}$. There are p^{2b-a} many choices and this gives the p^{2b-a} on the right hand side of (22). This choice then fixes $y_1 \bmod p^{2b-a}$ and hence $x_1 \equiv x_4 \equiv \xi' \pmod{p^{2b}}$ for some ξ'. Now we apply the definition of $I_1(X; 2b, b)$.

Broadly speaking the steps of this proof are

(1) Use translation invariance to shift 4 of the variables to be $0 \pmod{p^b}$.

(2) Since $x_2, x_3, x_5, x_6 \equiv 0 \pmod{p^b}$, we look at the quadratic equation mod p^{2b}.
(3) Find diagonal behavior in the remaining x_1 and x_4 variables at scale mod p^{2b}.

The analogue of translation invariance in decoupling is parabolic rescaling and so we have implemented Step 1 already in Lemma 5.6. Another way to view Lemma 5.6 is that we decouple the two copies of f_I into f_J which are at scale ν^{2b} all the while keeping the four copies of $f_{I(0,\nu^b)}$ on both sides. This corresponds to upgrading x_1 and x_4 from mod p^a knowledge to mod p^{2b} knowledge while gaining no new knowledge in the other remaining variables.

We now give our five proofs of Lemma 5.6. On a first read and to more easily compare the argument with the efficient congruencing argument in Section 6.1, it may be useful to set $\nu = 1/q$.

6.2. A first proof of Lemma 5.6. Our first proof will take advantage of the fact that both exponents on the left hand side of (17) are even and resembles a Córdoba–Fefferman argument (see for example [**9**, Section 3.2]). This proof is a modification of an argument that first appeared in [**16**, Lemma 2.7]. To prove Lemma 5.6, we prove Lemma 5.7 instead.

By definition, the left hand side of (18) is equal to

$$\int_{B(0,\nu^{-2b})} \sum_{J,J' \in P_{\nu^{2b}}(I)} f_J \overline{f_{J'}} |f_{I(0,\nu^b)}|^4. \tag{24}$$

As $(L \times \mathbb{Q}_q) \cap \bigcup_{K \in P_\delta} \theta_K = \{(\xi, \eta) \in \mathbb{Q}_q^2 : \xi \in L, |\eta - \xi^2| \leqslant \delta^2\}$, the definition of f_L then gives

$$f_L(x) = \int_{|t| \leqslant \delta^2} \int_L \widehat{f}(s, s^2 + t)\chi(sx_1 + s^2x_2)\chi(tx_2)\, ds\, dt$$

for any interval $L \subset \mathbb{Z}_q$. Expanding then gives that (24) is

$$\sum_{J,J' \in P_{\nu^{2b}}(I)} \int_{|t_1| \leqslant \delta^2, \dots, |t_6| \leqslant \delta^2} \int_{J \times I(0,\nu^b)^2 \times J' \times I(0,\nu^b)^2} (\cdots)_1 \int_{B(0,\nu^{-2b})} \chi((\cdots)_2)\, dx\, ds_1 \cdots ds_6\, dt_1 \cdots dt_6$$

where $s_1 \in J$, $s_4 \in J'$, $s_2, s_3, s_5, s_6 \in I(0, \nu^b)$,

$$(\cdots)_1 = \prod_{i=1}^{3} \widehat{f}(s_i, s_i^2 + t_i)\overline{\widehat{f}(s_{i+3}, s_{i+3}^2 + t_{i+3})},$$

and

$$(\cdots)_2 = \sum_{i=1}^{3}(s_i - s_{i+3})x_1 + (s_i^2 - s_{i+3}^2)x_2 + \sum_{i=1}^{3}(t_i - t_{i+3})x_2.$$

Since $|t_i| \leqslant \delta^2$, $|x_2| \leqslant \nu^{-2b}$, and $\delta \leqslant \nu^{2b}$, $|\sum_{i=1}^3 (t_i - t_{i+3})x_2| \leqslant 1$ and so $\chi(\sum_{i=1}^3 (t_i - t_{i+3})x_2) = 1$. Therefore

$$\int_{B(0,\nu^{-2b})} \chi((\cdots)_2)\, dx = \int_{B(0,\nu^{-2b})} \chi(\sum_{i=1}^{3}(s_i - s_{i+3})x_1 + (s_i^2 - s_{i-3}^2)x_2)\, dx$$
$$= \nu^{-4b} 1_{|\sum_{i=1}^3 s_i - s_{i+3}| \leqslant \nu^{2b}, |\sum_{i=1}^3 s_i^2 - s_{i+3}^2| \leqslant \nu^{2t}}$$

by Proposition 2.7. Since $\delta \leqslant \nu^{2b}$ and $s_2, s_3, s_5, s_6 \in I(0, \nu^b)$, if we have $|s_1^2 + s_2^2 + s_3^2 - s_4^2 - s_5^2 - s_6^2| \leqslant \nu^{2b}$, then

$$|s_1 + s_4||s_1 - s_4| \leqslant \nu^{2b}. \tag{25}$$

Notice that in this step, if we had set $\nu = 1/q$, then this is exactly (23). Recall that $s_1 \in J \subset I$, $s_4 \in J' \subset I$, and $d(I, 0) \geqslant q\nu$. For any $r \in I$, write $s_1 = r + s_1'$ and $s_4 = r + s_4'$ for some $|s_1'|, |s_4'| \leqslant \nu^a$. By the ultrametic triangle inequality, $|s_1 + s_4| = |2r + (s_1' + s_4')| \geqslant q\nu$ since $|2r| \geqslant q\nu > \nu^a \geqslant |s_1' + s_4'|$. Inserting this into (25) shows that we have $|s_1 - s_4| \leqslant (q\nu)^{-1}\nu^{2b}$ (corresponding to Step 3 in the efficient congruencing argument on Page 87) and hence we can insert the requirement that $d(J, J') \leqslant (q\nu)^{-1}\nu^{2b}$ into $\sum_{J,J'}$ in (24). Applying two instances of the Cauchy-Schwarz inequality gives that

$$\int_{B(0,\nu^{-2b})} \sum_{\substack{J,J' \in P_{\nu^{2b}}(I) \\ d(J,J') \leqslant (q\nu)^{-1}\nu^{2b}}} f_J \overline{f_{J'}} |f_{I(0,\nu^b)}|^4 \lesssim \nu^{-O(1)} \sum_{J \in P_{\nu^{2b}}(I)} \int_{B(0,\nu^{-2b})} |f_J|^2 |f_{I(0,\nu^b)}|^4 \tag{26}$$

which completes the proof.

6.3. A second proof of Lemma 5.6. Another way to interpret solution counting and the argument in Section 6.1 is to make use of the identity $\int f(x)\, dx = \widehat{f}(0)$ and ask under what conditions does 0 lie in the Fourier support of f. This was the point of view taken in [**8**]. Like the proof in Section 6.2, this second proof also makes use of the fact that 2 and 4 are even, however it takes more into account some geometric properties.

Like in (24), the left hand side of (17) is equal to

$$\sum_{J,J' \in P_{\nu^{2b}}(I)} \int_{\mathbb{Q}_q^2} f_J \overline{f_{J'}} |f_{I(0,\nu^b)}|^4 = \sum_{J,J' \in P_{\nu^{2b}}(I)} (\widehat{f_J} * \widehat{\overline{f_{J'}}} * \widehat{f_{I(0,\nu^b)}} * \widehat{f_{I(0,\nu^b)}} * \widehat{\overline{f_{I(0,\nu^b)}}} * \widehat{\overline{f_{I(0,\nu^b)}}})(0). \tag{27}$$

Observe that $\widehat{f_{I(0,\nu^b)}}$ and $\widehat{\overline{f_{I(0,\nu^b)}}}$ are supported in $\bigcup_{K \in P_\delta(I(0,\nu^b))} \theta_K \subset \{(\xi, \eta) : |\xi| \leqslant \nu^b, |\eta| \leqslant \nu^{2b}\}$. Partition this rectangle into squares $\{\square\}$ of side length ν^{2b}. Therefore (27) is equal to

$$\sum_{\square} \sum_{J,J' \in P_{\nu^{2b}}(I)} (\widehat{f_J} * \widehat{\overline{f_{J'}}} * (\widehat{f_{I(0,\nu^b)}} * \widehat{f_{I(0,\nu^b)}} * \widehat{\overline{f_{I(0,\nu^b)}}} * \widehat{\overline{f_{I(0,\nu^b)}}})1_{\square})(0). \tag{28}$$

Given a $\square$, we now need to decide for which $J, J' \in P_{\nu^{2b}}(I)$ is 0 contained in the support of $\widehat{f_J} * \widehat{\overline{f_{J'}}} * (\widehat{f_{I(0,\nu^b)}} * \widehat{f_{I(0,\nu^b)}} * \widehat{\overline{f_{I(0,\nu^b)}}} * \widehat{\overline{f_{I(0,\nu^b)}}})1_{\square}$. Since f_J is Fourier supported in $\bigcup_{K\in P_\delta(J)}\theta_K \subset \theta_J$ and similarly for $f_{J'}$, it suffices to ask for which $J, J' \in P_{\nu^{2b}}(I)$ is $0 \in \theta_J - \theta_{J'} + \square$. In fact, it suffices to ask something weaker. Recall the definition of τ_J in Section 3.1. As $\theta_J \subset \tau_J$, we will ask for which $J, J' \in P_{\nu^{2b}}(I)$ is $0 \in \tau_J - \tau_{J'} + \square$. Since $|J| = |J'| = \nu^{2b}$, τ_J and $\tau_{J'}$ are squares of side length ν^{2b}. Then $\tau_J - \tau_{J'} + \square$ is a square of side length ν^{2b} which contains 0. Therefore our question reduces to asking for which $J, J' \in P_{\nu^{2b}}(I)$ is such that

$$\tau_J - \tau_{J'} + \square = B(0, \nu^{2b}). \tag{29}$$

If (29) holds, then there exists an $s_J \in J$, $s_{J'} \in J'$ and $(b_1, b_2) \in \square$ such that

$$|s_J - s_{J'} + b_1| \leqslant \nu^{2b}$$
$$|s_J^2 - s_{J'}^2 + b_2| \leqslant \nu^{2b}.$$

We will only be interested in the second inequality. Since $\square \subset \{(\xi,\eta) : |\xi| \leqslant \nu^b, |\eta| \leqslant \nu^{2b}$, $|b_2| \leqslant \nu^{2b}$ and hence $|s_J^2 - s_{J'}^2| \leqslant \nu^{2b}$. Using that $s_J \in J \subset I$, $s_{J'} \subset J' \subset I$, $d(I, 0) \geqslant q\nu$ and the same reasoning as at the end of the proof of the previous section, we see $|s_J + s_{J'}| \geqslant q\nu$ and hence $|s_J - s_{J'}| \leqslant \nu^{2b}(q\nu)^{-1}$. But then by the ultrametric triangle inequality, $|s - s'| \leqslant \nu^{2b}(q\nu)^{-1}$ for all $s \in J$ and $s' \in J'$.

Therefore we have shown that given a $\square$, in order for 0 to lie in the support of $\widehat{f_J} * \widehat{\overline{f_{J'}}} * (\widehat{f_{I(0,\nu^b)}} * \widehat{f_{I(0,\nu^b)}} * \widehat{\overline{f_{I(0,\nu^b)}}} * \widehat{\overline{f_{I(0,\nu^b)}}})1_\square$, we cannot have that $d(J, J') > \nu^{2b}(q\nu)^{-1}$. Thus (28) is equal to

$$\sum_{\square} \sum_{\substack{J,J'\in P_{\nu^{2b}}(I)\\ d(J,J')\leqslant \nu^{2b}(q\nu)^{-1}}} \widehat{f_J} * \widehat{\overline{f_{J'}}} * (\widehat{f_{I(0,\nu^b)}} * \widehat{f_{I(0,\nu^b)}} * \widehat{\overline{f_{I(0,\nu^b)}}} * \widehat{\overline{f_{I(0,\nu^b)}}})1_\square.$$

Again using that $\widehat{f_{I(0,\nu^b)}} * \widehat{f_{I(0,\nu^b)}} * \widehat{\overline{f_{I(0,\nu^b)}}} * \widehat{\overline{f_{I(0,\nu^b)}}}$ is supported on the $\{\square\}$ and then finally applying $\int f(x)\,dx = \widehat{f}(0)$, shows that (27) is equal to

$$\int_{\mathbb{Q}_q^2} \sum_{\substack{J,J'\in P_{\nu^{2b}}(I)\\ d(J,J')\leqslant \nu^{2b}(q\nu)^{-1}}} f_J\overline{f_{J'}}|f_{I(0,\nu^b)}|^4$$

which by the same application of the Cauchy-Schwarz inequality as in (26) completes the proof.

6.4. A third proof of Lemma 5.6. The third proof of Lemma 5.6 relies importantly on that one of the exponents is a 2 and so suggests us to apply Plancherel. This argument first appeared in [**19**, Proposition 19] and is similar to the argument in [**12**] which we will present in the next section. An argument similar to this one was used to prove decoupling for fractal sets on the parabola in [**7**]. By Plancherel, the left hand side of (17) is

$$\int_{\mathbb{Q}_q^2} \Big|\sum_{J\in P_{\nu^{2b}}(I)} f_J f_{I(0,\nu^b)}^2\Big|^2 = \int_{\mathbb{Q}_q^2} \Big|\sum_{J\in P_{\nu^{2b}}(I)} \widehat{f_J} * \widehat{f_{I(0,\nu^b)}} * \widehat{f_{I(0,\nu^b)}}\Big|^2. \tag{30}$$

If we could show that the $\{\widehat{f_J} * \widehat{f_{I(0,\nu^b)}} * \widehat{f_{I(0,\nu^b)}}\}_{J\in P_{\nu^{2b}}(I)}$ are almost pairwise orthogonal, then combining this with another application of Plancherel in (30) would finish the proof.

Inspired by that we only looked at the quadratic equation in Section 6.1, we only look at the vertical projection of the supports of $\widehat{f_J} * \widehat{f_{I(0,\nu^b)}} * \widehat{f_{I(0,\nu^b)}}$. Since

$\widehat{f_{I(0,\nu^b)}}$ is supported in the strip $\{(\xi,\eta)\in\mathbb{Q}_q^2 : |\eta|\leqslant \nu^{2b}\}$, $\widehat{f_{I(0,\nu^b)}} * \widehat{f_{I(0,\nu^b)}}$ is also supported in the strip $\{(\xi,\eta)\in\mathbb{Q}_q^2 : |\eta|\leqslant \nu^{2b}\}$.

Next, $\widehat{f_J}$ is supported in $\theta_J \subset \tau_J \subset S_J := \{(\xi,\eta)\in\mathbb{Q}_q^2 : |\eta - s_J^2|\leqslant \nu^{2b}\}$ for any $s_J\in J$. The ultrametric triangle inequality then shows that $\widehat{f_J} * \widehat{f_{I(0,\nu^b)}} * \widehat{f_{I(0,\nu^b)}}$ is supported in the strip S_J. Now we claim that the $\{S_J\}$ are almost pairwise disjoint. Suppose there was a J and J' such that $S_J\cap S_{J'}\neq\varnothing$. Then the ultrametric triangle inequality gives that $|s_{J'}^2 - s_J^2|\leqslant \nu^{2b}$. But just like as in the argument in Sections 6.2 and 6.3, since $J, J'\subset I$ and $d(I,0)\geqslant q\nu$, it follows that $|s_J + s_{J'}|\geqslant q\nu$ and hence $|s_J - s_{J'}|\leqslant \nu^{2b}(q\nu)^{-1}$ and $d(J,J')\leqslant \nu^{2b}(q\nu)^{-1}$. Therefore (30) is equal to

$$\int_{\mathbb{Q}_q^2}\sum_{\substack{J,J'\in P_{\nu^{2b}}(I)\\ d(J,J')\leqslant \nu^{2b}(q\nu)^{-1}}} \widehat{f_J} * \widehat{f_{I(0,\nu^b)}} * \widehat{f_{I(0,\nu^b)}}\overline{\widehat{f_{J'}} * \widehat{f_{I(0,\nu^b)}} * \widehat{f_{I(0,\nu^b)}}}$$

which by the same application of the Cauchy-Schwarz inequality as in (26) gives that this is

$$\lesssim \nu^{-O(1)}\int_{\mathbb{Q}_q^2}\sum_{J\in P_{\nu^{2b}}(I)} |\widehat{f_J} * \widehat{f_{I(0,\nu^b)}} * \widehat{f_{I(0,\nu^b)}}|^2.$$

Applying Plancherel then finishes the proof.

6.5. A fourth proof of Lemma 5.6. In this fourth proof, we will rely on that the exponent 2 suggests the use of Plancherel, much like in the third proof. However to avoid using that 4 is even, we will instead rely on the locally constant property. This proof is essentially the argument in [**12**] in the case of the parabola.

To prove Lemma 5.6, we prove (18) instead. Partition $B(0,\nu^{-2b})$ into translates of T_{ν^b}, that is, vertical $\nu^{-b}\times\nu^{-2b}$ rectangles $\{\square\}$. Such a partition exists by Lemma 3.3 (in fact the proof of Lemma 3.3 gives an explicit construction). Therefore it suffices to show that for each $\square$, we have

$$\int_\square |f_I|^2|f_{I(0,\nu^b)}|^4 \lesssim \nu^{-O(1)}\sum_{J\in P_{\nu^{2b}}(I)}\int_\square |f_J|^2|f_{I(0,\nu^b)}|^4 \tag{31}$$

for all f Fourier supported in $\bigcup_{K\in P_\delta}\theta_K$. Since we want (31) to be true for all f Fourier supported in $\bigcup_{K\in P_\delta}\theta_K$, it suffices to prove prove (31) when $\square = T_{\nu^b}$. That is, we just need to show that

$$\int_{T_{\nu^b}} |f_I|^2|f_{I(0,\nu^b)}|^4 \lesssim \nu^{-O(1)}\sum_{J\in P_{\nu^{2b}}(I)}\int_{T_{\nu^b}} |f_J|^2|f_{I(0,\nu^b)}|^4 \tag{32}$$

for all f Fourier supported in $\bigcup_{K\in P_\delta}\theta_K$. By the locally constant property, $|f_{I(0,\nu^b)}|$ is constant on any translate of T_{ν^b}. Thus to prove (32) it suffices to just show that

$$\begin{aligned}\int_{|x|\leqslant\nu^{-b},|y|\leqslant\nu^{-2b}} &|f_I(x,y)|^2\,dx\,dy\\ &\lesssim \nu^{-O(1)}\sum_{J\in P_{\nu^{2b}}(I)}\int_{|x|\leqslant\nu^{-b},|y|\leqslant\nu^{-2b}}|f_J(x,y)|^2\,dx\,dy.\end{aligned}\tag{33}$$

Since in the efficient congruencing argument, we only looked at the quadratic equation, this suggests in (33) that we freeze the x variable and look only at (33) in the

y variable. Thus we need to show that for every x_0 such that $|x_0| \leqslant \nu^{-b}$, we have

$$\int_{|y|\leqslant\nu^{-2b}} |f_I(x_0,y)|^2\,dy \leqslant \sum_{J\in P_{\nu^{2b}}(I)} \int_{|y|\leqslant\nu^{-2b}} |f_J(x_0,y)|^2\,dy. \tag{34}$$

Rewriting the left hand side of (34) and applying Plancherel, we obtain

$$\begin{aligned}\int_{\mathbb{Q}_q} &|\sum_{J\in P_{\nu^{2b}}(I)} f_J(x_0,y)1_{|y|\leqslant\nu^{-2b}}|^2\,dy\\ &= \int_{\mathbb{Q}_q} |\sum_{J\in P_{\nu^{2b}}(I)} \widetilde{f_J}(x_0,\eta) * \nu^{-2b}1_{|\eta|\leqslant\nu^{2b}}|^2\,d\eta\end{aligned} \tag{35}$$

where here we use $\widetilde{f}$ to just denote the Fourier transform in the second variable. Let $F_{x_0}(\eta) := \int_{\mathbb{Q}_q} \widehat{f}(\xi,\eta)\chi(\xi x_0)1_{\theta_J}(\xi,\eta)\,d\xi$. Then $f_J(x_0,y) = \int_{\mathbb{Q}_q} F_{x_0}(\eta)\chi(\eta y)\,d\eta$ and so $\widetilde{f_J}(x_0,y) = F_{x_0}(y)$ which is supported in the projection of θ_J onto the y-axis, but in Section 6.4, we saw that such a projection is contained in S_J. By the ultametric triangle inequality, the convolution of $\widetilde{f_J}(x_0,\eta)$ with $1_{|\eta|\leqslant\nu^{2b}}$ does not change its support. Since from Section 6.4, we saw that $S_J \cap S_{J'} = \varnothing$ if $d(J,J') > \nu^{2b}(q\nu)^{-1}$, by the same argument as at the end of Section 6.4, it follows that (35) is

$$\lesssim \nu^{-O(1)} \int_{\mathbb{Q}_q} \sum_{J\in P_{\nu^{2b}}(I)} |\widetilde{f_J}(x_0,\eta) * \nu^{-2b}1_{|\eta|\leqslant\nu^{2b}}|^2\,d\eta.$$

Undoing Plancherel then gives the right hand side of (34) which then completes the proof.

6.6. A fifth proof of Lemma 5.6. The previous four proofs all rely on the fact that $\int_{\mathbb{Q}_q^2} f_J\overline{f_{J'}}|f_{I(0,\nu^b)}|^4 = 0$ if $d(J,J') > (q\nu)^{-1}\nu^{2b}$ or some variant. Our final proof is qualitatively different in that it does not go through this step. The proof of Lemma 5.6 given here is in the style of Bourgain and Demeter's proof of decoupling for the paraboloid [**3, 4**] specialized to the case of the parabola. The Bourgain and Demeter argument uses a different bilinearization than in (14), but the main ideas are similar in that this proof makes use of bilinear Kakeya and wavepacket decomposition. This argument (in the real setting) first appeared in [**16**, Section 3].

To prove Lemma 5.6, we first reduce to proving the $a = b$ case.

LEMMA 6.1. *If* $1 \leqslant a \leqslant b$ *and* $|I| = \nu^a$, *then*

$$\int_{\mathbb{Q}_q^2} |f_I|^2|f_{I(0,\nu^b)}|^4 \leqslant \sum_{J\in P_{\nu^b}(I)} \int_{\mathbb{Q}_q^2} |f_J|^2|f_{I(0,\nu^b)}|^4 \tag{36}$$

for all f *Fourier supported in* $\bigcup_{K\in P_\delta} \theta_K$.

PROOF. By partitioning $\mathbb{Q}_q^2$ into squares of side length ν^{-b} and since we want (36) to be true for all f Fourier supported in $\bigcup_{K\in P_\delta}\theta_K$, it suffices to show that

$$\int_{B(0,\nu^{-b})} |f_I|^2|f_{I(0,\nu^b)}|^4 \leqslant \sum_{J\in P_{\nu^b}(I)} \int_{B(0,\nu^{-b})} |f_J|^2|f_{I(0,\nu^b)}|^4.$$

Since $|f_{I(0,\nu^b)}|$ is constant on T_{ν^b} and $B(0,\nu^{-b})\subset T_{\nu^b}$, it suffices to show that

$$\int_{\mathbb{Q}_q^2}|\sum_{J\in P_{\nu^b}(I)} f_J 1_{B(0,\nu^{-b})}|^2 \leqslant \sum_{J\in P_{\nu^b}(I)}\int_{\mathbb{Q}_q^2}|f_J1_{B(0,\nu^{-b})}|^2.$$

But this is just an application of Plancherel since $f_J1_{B(0,\nu^{-b})}$ is Fourier supported on $\theta_J+B(0,\nu^b)$ and by Lemma 3.2, this is equal to θ_J. This completes the proof of Lemma 6.1. □

Therefore to prove Lemma 5.7 (and hence Lemma 5.6), it suffices to assume that $a=b$. To do this, we first show an intermediate inequality that is an application of wavepacket decomposition and bilinear Kakeya (and is also related to the ball inflation step in [**4**], see [**16**, Lemma 3.4]):

$$\fint_{B(0,\nu^{-2b})}|f_I|^2|f_{I(0,\nu^b)}|^4 \lesssim \nu^{-O(1)}\Big(\fint_{B(0,\nu^{-2b})}|f_I|^2\Big)\Big(\fint_{B(0,\nu^{-2b})}|f_{I(0,\nu^b)}|^4\Big). \tag{37}$$

where $|I|=\nu^b$, $d(I,0)\geqslant q\nu$, f is Fourier supported in $\bigcup_{K\in P_\delta}\theta_K$ and $\fint_B := \frac{1}{|B|}\int_B$.

Partition $B(0,\nu^{-2b})$ by translates of $T_{0,I}$ and call this collection $\mathcal{T}(I)$. Similarly, partition $B(0,\nu^{-2b})$ by translates of $T_{0,I(0,\nu^b)}=T_{\nu^b}$ and call this collection $\mathcal{T}(I(0,\nu^b))$. Since f_I is Fourier supported in θ_I and $f_{I(0,\nu^b)}$ is Fourier supported in $\theta_{I(0,\nu^b)}$, from wavepacket decomposition, we can write $|f_I|^2=\sum_{T\in\mathbb{T}(I)}c_T^21_T$ and $|f_{I(0,\nu^b)}|^4=\sum_{T'\in\mathcal{T}(I(0,\nu^b))}c_{T'}^41_{T'}$ for some nonnegative constants c_T and $c_{T'}$. Thus the left hand side of (37) becomes

$$\sum_{\substack{T\in\mathcal{T}(I)\\ T'\in\mathcal{T}(I(0,\nu^b))}} c_T^2c_{T'}^4\frac{|T\cap T'|}{|B(0,\nu^{-2b})|}. \tag{38}$$

By Lemma 3.6,

$$\frac{|T\cap T'|}{|B(0,\nu^{-2b})|}\leqslant\frac{\nu^{-2b}/d(I,0)}{\nu^{-4b}}\leqslant\nu^{2b}(q\nu)^{-1}.$$

Therefore (38) is

$$\leqslant (q\nu)^{-1}\sum_{\substack{T\in\mathcal{T}(I)\\ T'\in\mathcal{T}(I(0,\nu^b))}} c_T^2c_{T'}^4\nu^{2b}=(q\nu)^{-1}\Big(\nu^b\sum_{T\in\mathcal{T}(I)}c_T^2\Big)\Big(\nu^b\sum_{T'\in\mathcal{T}(I')}c_{T'}^4\Big)$$

which is exactly the right hand side of (37). Thus we have shown that

$$\int_{B(0,\nu^{-2b})}|f_I|^2|f_{I(0,\nu^b)}|^4 \lesssim \nu^{-O(1)}\Big(\fint_{B(0,\nu^{-2b})}|f_I|^2\Big)\Big(\int_{B(0,\nu^{-2b})}|f_{I(0,\nu^b)}|^4\Big).$$

We next show that

$$\fint_{B(0,\nu^{-2b})}|f_I|^2\leqslant\sum_{J\in P_{\nu^{2b}}(I)}\fint_{B(0,\nu^{-2b})}|f_J|^2. \tag{39}$$

This would finish the proof of (18) since $|J|=\nu^{2b}$ and so $|f_J|$ is constant on $B(0,\nu^{-2b})$ by Corollary 3.5 and hence

$$\Big(\fint_{B(0,\nu^{-2b})}|f_J|^2\Big)\Big(\int_{B(0,\nu^{-2b})}|f_{I(0,\nu^b)}|^4\Big)=\int_{B(0,\nu^{-2b})}|f_J|^2|f_{I(0,\nu^b)}|^4.$$

Thus it remains to show (39). By Plancherel,

$$\int_{B(0,\nu^{-2b})} |f_I|^2 = \int_{\mathbb{Q}_q} |\sum_{J\in P_{\nu^{2b}}(I)} f_J 1_{B(0,\nu^{-2b})}|^2$$
$$= \int_{\mathbb{Q}_q} |\sum_{J\in P_{\nu^{2b}}(I)} \widehat{f_J} * \widehat{1}_{B(0,\nu^{-2b})}|^2 = \sum_{J\in P_{\nu^{2b}}(I)} \int_{B(0,\nu^{-2b})} |f_J|^2.$$

where in the last equality we have used that the support of $\widehat{1}_{B(0,\nu^{-2b})}$ is $B(0,\nu^{2b})$ and since $\theta_J + B(0,\nu^{2b}) = \theta_J$, each term in the $\sum_J$ are orthogonal.

Acknowledgments

This expository paper evolved from the slides of a talk the author gave on March 26, 2022 at the AMS Spring Central Sectional Meeting and also the notes for a series of lectures given by the author from April to May 2022 at the National Center for Theoretical Science (NCTS) in Taipei, Taiwan. A recording of this lecture series can be viewed at `https://youtu.be/_ML8jAu5cqo&list=PLQZfZKhc0kiAXeVcTAnYT9_YwpQ13SMB5`. The author would like to thank NCTS for the kind hospitality during his visit. The author would also like to thank the anonymous referee for their comments and suggestions in improving the exposition.

References

[1] Kirsti D. Biggs, *Efficient congruencing in ellipsephic sets: the quadratic case*, Acta Arith. **200** (2021), no. 4, 331–348, DOI 10.4064/aa191216-8-2. MR4331907

[2] Jean Bourgain, *Fourier transform restriction phenomena for certain lattice subsets and applications to nonlinear evolution equations. I. Schrödinger equations*, Geom. Funct. Anal. **3** (1993), no. 2, 107–156, DOI 10.1007/BF01896020. MR1209299

[3] Jean Bourgain and Ciprian Demeter, *The proof of the l^2 decoupling conjecture*, Ann. of Math. (2) **182** (2015), no. 1, 351–389, DOI 10.4007/annals.2015.182.1.9. MR3374964

[4] Jean Bourgain and Ciprian Demeter, *A study guide for the l^2 decoupling theorem*, Chinese Ann. Math. Ser. B **38** (2017), no. 1, 173–200, DOI 10.1007/s11401-016-1066-1. MR3592159

[5] Jean Bourgain, Ciprian Demeter, and Larry Guth, *Proof of the main conjecture in Vinogradov's mean value theorem for degrees higher than three*, Ann. of Math. (2) **184** (2016), no. 2, 633–682, DOI 10.4007/annals.2016.184.2.7. MR3548534

[6] Jean Bourgain and Larry Guth, *Bounds on oscillatory integral operators based on multilinear estimates*, Geom. Funct. Anal. **21** (2011), no. 6, 1239–1295, DOI 10.1007/s00039-011-0140-9. MR2860188

[7] Alan Chang, Jaume de Dios Pont, Rachel Greenfeld, Asgar Jamneshan, Zane Kun Li, and José Madrid, *Decoupling for fractal subsets of the parabola*, Math. Z. **301** (2022), no. 2, 1851–1879, DOI 10.1007/s00209-021-02950-0. MR4418339

[8] Brian Cook, Kevin Hughes, Zane Kun Li, Akshat Mudgal, Olivier Robert, and Po-Lam Yung, *A decoupling interpretation of an old argument for Vinogradov's mean value theorem*, To appear in Mathematika, `arXiv:2207.01097`, 2022.

[9] Ciprian Demeter, *Fourier restriction, decoupling, and applications*, Cambridge Studies in Advanced Mathematics, vol. 184, Cambridge University Press, Cambridge, 2020, DOI 10.1017/9781108584401. MR3971577

[10] Shaoming Guo, Zane Kun Li, and Po-Lam Yung, *Improved discrete restriction for the parabola*, To appear in *Mathematical Research Letters*, `arXiv:2103.09795`, 2021.

[11] Shaoming Guo, Zane Kun Li, and Po-Lam Yung, *A bilinear proof of decoupling for the cubic moment curve*, Trans. Amer. Math. Soc. **374** (2021), no. 8, 5405–5432, DOI 10.1090/tran/8363. MR4293776

[12] Shaoming Guo, Zane Kun Li, Po-Lam Yung, and Pavel Zorin-Kranich, *A short proof of ℓ^2 decoupling for the moment curve*, Amer. J. Math. **143** (2021), no. 6, 1983–1998, DOI 10.1353/ajm.2021.0048. MR4349137

[13] Larry Guth, Dominique Maldague, and Hong Wang, *Improved decoupling for the parabola*, To appear in J. Eur. Math. Soc. (JEMS), arXiv:2009.07953, 2022.

[14] D. R. Heath-Brown, *The cubic case of Vinogradov's mean value theorem*, Essent. Number Theory **1** (2022), no. 1, 1–12, DOI 10.2140/ent.2022.1.1. MR4573248

[15] Kelsey Houston-Edwards, *An infinite universe of number systems*. https://www.quantamagazine.org/how-the-towering-p-adic-numbers-work-20201019/, 2020.

[16] Zane Kun Li, *An l^2 decoupling interpretation of efficient congruencing: the parabola*, Rev. Mat. Iberoam. **37** (2021), no. 5, 1761–1802, DOI 10.4171/rmi/1248. MR4276295

[17] Lillian B. Pierce, *The Vinogradov mean value theorem [after Wooley, and Bourgain, Demeter and Guth]*, Astérisque **407** (2019), Exp. No. 1134, 479–564, DOI 10.24033/ast. Séminaire Bourbaki. Vol. 2016/2017. Exposés 1120–1135. MR3939285

[18] Mitchell H. Taibleson, *Fourier analysis on local fields*, Princeton University Press, Princeton, N.J.; University of Tokyo Press, Tokyo, 1975. MR0487295

[19] Terence Tao, *247B, Notes 2: Decoupling theory*, https://terrytao.wordpress.com/2020/04/13/247b-notes-2-decoupling-theory/.

[20] Terence Tao, *254A, Notes 5: Bounding exponential sums and the zeta function*, https://terrytao.wordpress.com/2015/02/07/254a-notes-5-bounding-exponential-sums-and-the-zeta-function/.

[21] Christoph Thiele, *Wave packet analysis*, CBMS Regional Conference Series in Mathematics, vol. 105, Published for the Conference Board of the Mathematical Sciences, Washington, DC; by the American Mathematical Society, Providence, RI, 2006, DOI 10.1090/cbms/105. MR2199086

[22] Robert C. Vaughan, *The Hardy–Littlewood method*, 2nd ed., Cambridge Tracts in Mathematics, vol. 125, Cambridge University Press, Cambridge, 1997, DOI 10.1017/CBO9780511470929. MR1435742

[23] Vasiliĭ S. Vladimirov, Igor V. Volovich, and Evgeny I. Zelenov, *p-adic analysis and mathematical physics*, Series on Soviet and East European Mathematics, vol. 1, World Scientific Publishing Co., Inc., River Edge, NJ, 1994, DOI 10.1142/1581. MR1288093

[24] Trevor D. Wooley, *Translation invariance, exponential sums, and Waring's problem*, Proceedings of the International Congress of Mathematicians—Seoul 2014. Vol. II, Kyung Moon Sa, Seoul, 2014, pp. 505–529. MR3728625

[25] Trevor D. Wooley, *The cubic case of the main conjecture in Vinogradov's mean value theorem*, Adv. Math. **294** (2016), 532–561, DOI 10.1016/j.aim.2016.02.033. MR3479572

[26] Trevor D. Wooley, *Nested efficient congruencing and relatives of Vinogradov's mean value theorem*, Proc. Lond. Math. Soc. (3) **118** (2019), no. 4, 942–1016, DOI 10.1112/plms.12204. MR3938716

Department of Mathematics, North Carolina State University, Raleigh, North Carolina, 27695

Email address: zkli@ncsu.edu

Contemporary Mathematics
Volume **792**, 2024
https://doi.org/10.1090/conm/792/15895

On a free Schrödinger solution studied by Barceló–Bennett–Carbery–Ruiz–Vilela

Xiumin Du, Yumeng Ou, Hong Wang, and Ruixiang Zhang

ABSTRACT. We present a free Schrödinger solution studied by Barceló–Bennett–Carbery–Ruiz–Vilela and show why it can be viewed as a sharp example for the recently discovered refined decoupling theorem.

1. Introduction

Let $d \geq 2$. Consider the free Schrödinger equation:

$$\begin{cases} iu_t - \Delta_x(u) = 0, \\ u(x,0) = f(x) \end{cases} \tag{1.1}$$

for $(x,t) \in \mathbb{R}^{d-1} \times \mathbb{R}$. By taking the Fourier transform of both sides, we know $\operatorname{supp} \hat{u} \subset \widetilde{P^{d-1}}$. Here $\widetilde{P^{d-1}}$ is the paraboloid:

$$\widetilde{P^{d-1}} = \{\xi_d = |\xi'|^2\}, \quad \xi = (\xi', \xi_d) := (\xi_1, \ldots, \xi_{d-1}, \xi_d) \in \mathbb{R}^d.$$

Because of the above property, such functions u are closely related to the *Fourier restriction theory* and have been extensively studied by Fourier analysts. In light of the Littlewood-Paley decomposition, people are often interested in functions g on $\mathbb{R}^n$ such that $\operatorname{supp} \hat{g}$ is in the *truncated paraboloid*

$$P^{d-1} = \{\xi_d = |\xi'|^2, \, |\xi_j| \leq 1, \forall 1 \leq j \leq d-1\}.$$

Fourier analysts are then interested in $L^p \to L^q$ estimates of such g on certain subsets of $\mathbb{R}^d$, and a great amount of related recent progress has been made.

In this note, we first review an example of such a function g studied by Barceló–Bennett–Carbery–Ruiz–Vilela in [**1**]. Next, we present a recent result known as "refined decoupling" (proved independently by Guth–Iosevich–Ou–Wang [**7**] and Du–Zhang). Refined decoupling has seen powerful applications in recent years such as in the Falconer distance problem [**5**, **7**] and small cap decouplings [**4**]. It would thus be interesting to know various sharp examples for this estimate. In this note, we show that Barceló-Bennett–Carbery–Ruiz–Vilela's free Schrödinger solutions are

2020 *Mathematics Subject Classification.* Primary 42B20.

The first author was supported by NSF DMS-2107729 (transferred from DMS-1856475) and Sloan Research Fellowship.

The second author was supported by NSF DMS-2142221 and NSF DMS-2055008.

The third author was supported by NSF DMS-2141426.

The fourth author was supported by NSF DMS-2207281 (transferred from DMS-1856541).

always (almost) sharp examples for refined decoupling. Moreover, we show in the end that this example is also sharp for an L^2 estimate used as a key step in many recent arguments studying the Falconer distance problem.

REMARK 1.1. Historically, [**1**] introduced this example and generalizations to provide useful test cases for L^2-average decay estimates of Fourier transforms of fractal measures. Therefore, it is perhaps not surprising that the example may be relevant for testing against other related results such as refined decoupling.

REMARK 1.2. In [**3**], inequality (4) is essentially a restatement of refined decoupling and it was remarked (Remark 1.3) that Knapp examples make refined decoupling (almost) sharp too. To put [**3**] into historical context, Guth [**6**] provided another different example capturing limits of decoupling. Guth's example was worked out carefully in [**3**] to show sharpness of their study of Mizohata-Takeuchi conjecture using refined decoupling estimate.

2. Barceló–Bennett–Carbery–Ruiz–Vilela's free Schrödinger solution

Let $0 < \sigma < 1/2$ and $R > 1$ be fixed parameters. Let $\mathrm{d}\omega$ be the hypersurface measure on P^{d-1}. Barceló–Bennett–Carbery–Ruiz–Vilela's free Schrödinger solution is a function g such that

$$\hat{g}(\xi) = h(\xi)\,\mathrm{d}\omega$$

where

$$h(\xi) = \sum_{\substack{l_1,\dots,l_{d-1}\in\mathbb{Z},\\ 1\le l_1,\dots,l_{d-1}<R^\sigma}} 1_{(l_1R^{-\sigma}-R^{-1},l_1R^{-\sigma}+R^{-1})\times\cdots\times(l_{d-1}R^{-\sigma}-R^{-1},l_{d-1}R^{-\sigma}+R^{-1})}.$$

We now state the most relevant properties of the above function g here. By elementary computations, one can check that $|g| \sim R^{(d-1)(\sigma-1)}$ at all points of the form $(n_1R^\sigma, n_2R^\sigma, \dots, n_{d-1}R^\sigma, n_dR^{2\sigma})$ inside the ball B_{c_dR} of radius c_dR, where $n_j \in \mathbb{Z}$ and $c_d > 0$ is a small constant only depending on the ambient dimension. Note that $R^{(d-1)(\sigma-1)}$ is comparable to $\|g\|_\infty$ by triangle inequality. Moreover, $|g| \sim R^{(d-1)(\sigma-1)}$ inside a ball of radius $\sim_d 1$ around every point above by a similar computation. We refer the reader to [**1**] for more detailed justification of these facts.

3. Wave packet decomposition

In order to introduce the refined decoupling inequality, we first briefly recall the *wave packet decomposition*, a standard tool for analyzing functions with Fourier support in P^{d-1}. Here, we present (the rescaled version of) the wave packet decomposition used in [**7**].

Fix a parameter $R > 1$. Decompose P^{d-1} into pieces θ such that the projection of each θ onto the hyperplane of the first $d-1$ coordinates is a square of side length $R^{-\frac{1}{2}}$. By elementary differential geometry, each θ is contained in a box of dimensions $R^{-\frac{1}{2}} \times \cdots \times R^{-\frac{1}{2}} \times R^{-1}$. Pick such a box and let T_θ be its dual box centered at the origin. Note that T_θ is roughly a tube of thickness $R^{\frac{1}{2}}$ and length R. Let $B_R \subset \mathbb{R}^d$ be the ball centered at the origin with radius R. Tile B_R by translations of T_θ and call this family $\mathbb{T}_\theta$.

For a function v whose Fourier support is in $P^{d-1} \subset \mathbb{R}^d$, one can decompose

$$v = \sum_{\theta, T:\, T \in \mathbb{T}_\theta} v_{\theta,T}$$

inside B_R such that:

- Each $\widehat{v_{\theta,T}}$ is supported in a box of dimensions $\sim R^{-\frac{1}{2}} \times \cdots \times R^{-\frac{1}{2}} \times R^{-1}$ containing θ.
- Each $v_{\theta,T}$ (known as a wave packet) is morally supported in T and rapidly decays outside of it.
- Each $|v_{\theta,T}|$ is morally a constant on T and we will call this constant the *magnitude* of $v_{\theta,T}$.
- Different $v_{\theta,T}$ are morally L^2-orthogonal on every $R^{\frac{1}{2}}$-ball. This property is known as *local orthogonality* and is very useful in Fourier restriction type problems. We do not need its detailed description here.

For each θ, we also define

$$v_\theta = \sum_{T \in \mathbb{T}_\theta} v_{\theta,T}.$$

4. Refined decoupling and its sharpness

4.1. The refined decoupling theorem. We can now state the refined decoupling theorem:

THEOREM 4.1 (Refined decoupling [**7**]). *Suppose* $\operatorname{supp} \hat{v} \subset P^{d-1}$ *and* $R > 1$. *Suppose that in the wave packet decomposition of* v *in* B_R*, every two wave packets have comparable magnitudes. Let* $X \subset B_R$ *such that each* $x \in X$ *hits the essential support of* $\leq M$ *wave packets, then for* $p = \frac{2(d+1)}{d-1}$,

$$\|v\|_{L^p(X)} \lesssim_\varepsilon R^\varepsilon M^{\frac{1}{2}-\frac{1}{p}} \left(\sum_\theta \|v_\theta\|^p_{L^p(w_{B_R})} \right)^{\frac{1}{p}}. \tag{4.1}$$

Here we have a weight w_{B_R} included on the right hand side for technical reasons. It behaves like 1_{B_R} but has a rapidly decaying tail. Morally one can think of $\|\cdot\|_{L^p(w_{B_R})}$ as $\|\cdot\|_{L^p(B_R)}$. Theorem 4.1 is named refined decoupling because it is a refinement of the celebrated Bourgain–Demeter decoupling theorem for paraboloids [**2**].

We remark that the assumption that all wave packets have comparable magnitudes in the theorem is usually harmless. In applications, one can usually reduce a general situation to this case by dyadic pigeonholing.

In [**7**], Theorem 4.1 is one of the central ingredients the authors use to make progress on the Falconer distance conjecture in $\mathbb{R}^2$. The theorem is also useful in other problems of similar flavors such as Schrödinger maximal function estimates.

4.2. An almost sharp example for Theorem 4.1. The function g we discussed in §2 is an (almost) sharp example for Theorem 4.1, as we explain below.

Note that for the function g, if we take X to be the 1-neighborhood of $\{(n_1 R^\sigma, n_2 R^\sigma, \ldots, n_{d-1} R^\sigma, n_d R^{2\sigma}) : n_1, \ldots, n_d \in \mathbb{Z}\} \bigcap B_R^d$, then since the Fourier support of g only intersects $\sim R^{(d-1)\sigma}$ θ's and that the supports of wave packets from one θ are essentially nonoverlapping, the relevant M is $\lessapprox R^{(d-1)\sigma}$. In fact, since g attains almost its maximal possible value at each $(n_1 R^\sigma, n_2 R^\sigma, \ldots, n_{d-1} R^\sigma, n_d R^{2\sigma})$, one can further see that M is indeed $\sim R^{(d-1)\sigma}$, but even without this stronger observation one can still see sharpness from the computation below.

Recall that $p = \frac{2(d+1)}{d-1}$. Because of the property that g is almost the largest possible at each $(n_1R^\sigma, n_2R^\sigma, \ldots, n_{d-1}R^\sigma, n_dR^{2\sigma})$, as discussed in §2, we see that

$$\|g\|_{L^p(X)} \sim R^{(d-1)(\sigma-1)}|X|^{1/p} \sim R^{(d-1)(\sigma-1)}R^{\frac{d-(d+1)\sigma}{p}} \sim R^{\frac{d-1}{2}\sigma - \frac{(d-1)(d+2)}{2(d+1)}}. \tag{4.2}$$

Let us look at the right hand side of (4.1). Each g_θ is supported in a ball of radius $\sim R^{-1}$, so one can see that each $|g_\theta|$ is $\sim R^{-(d-1)}$ on a ball of radius $\sim R$ centered at the origin and has the same value on the whole space as an upper bound. Hence each

$$\|g_\theta\|_{L^p(w_{B_R})} \sim R^{-(d-1)+\frac{d}{p}} \sim R^{-\frac{(d-1)(d+2)}{2(d+1)}}$$

and the right hand side of (4.1) for $v = g$ is

$$\sim R^\varepsilon \cdot R^{\frac{(d-1)\sigma}{d+1}} \cdot R^{\frac{(d-1)\sigma}{p}} \cdot R^{-\frac{(d-1)(d+2)}{2(d+1)}} \sim R^{\frac{d-1}{2}\sigma - \frac{(d-1)(d+2)}{2(d+1)}+\varepsilon}. \tag{4.3}$$

We see that the right hand sides of (4.2) and (4.3) match except for the R^ε-loss, showing that g is an almost sharp example for Theorem 4.1.

Since the decoupling theorem for the paraboloid by Bourgain–Demeter is weaker than Theorem 4.1, we see that this function g is also sharp for Bourgain–Demeter's decoupling theorem.

As mentioned above, the refined decoupling Theorem 4.1 has other consequences that are useful in problems in geometric measure theory such as the Falconer distance problem. For example, by dyadic pigeonholing and Hölder's ineuqality, it implies the following corollary:

COROLLARY 4.2. *Let $R > 1$ and $\alpha > 0$. Suppose a set $Y \subset B_R$ is a union of lattice 1-cubes with the following "fractal structure (between scale 1 and R)":*

$$|B_r \bigcap Y| \lesssim r^\alpha, \quad \forall 1 \le r \le R, \forall B_r \subset B_R.$$

Suppose $\operatorname{supp} \hat{v} \subset P^{d-1}$ with $\hat{v} = \varphi \, d\omega$ such that in the wave packet decomposition of v in B_R, each wave packet has its essential support hits a $\lesssim R^{-\frac{(d-1)}{2}}$ fraction of unit cubes in Y. Then

$$\|v\|_{L^2(Y)} \lesssim_\varepsilon R^{\frac{1}{d+1}(\alpha - \frac{d-1}{2})+\varepsilon}\|\varphi\|_{L^2(d\omega)}. \tag{4.4}$$

By a similar computation, one can see that if we take $\alpha = d - (d+1)\sigma$ and v to be the function g in §2, one has again an almost sharp example for Corollary 4.2. In this example, both sides of (4.4) are comparable to or close to $R^{\frac{d-3}{2}\sigma - \frac{d-2}{2}}$.

Theorem 4.1 and various versions of Corollary 4.2 were used in **[7]** and later works such as **[5]** to make progress towards the Falconer distance conjecture. The sharpness of these two propositions showed in this section suggests that to make further progress beyond e.g. **[7]**, one either has to sharpen other components in these two papers or to design new approaches.

References

[1] Juan A. Barceló, Jonathan M. Bennett, Anthony Carbery, Alberto Ruiz, and María C. Vilela, *Some special solutions of the Schrödinger equation*, Indiana Univ. Math. J. **56** (2007), no. 4, 1581–1593, DOI 10.1512/iumj.2007.56.3016. MR2354692

[2] Jean Bourgain and Ciprian Demeter, *The proof of the l^2 decoupling conjecture*, Ann. of Math. (2) **182** (2015), no. 1, 351–389, DOI 10.4007/annals.2015.182.1.9. MR3374964

[3] Anthony Carbery, Marina Iliopoulou, and Hong Wang, *Some sharp inequalities of Mizohata-Takeuchi-type*, Preprint, arXiv:2302.11877, (2023).

[4] Ciprian Demeter, Larry Guth, and Hong Wang, *Small cap decouplings*, Geom. Funct. Anal. **30** (2020), no. 4, 989–1062, DOI 10.1007/s00039-020-00541-5. With an appendix by D. R. Heath-Brown. MR4153908

[5] Xiumin Du, Alex Iosevich, Yumeng Ou, Hong Wang, and Ruixiang Zhang, *An improved result for Falconer's distance set problem in even dimensions*, Math. Ann. **380** (2021), no. 3-4, 1215–1231, DOI 10.1007/s00208-021-02170-1. MR4297185

[6] Larry Guth, *An enemy scenario in restriction theory*, joint talk for AIM Research Community *Fourier restriction conjecture and related problems* and HAPPY network (2022), https://www.youtube.com/watch?v=x-DET83UjFg, (2022).

[7] Larry Guth, Alex Iosevich, Yumeng Ou, and Hong Wang, *On Falconer's distance set problem in the plane*, Invent. Math. **219** (2020), no. 3, 779–830, DOI 10.1007/s00222-019-00917-x. MR4055179

Department of Mathematics, Northwestern University, Evanston, Ilinois 60208
Email address: xdu@northwestern.edu

Department of Mathematics, University of Pennsylvania, Philadelphia, Pennsylvania 19104
Email address: yumengou@sas.upenn.edu

Department of Mathematics, University of California, Los Angeles, Los Angeles, California 90095
Email address: hongwang@math.ucla.edu

Department of Mathematics, University of California, Berkeley, Berkeley, California 94720
Email address: ruixiang@berkeley.edu

Contemporary Mathematics
Volume **792**, 2024
https://doi.org/10.1090/conm/792/15896

Some results for the Szegő and Bergman projections on planar domains

Nathan A. Wagner

ABSTRACT. The purpose of this note is to prove some boundedness/compactness results of a harmonic analysis flavor for the Bergman and Szegő projections on certain classes of planar domains using conformal mappings. In particular, we prove weighted estimates for the projections, provide quantitative L^p estimates and a specific example of such estimates on a domain with a sharp p range, and show that the "difference" of the Bergman and Szegő projections is compact at the endpoints $p = 1, \infty$ for domains with sufficient smoothness. We also pose some open questions that naturally arise from our investigation.

1. Introduction

Let $\mathbb{D}$ denote the unit disc and $\partial\mathbb{D}$ the unit circle. Let $A^2(\mathbb{D})$ denote the Bergman space of square-integrable holomorphic functions on the unit disc and $H^2(\partial\mathbb{D})$ denote the Hardy space on the unit circle, which is the closed subspace of $L^2(\partial\mathbb{D})$ consisting of functions whose Fourier series are of the form $f(e^{\mathrm{i}\theta}) = \sum_{n=0}^{\infty} a_n e^{in\theta}$. It is well-known that every function in $H^2(\partial\mathbb{D})$ can be identified with a holomorphic function in $\mathbb{D}$ with square-summable Taylor coefficients (equivalently, belonging to the space $H^2(\mathbb{D})$) via the Poisson extension, and conversely, the boundary values of any such holomorphic function are well-defined as nontangential limits almost everywhere on $\partial\mathbb{D}$ and belong to $H^2(\partial\mathbb{D})$. The Bergman projection $\mathcal{B}_0$ is the orthogonal projection from $L^2(\mathbb{D})$ to $A^2(\mathbb{D})$, and can be expressed as the following integral for $f \in L^2(\mathbb{D})$ and $z \in \mathbb{D}$:

$$\mathcal{B}_0 f(z) = \int_{\mathbb{D}} \frac{f(w)}{(1-\overline{w}z)^2}\, dA(w),$$

where dA denotes (normalized) Lebesgue area measure on $\mathbb{D}$. On the other hand, the Szegő projection $\mathcal{S}_0$ is the orthogonal projection from $L^2(\partial\mathbb{D})$ to $H^2(\partial\mathbb{D})$, and has the following integral representation for $f \in L^2(\partial\mathbb{D})$ and $\zeta \in \partial\mathbb{D}$:

$$\mathcal{S}_0 f(\zeta) = \lim_{r\to 1^-} \int_0^{2\pi} \frac{f(e^{\mathrm{i}\theta})}{(1-e^{-\mathrm{i}\theta} r\zeta)}\, d\tilde{\theta}, \quad d\tilde{\theta} = \frac{d\theta}{2\pi}.$$

2020 *Mathematics Subject Classification.* Primary 43A99; Secondary 30E20, 30H10, 30H20, 30C20.

The research that led to this manuscript was partially completed when the author was supported by an NSF GRF grant (DGE-1745038).

Conformal mapping then allows us to express the Bergman and Szegő projections on other planar domains in terms of the projections on the unit disc/circle. Let $\Omega \subsetneq \mathbb{C}$ be a simply connected domain. When we consider the Szegő projection, we will make the additional assumption that Ω is bounded and has rectifiable boundary. For such a domain, we know by the Riemann mapping theorem that there exists a biholomorphic map $\varphi : \mathbb{D} \to \Omega$. In particular, if Ω is also assumed to be bounded and rectifiable, we know that φ extends to a continuous bijective mapping of the closures $\bar{\mathbb{D}} \to \bar{\Omega}$ and φ' has nontangential boundary limits almost everywhere. Moreover, if we identify φ' with its boundary values, it is known $\varphi' \in H^1(\partial\mathbb{D})$ (see [**19**]). When we refer to φ' in the case of the Szegő projection, we will make this identification without further comment.

The Bergman projection $\mathcal{B}$ is again defined to be the orthogonal projection from $L^2(\Omega)$ to $A^2(\Omega)$, where $A^2(\Omega)$ is the Bergman space of square-integrable functions on Ω. Having fixed a conformal map φ, define a map τ from $L^2(\Omega)$ to $L^2(\mathbb{D})$ as follows:

$$\tau(f) := (f \circ \varphi) \cdot (\varphi')$$

and notice that τ is a surjective isometry. Then, using a transformation law for the Bergman kernel, one can show the Bergman projection $\mathcal{B}$ can be expressed

$$\mathcal{B} = \tau^{-1}\mathcal{B}_0\tau. \tag{1.1}$$

Similarly, the Szegő projection $\mathcal{S}$ can be defined as the orthogonal projection from $L^2(\partial\Omega)$ to $H^2(\partial\Omega)$, where $H^2(\partial\Omega)$ is an appropriate notion of Hardy space. This definition can be made more explicit by conformal mapping: analogously define a surjective isometry $\tau_{1/2}$ from $L^2(\partial\Omega)$ to $L^2(\partial\mathbb{D})$ as follows:

$$\tau_{1/2}(f)(e^{i\theta}) := f(\varphi(e^{i\theta})) \cdot (\varphi'(e^{i\theta}))^{1/2}.$$

We remark that as a consequence of the fact that φ' is nonvanishing on $\mathbb{D}$, there exists a holomorphic function ϕ on $\mathbb{D}$ such that $\phi^2 = \varphi'$. Moreover, ϕ has nontangential boundary limits almost everywhere (in fact, $\phi \in H^2(\mathbb{D})$) by virtue of the fact that $\varphi' \in H^1(\mathbb{D})$. Therefore, we may take $(\varphi'(e^{i\theta}))^{1/2} = \phi(e^{i\theta})$ for almost every $\theta \in [0, 2\pi)$, and this consistently defines a $L^2(\partial\mathbb{D})$ function. Then the Szegő projection on $L^2(\partial\Omega)$ may be defined

$$\mathcal{S} = \tau_{1/2}^{-1}\mathcal{S}_0\tau_{1/2}. \tag{1.2}$$

While the L^2 regularity of both projection operators is immediate from their definitions, the L^p regularity is highly dependent on the boundary geometry of the domain. The technique of relating the projections to operators on the disk via conformal mapping was used by Lanzani and Stein to study the L^p regularity of the Bergman and Szegő projections in [**19**] on several classes of planar domains including graph domains, local graph domains, domains with the vanishing chord-arc property, and Lipschitz domains. In some of these cases (local graph, graph, and Lipschitz) they show that the Szegő/Bergman projections are bounded on L^p for a limited range $p_0 < p < p_0' = \frac{p_0}{p_0-1}$ where $p_0 > 1$, and they additionally demonstrate that the range cannot be improved in the Lipschitz case for the Szegő projection. This behavior of the projections stands in stark contrast to the smooth boundary case, or even the case when the boundary is of class C^1 or convex. In these situations, the L^p boundedness for $1 < p < \infty$ can be recovered (see, for example [**5**, Theorem 2.1], [**25, 26**]). However, in the case that Ω has the vanishing chord-arc property, Lanzani and Stein proved that the projections are bounded on L^p for the full range as well, which strengthens the C^1 result.

It turns out that the *unweighted* L^p regularity of either projection on Ω or $\partial\Omega$ is equivalent to the boundedness on a *weighted* L^p space of the corresponding projection on the unit disc/circle. The boundedness on the unit disc/circle can then be deduced from known weighted theory; in particular, the Szegő projection is bounded on $L^p_\omega(\partial\mathbb{D})$ if and only if $\omega \in A_p(\partial\mathbb{D})$ (see [**21**]), and the Bergman projection is bounded on $L^p_\omega(\mathbb{D})$ if and only if $\omega \in A_p^+(\mathbb{D})$ (see [**5**]). These weight classes are defined precisely in Section 2. We also mention that this conformal mapping technique was used by Lanzani and Koenig [**18**] to study mapping properties of the difference operator $\mathcal{BP} - \mathcal{PS}$, where $\mathcal{P}$ denotes the Poisson extension operator.

In this note, we prove a collection of results, including weighted estimates and compactness, for the Bergman and Szegő projections and related operators on planar domains. The unifying principle in our exploration is the use of conformal mapping techniques, in particular equations (1.1) and (1.2) and related identities to prove these harmonic analysis theorems. First, we provide sufficient conditions on a weight ω for the Bergman or Szegő projection to be bounded on L^p_ω for a general planar domain. In specific cases, these conditions translate to membership in an expected weight class that is the intersection of a Muckenhoupt class with a reverse Hölder class. We also use conformal mapping to provide a quantitative upper estimate on the operator norms of the projections on unweighted L^p spaces and work out a specific example where we also provide a lower bound. Finally, we study the compactness of the difference operator $\mathcal{BP} - \mathcal{PS}$. Throughout the paper, we list several open questions as well.

This paper is organized as follows. In Section 2, we prove the weighted L^p estimates for the projection operators (Theorems 2.1 and 2.2) and note several applications to specific domain classes (Corollaries 2.7, 2.8, 2.9, and 2.10). In Section 3, we show how the conformal techniques lead to quantitative (unweighted) L^p norm estimates for the projections and work out an example where we can obtain concrete upper and lower bounds (Theorem 3.5). Finally, in Section 4, we prove endpoint compactness results for the difference operator $\mathcal{BP} - \mathcal{PS}$ (Theorem 4.4).

The author would like to thank Brett Wick, Cody Stockdale, and Walton Green for several helpful comments. He would also like to express his gratitude to the anonymous referee, whose suggestions led to many improvements in the paper.

2. Weighted estimates

In this section, we prove general weighted estimates for the Bergman and Szegő projections. As a consequence, we deduce specific results on certain classes of planar domains. To summarize, we establish that for these domain classes, if the Szegő or Bergman projection is bounded on L^p for a limited range of p, then the operator extends boundedly in the same range to the weighted space $L^p_\omega(\Omega)$, provided the pullback of the weight ω to the unit disc belongs to an appropriate Muckenhoupt class intersected with a reverse Hölder class. In particular, see Subsection 2.2.

Weighted theory for operators that are bounded on a limited range of L^p spaces has been studied by several authors. Auscher and Martell proved that for a certain class of operators $\mathcal{T}$ that are bounded on a limited range $p_0 < p < q_0$ of L^p spaces, weighted L^p estimates can be obtained [**1**]. In particular, if $\omega \in A_{\frac{p}{p_0}} \cap \mathrm{RH}_{\left(\frac{q_0}{p}\right)'}$, then $\mathcal{T}$ maps $L^p(\omega)$ to $L^p(\omega)$. Bernicot, Frey, and Petermichel later elaborated on this result to provide a sharp weighted estimate using techniques of sparse bounds [**7**]. The weight classes we consider below are inspired by these results.

Recall that a weight is a locally integrable function that is positive almost everywhere. For $1 < p < \infty$, we say a weight ω on $\partial\mathbb{D}$ belongs to $A_p(\partial\mathbb{D})$ if the following quantity is finite:

$$[\omega]_{A_p} := \sup_I \frac{1}{|I|}\int_I \omega \, d\tilde{\theta} \left(\frac{1}{|I|}\int_I \omega^{-\frac{1}{p-1}} \, d\tilde{\theta}\right)^{p-1},$$

where I is an arc on the boundary of $\mathbb{D}$, $d\tilde{\theta}$ denotes normalized arc length measure, and $|I| = \int_I d\tilde{\theta}$ denotes the (normalized) arc length of I. On the other hand, for $1 < q < \infty$ we say that ω lies in the reverse Hölder class $\mathrm{RH}_q(\partial\mathbb{D})$ if

$$[\omega]_{\mathrm{RH}_q(\partial\mathbb{D})} := \sup_I \frac{\left(\frac{1}{|I|}\int_I \omega^q \, d\tilde{\theta}\right)^{1/q}}{\frac{1}{|I|}\int_I \omega \, d\tilde{\theta}} < \infty.$$

Obviously, the normalization is not important in the definition of these weight classes; it is simply for convenience.

Analogously for the Bergman projection, we define the A_p^+ class of weights with respect to certain Carleson regions. For $z_0 \in \mathbb{C}$, let $D_r(z_0) = \{z : |z - z_0| < r\}$ be the (Euclidean) disc of radius r centered at z_0. We say a weight ω on $\mathbb{D}$ belongs to $A_p^+(\mathbb{D})$ if the following quantity is finite:

$$[\omega]_{A_p^+} := \sup_{\substack{z\in\partial\mathbb{D}\\ r>0}} \frac{1}{|D_r(z)\cap\mathbb{D}|}\int_{D_r(z)\cap\mathbb{D}} \omega \, dA \left(\frac{1}{|D_r(z)\cap\mathbb{D}|}\int_{D_r(z)\cap\mathbb{D}} \omega^{-\frac{1}{p-1}} \, dA\right)^{p-1},$$

where dA denotes (normalized) planar Lebesgue measure and $|D_r(z)| = A(D_r(z))$.

Analogous to our earlier definition, for $1 < q < \infty$ we say that $\omega \in \mathrm{RH}_q(\mathbb{D})$ if

$$[\omega]_{\mathrm{RH}_q(\mathbb{D})} := \sup_{\substack{z\in\partial\mathbb{D}\\ r>0}} \frac{\left(\frac{1}{|D_r(z)\cap\mathbb{D}|}\int_{D_r(z)\cap\mathbb{D}} \omega^q \, dA\right)^{1/q}}{\frac{1}{|D_r(z)\cap\mathbb{D}|}\int_{D_r(z)\cap\mathbb{D}} \omega \, dA} < \infty.$$

We remark in passing that a subclass of A_p^+ weights (more precisely, a limiting class when $p = \infty$ that satisfy an additional regularity condition) that belong to $\mathrm{RH}_q(\mathbb{D})$ were the main object of study in [**2**].

We have the following two general theorems.

THEOREM 2.1. *Suppose $\Omega \subset \mathbb{C}$ is a bounded, simply connected domain with rectifiable boundary and $1 < p_0 < 2$. Suppose there exists a biholomorphic mapping $\varphi : \bar{\mathbb{D}} \to \bar{\Omega}$ that has the property that $|\varphi'|^{\left(1-\frac{p}{2}\right)\left(\frac{p_0}{p(p_0-1)}\right)} \in A_2(\partial\mathbb{D})$ for $p_0 < p < \frac{p_0}{p_0-1}$. If $p_0 < p < \frac{p_0}{p_0-1}$ and $\omega \circ \varphi \in A_{\frac{p}{p_0}}(\partial\mathbb{D}) \cap \mathrm{RH}_{\left(\frac{p_0'}{p}\right)'}(\partial\mathbb{D})$, then the Szegő projection*

$\mathcal{S}$ extends to a bounded operator on $L^p_\omega(\partial\Omega)$. Moreover,

$$\|\mathcal{S}\|_{L^p_\omega(\partial\Omega)\to L^p_\omega(\partial\Omega)}$$
$$\lesssim C_p\left([\omega\circ\varphi]_{A_{\frac{p}{p_0}}}\times[\omega\circ\varphi]_{\mathrm{RH}_{\left(\frac{p_0'}{p}\right)'}}\times\left[|\varphi'|^{(1-\frac{p}{2})\left(\frac{p_0}{p(p_0-1)}\right)}\right]_{A_2}^{\frac{p(p_0-1)}{p_0}}\right)^{\max\{1,\frac{1}{p-1}\}},$$

where $C_p=\max\{p,\frac{1}{p-1}\}$ and the implicit constant is independent of p and Ω.

THEOREM 2.2. *Suppose $\Omega\subsetneq\mathbb{C}$ is a simply connected domain. Suppose there exists a biholomorphic mapping $\varphi:\mathbb{D}\to\Omega$ that has the property that $|\varphi'|^{(2-p)\left(\frac{p_0}{p(p_0-1)}\right)}\in A_2^+(\mathbb{D})$ for $p_0<p<\frac{p_0}{p_0-1}$. If $p_0<p<\frac{p_0}{p_0-1}$ and $\omega\circ\varphi\in A^+_{\frac{p}{p_0}}(\mathbb{D})\cap\mathrm{RH}_{\left(\frac{p_0'}{p}\right)'}(\mathbb{D})$, then the Bergman projection $\mathcal{B}$ is bounded on $L^p_\omega(\Omega)$. Moreover,*

$$\|\mathcal{B}\|_{L^p_\omega(\Omega)\to L^p_\omega(\Omega)}$$
$$\lesssim C_p\left([\omega\circ\varphi]_{A^+_{\frac{p}{p_0}}}\times[\omega\circ\varphi]_{\mathrm{RH}_{\left(\frac{p_0'}{p}\right)'}}\times\left[|\varphi'|^{(2-p)\left(\frac{p_0}{p(p_0-1)}\right)}\right]_{A_2^+}^{\frac{p(p_0-1)}{p_0}}\right)^{\max\{1,\frac{1}{p-1}\}},$$

where $C_p=\max\{p,\frac{1}{p-1}\}$ and the implicit constant is independent of p and Ω.

We denote arc length measure on $\partial\Omega$ by $d\sigma$. For convenience in what follows, for a locally integrable function f we let

$$\fint_I f\,d\tilde{\theta}=\frac{1}{|I|}\int_I f\,d\tilde{\theta}.$$

We use analogous notation for averages in the Bergman case when the integrals are taken over a domain in $\mathbb{C}$ with respect to Lebesgue measure dA in the plane.

2.1. Proofs of main theorems. These theorems are established by transferring the problem to the unit disc via the following two lemmas. In what follows, $\varphi:\bar{\mathbb{D}}\to\bar{\Omega}$ always denotes the (fixed) Riemann map.

LEMMA 2.3. *Let $1\le p<\infty$. The boundedness of $\mathcal{S}$ on $L^p_\omega(\partial\Omega)$ is equivalent to the boundedness of $\mathcal{S}_0$ on $L^p_{(\omega\circ\varphi)\nu}(\partial\mathbb{D})$ where $\nu(e^{i\theta})=|\varphi'(e^{i\theta})|^{1-\frac{p}{2}}$. Therefore, for $1<p<\infty$, the boundedness of $\mathcal{S}$ on $L^p_\omega(\partial\Omega)$ is equivalent to the condition $(\omega\circ\varphi)\nu\in A_p(\partial\mathbb{D})$.*

PROOF. To prove the first statement, it suffices to show that $\tau_{1/2}$ is also a surjective isometry from $L^p_\omega(\partial\Omega)$ to $L^p_{(\omega\circ\varphi)\nu}(\partial\mathbb{D})$. It is well-known (see [**19**, pp. 66]) that $d\sigma(\zeta)=|\varphi'(e^{i\theta})|d\theta$ where $\zeta=\varphi(e^{i\theta})$, so we may apply a change of variable. We compute, for a function $f\in L^p_\omega(\partial\Omega)$:

$$\begin{aligned}\int_{\partial\Omega}|f|^p\omega\,d\sigma &= \int_{\partial\mathbb{D}}|f(\varphi)|^p(\omega\circ\varphi)|\varphi'|\,d\theta\\ &= \int_{\partial\mathbb{D}}|f(\varphi)|^p|\varphi'|^{p/2}(\omega\circ\varphi)|\varphi'|^{1-\frac{p}{2}}\,d\theta\\ &= \int_{\partial\mathbb{D}}|\tau_{1/2}(f)|^p(\omega\circ\varphi)|\varphi'|^{1-\frac{p}{2}}\,d\theta\end{aligned}$$

which establishes that $\tau_{1/2}$ is isometric. Surjectivity is obvious.

The second statement is an immediate consequence of the first by standard facts about the Szegő projection on the unit disc (see [**19**, pp. 66] or [**21**]). □

The second lemma is completely analogous to the first.

LEMMA 2.4. *The boundedness of $\mathcal{B}$ on $L^p_\omega(\Omega)$ is equivalent to the boundedness of $\mathcal{B}_0$ on $L^p_{(\omega\circ\varphi)\nu}(\mathbb{D})$ where $\nu(z) = |\varphi'(z)|^{2-p}$. Therefore, the boundedness of $\mathcal{B}$ on $L^p_\omega(\Omega)$ is equivalent to the condition $(\omega\circ\varphi)\nu \in A_p^+(\mathbb{D})$.*

PROOF. The proof is virtually identical to the proof of the preceding lemma with minor changes. In particular, the map $\tau_{1/2}$ needs to be replaced by the map τ and the integration now takes places on $\mathbb{D}$.

□

The following two propositions will enable us to prove Theorems 2.1 and 2.2.

PROPOSITION 2.5. *Let $1 < p_0 < p < \frac{p_0}{p_0-1}$ and suppose $(\omega\circ\varphi) \in A_{\frac{p}{p_0}}(\partial\mathbb{D}) \cap RH_{\left(\frac{p_0'}{p}\right)'}(\partial\mathbb{D})$. Let $\nu(e^{i\theta}) = |\varphi'(e^{i\theta})|^{1-\frac{p}{2}}$. If $\nu^{\frac{p_0}{p(p_0-1)}} \in A_2(\partial\mathbb{D})$, then $(\omega\circ\varphi)\nu \in A_p(\partial\mathbb{D})$ and*

$$[(\omega\circ\varphi)\nu]_{A_p} \le [\omega\circ\varphi]_{A_{\frac{p}{p_0}}} \times [\omega\circ\varphi]_{RH_{\left(\frac{p_0'}{p}\right)'}} \times \left[\nu^{\frac{p_0'}{p}}\right]_{A_2}^{\frac{p}{p_0'}}.$$

PROOF. We must show

$$\sup_I \fint_I (\omega\circ\varphi)\nu\, d\tilde\theta \left(\fint_I ((\omega\circ\varphi)\nu)^{-\frac{1}{p-1}}\, d\tilde\theta\right)^{p-1} \le [\omega\circ\varphi]_{A_{\frac{p}{p_0}}} \times [\omega\circ\varphi]_{\mathrm{RH}_{\left(\frac{p_0'}{p}\right)'}} \times \left[\nu^{\frac{p_0'}{p}}\right]_{A_2}^{\frac{p}{p_0'}}.$$

Applying Hölder's inequality with exponents $r = \frac{p_0'}{p}$ and r' and using the fact that $(\omega\circ\varphi) \in \mathrm{RH}_{r'}(\partial\mathbb{D})$, we obtain

$$\begin{aligned} \fint_I (\omega\circ\varphi)\nu\, d\tilde\theta &\le \left(\fint_I (\omega\circ\varphi)^{r'}\, d\tilde\theta\right)^{\frac{1}{r'}} \left(\fint_I \nu^r\, d\tilde\theta\right)^{\frac{1}{r}} \\ &\le [\omega\circ\varphi]_{\mathrm{RH}_{r'}} \left(\fint_I (\omega\circ\varphi)\, d\tilde\theta\right) \left(\fint_I \nu^r\, d\tilde\theta\right)^{\frac{1}{r}}. \end{aligned}$$

On the other hand, let $s = \frac{p}{p_0}$. Let $t = \frac{(p-1)}{s-1}$ and note $t > 1$. Then, applying Hölder's inequality with exponents t and $t' = \frac{p_0(p-1)}{p(p_0-1)} = (p-1)r$, we obtain

$$\left(\fint_I ((\omega\circ\varphi)\nu)^{-\frac{1}{p-1}}\, d\tilde\theta\right)^{p-1} \le \left(\fint_I (\omega\circ\varphi)^{-\frac{1}{s-1}}\, d\tilde\theta\right)^{s-1} \left(\fint_I \nu^{-r}\, d\tilde\theta\right)^{\frac{1}{r}}.$$

Putting these facts together, we then get

$$
\begin{aligned}
&\fint_I (\omega\circ\varphi)\nu\, d\tilde\theta \left(\fint_I ((\omega\circ\varphi)\nu)^{-\frac{1}{p-1}}\, d\tilde\theta\right)^{p-1}\\
\leq &[\omega\circ\varphi]_{\mathrm{RH}_{r'}} \left(\fint_I (\omega\circ\varphi)\, d\tilde\theta\right)\left(\fint_I \nu^r\, d\tilde\theta\right)^{\frac{1}{r}}\left(\fint_I (\omega\circ\varphi)^{-\frac{1}{s-1}}\, d\tilde\theta\right)^{s-1}\left(\fint_I \nu^{-r}\, d\tilde\theta\right)^{\frac{1}{r}}\\
\leq &[\omega\circ\varphi]_{A_s(\partial\mathbb{D})}\times[\omega\circ\varphi]_{\mathrm{RH}_{r'}}\times[\nu^r]^{\frac{1}{r}}_{A_2(b\mathbb{D})}
\end{aligned}
$$

as required. □

Proposition 2.6. *Let $1 < p_0 < p < \frac{p_0}{p_0-1}$ and suppose $(\omega\circ\varphi) \in A^+_{\frac{p}{p_0}}(\mathbb{D}) \cap RH_{\left(\frac{p_0'}{p}\right)'}(\mathbb{D})$. Let $\nu(z) = |\varphi'(z)|^{2-p}$. If $\nu^{\frac{p_0}{p(p_0-1)}} \in A^+_2(\mathbb{D})$, then $(\omega\circ\varphi)\nu \in A^+_p(\mathbb{D})$ and*

$$[(\omega\circ\varphi)\nu]_{A^+_p} \leq [\omega\circ\varphi]_{A^+_{\frac{p}{p_0}}} \times [\omega\circ\varphi]_{RH_{\left(\frac{p_0'}{p}\right)'}} \times \left[\nu^{\frac{p_0'}{p}}\right]^{\frac{p}{p_0'}}_{A^+_2}.$$

Proof. The proof follows exactly the same way as Proposition 2.5. □

Proof of Theorem 2.1. The theorem will follow from Lemma 2.3 and Proposition 2.5 once we have deduced the following weighted norm estimates for the Szegő projection on the circle:

$$\|\mathcal{S}_0\|_{L^p_\omega(\partial\mathbb{D})\to L^p_\omega(\partial\mathbb{D})} \lesssim C_p[\omega]^{\max\{1,\frac{1}{p-1}\}}_{A_p(\partial\mathbb{D})}, \quad C_p = \max\{p, \frac{1}{p-1}\}.$$

This estimate can be deduced as follows. We will begin by proving it for $p = 2$, and then we will extend the estimate to all $p \in (1,\infty)$ via extrapolation. It is also probably possible in some way to deduce this bound from the case of the Hilbert transform on the line, but we argue directly instead. First, note that the Szegő projection on the circle has kernel (with respect to normalized arc length measure $d\tilde\theta$)

$$K(z,w) = \frac{1}{1-\overline{w}z}$$

in the precise sense that if $f \in L^2(\partial\mathbb{D})$ we have, for almost every $z \in \partial\mathbb{D}\setminus \mathrm{supp} f$:

$$\mathcal{S}_0 f(z) = \int_{\partial\mathbb{D}} \frac{f(w)}{1-\overline{w}z}\, d\tilde\theta(w).$$

This fact is a well-known consequence of the Plemelj jump formula.

Moreover, we see that $\mathcal{S}_0$ can be viewed as a Calderón–Zygmund operator on the circle $\partial\mathbb{D}$ viewed as a space of homogeneous type with respect to the measure $d\tilde\theta$ and arc length metric

$$d(z,w) = \min\{|\theta-\phi|, 2\pi - |\theta-\phi| : z = e^{\mathrm{i}\theta}, w = e^{\mathrm{i}\phi}, \theta,\phi \in [0,2\pi)\}$$

We remark in passing that one could also take the Euclidean metric by viewing z, w as living in the ambient space $\mathbb{C}$, and this would have no serious impact on the proof. Then, $\mathcal{S}_0$ is obviously bounded on $L^2(\partial\mathbb{D})$, and the size and smoothness estimates on its kernel then follow from direct calculation, which we omit. The sharp weighted estimates on L^p for $p = 2$ then follow from the A_2 theorem on spaces of homogeneous type (see [**4**, Theorem 1.1]), which states that if T is a Calderón-Zygmund operator on a space of homogeneous type and w is an A_2 weight, then

$$\|T\|_{L^2(w)\to L^2(w)} \lesssim [w]_{A_2}. \tag{2.1}$$

Given $\zeta \in \partial\mathbb{D}$ and $0 < r < 2\pi$, let $I(\zeta, r) = \{\xi \in \partial\mathbb{D} : d(\xi,\zeta) < r\}$. Let $\mathfrak{M}_{\partial\mathbb{D}}$ be the Hardy-Littlewood Maximal function defined

$$\mathfrak{M}_{\partial\mathbb{D}} f(z) = \sup_{\substack{\zeta\in\partial\mathbb{D} \\ r\in(0,2\pi): \\ z\ni I(\zeta,r)}} \frac{1}{|I(\zeta,r)|}\int_{I(\zeta,r)} |f|\, d\tilde{\theta}.$$

A line-by-line imitation of the Euclidean extrapolation proof (replace $\mathbb{R}^n$ by $\partial\mathbb{D}$, etc) in [**8**, Theorem 3.22], together with (2.1) yields the estimates

$$\int_{\partial\mathbb{D}} |\mathcal{S}_0 f|^p \omega\, d\theta \lesssim \|\mathfrak{M}_{\partial\mathbb{D}}\|^{2-p}_{L^p_\omega(\partial\mathbb{D})\to L^p_\omega(\partial\mathbb{D})} [\omega]_{A_p} \left(\int_{\partial\mathbb{D}} |f|^p\omega\, d\theta\right), \quad 1<p<2 \tag{2.2}$$

and

$$\int_{\partial\mathbb{D}} |\mathcal{S}_0 f|^p \omega\, d\theta \lesssim \|\mathfrak{M}_{\partial\mathbb{D}}\|^{\frac{p-2}{p-1}}_{L^p_{\omega'}(\partial\mathbb{D})\to L^p_{\omega'}(\partial\mathbb{D})} [\omega]^{\frac{1}{p-1}}_{A_p} \left(\int_{\partial\mathbb{D}} |f|^p\omega\, d\mu\right), \tag{2.3}$$

$$\omega' = \omega^{1-p'}, \quad 2<p<\infty,$$

where $\frac{1}{p}+\frac{1}{p'}=1$ and the implicit constants are independent of the exponent p and weight ω.

Finally, the following estimate for the maximal function (which is also valid on any space of homogeneous type) is given in [**16**, Proposition 7.13]:

$$\|\mathfrak{M}_{\partial\mathbb{D}}\|_{L^p_\omega(\partial\mathbb{D})\to L^p_\omega(\partial\mathbb{D})} \lesssim (p')^{1/p} p^{1/p'} [\omega]^{\frac{1}{p-1}}_{A_p}, 1<p<\infty. \tag{2.4}$$

The result then follows by combining (2.2), (2.3), and (2.4).

We also refer the reader to [**10**] for a similar estimate in the Euclidean setting of $\mathbb{R}^n$. □

Proof of Theorem 2.2. The theorem follows from Lemma 2.4, Proposition 2.6, and the fact that $\|\mathcal{B}_0\|_{L^p_\omega(\mathbb{D})} \lesssim C_p[\omega]^{\max\{1,\frac{1}{p-1}\}}_{A^+_p(\mathbb{D})}$. This last fact can be found in [**24**, Theorem 2] (the quantative dependence on p is implicit in the proof but is also stated explicitly in [**15**, Theorem 1.2]). In the notation of the reference, we take $b=0$ and $u=\omega$, $P_0=\mathcal{B}_0$, and $B_{p,0}=A^+_p(\mathbb{D})$. We remark that here the quantitative estimate is a direct consequence of a "sparse domination" for the Bergman projection, and does not involve singular integral theory. Indeed, the Bergman projection is easier to deal with than the Szegő projection because it is not a true singular integtal operator; cancellation does not play a role in its boundedness and only the size of its kernel matters. See also [**23**] for a similar result on the upper-half plane. □

2.2. Applications. We now present several applications of Theorems 2.1 and 2.2. We consider in particular local graph domains and bounded Lipschitz domains. We say Ω is a graph domain if there exists an angle $\alpha > 0$ and a continuous function $\gamma : \mathbb{R}\to\mathbb{R}$ so that if $z = x + iy$,

$$\Omega = \{e^{i\alpha} z : y > \gamma(x)\}.$$

Similarly, we say Ω is a local graph domain if there exist finitely many graph domains $\Omega_1, \ldots, \Omega_n$ together with arcs on the boundary $\Gamma_1, \ldots, \Gamma_n$ covering $\partial\Omega$

and discs D_j so that $\Gamma_j \subset D_j$ and $\Omega \cap D_j = \Omega_j \cap D_j$ for $1 \leq j \leq n$. If each γ_j is a Lipschitz map with Lipschitz constant M_j, we further say that Ω is a bounded Lipschitz domain with Lipschitz constant $M = \max_{1 \leq j \leq n} M_j$.

COROLLARY 2.7 *Let Ω be a bounded, simply connected local graph domain. Then the Szegő projection $\mathcal{S}$ is bounded on $L^p_\omega(\partial\Omega)$ for $\frac{4}{3} < p < 4$ and all ω satisfying $(\omega \circ \varphi) \in A_{\frac{3p}{4}}(\partial\mathbb{D}) \cap RH_{(\frac{4}{p})'}(\partial\mathbb{D})$.*

COROLLARY 2.8. *Let Ω be a bounded Lipschitz domain with Lipschitz constant M and let $p_M = 2(\frac{\pi}{2\arctan M} + 1)$. Then the Szegő projection $\mathcal{S}$ is bounded on $L^p_\omega(\partial\Omega)$ for $p'_M < p < p_M$ and all ω satisfying $(\omega \circ \varphi) \in A_{\frac{p}{p'_M}}(\partial\mathbb{D}) \cap RH_{(\frac{p_M}{p})'}(\partial\mathbb{D})$.*

COROLLARY 2.9. *Let Ω be a bounded, simply connected local graph domain. Then the Bergman projection $\mathcal{B}$ is bounded on $L^p_\omega(\Omega)$ for $\frac{4}{3} < p < 4$ and all ω satisfying $(\omega \circ \varphi) \in A^+_{\frac{3p}{4}}(\mathbb{D}) \cap RH_{(\frac{4}{p})'}(\mathbb{D})$.*

COROLLARY 2.10. *Let Ω be a bounded Lipschitz domain with Lipschitz constant M and let $p_M = 2(\frac{\pi}{2\arctan M} + 1)$. Then the Bergman projection $\mathcal{B}$ is bounded on $L^p_\omega(\Omega)$ for $p'_M < p < p_M$ and all ω satisfying $(\omega \circ \varphi) \in A^+_{\frac{p}{p'_M}}(\mathbb{D}) \cap RH_{(\frac{p_M}{p})'}(\mathbb{D})$.*

All of these corollaries are deduced via Theorem 2.1 or 2.2 in the same way; by verifying that $|\varphi'|^{(1-\frac{p}{2})(\frac{p_0}{p(p_0-1)})}$ (or its square) is an $A_2(\partial\mathbb{D})$ or (respectively, $A^+_2(\mathbb{D})$) weight. The details can be found in [**19**], but we recall relevant facts for completeness. If Ω is any local graph domain, it is shown in [**19**, Proposition 2.2] that $|\varphi'|^r \in A_2(\partial\mathbb{D})$ for $0 \leq r < 1$ (and thus actually for $-1 < r < 1$). Note if $p_0 = \frac{4}{3}$ and $p \in (\frac{4}{3}, 4)$, we have $(1 - \frac{p}{2})(\frac{p_0}{p(p_0-1)}) = \frac{4}{p} - 2$, and $-1 < \frac{4}{p} - 2 < 1$ for p in this range. The argument for when Ω is a bounded Lipschitz domain is basically identical and follows from [**19**, Corollary 2.3].

On the other hand, for the same domains when considering the Bergman projection, Lanzani and Stein show that φ' is an outer function on $\mathbb{D}$ [**19**, Proposition 4.1]. A result of Békollè [**5**, Theorem 1.2] is then invoked which states if φ' is outer, $\alpha \in \mathbb{R}$, and $|\varphi'(e^{i\theta})|^\alpha \in A_p(\partial\mathbb{D})$, then $|\varphi'(z)|^{2\alpha} \in A^+_p(\mathbb{D})$. Thus, the two corresponding results for the Bergman projection follow from the arguments above pertaining to the Szegő projection.

REMARK 2.11. Notice that the conditions on the weight for boundedness in all of these results are, of course, in terms of its pullback to the unit disc. It is unclear whether these weights can be described in a conformal-free way that is intrinsic to the domain Ω, especially in the case of a domain with nonsmooth boundary.

3. An upper bound and a quantitative example

3.1. Upper quantitative bound. One can use similar methods as above to obtain quantitative estimates on the norm of the unweighted Szegő or Bergman projection. We have the following result:

THEOREM 3.1. *Let $\Omega \subset \mathbb{C}$ be a simply connected, bounded domain with rectifiable boundary. Let $\varphi : \bar{\mathbb{D}} \to \bar{\Omega}$ be a conformal mapping. The following estimate holds for the Szegő projection $\mathcal{S}$:*

$$\|\mathcal{S}\|_{L^p(\partial\Omega)\to L^p(\partial\Omega)} \lesssim C_p \left[|\varphi'|^{1-\frac{p}{2}}\right]_{A_p}^{\max\{1,\frac{1}{p-1}\}},$$

where $C_p = \max\{p, \frac{1}{p-1}\}$ *and the implicit constant is independent of* p *and* Ω.

PROOF OF THEOREM 3.1. First, note that the proof of Lemma 2.3 demonstrates that

$$\|\mathcal{S}\|_{L^p(\partial\Omega)\to L^p(\partial\Omega)} = \|\mathcal{S}_0\|_{L^p_\omega(\partial\mathbb{D})\to L^p_\omega(\partial\mathbb{D})},$$

where $\omega = |\varphi'|^{1-\frac{p}{2}}$. The upper bound then follows from noting that the Szegő projection on the unit circle is a Calderón-Zygmund operator on a space of homogeneous type as we argued in the proof of Theorem 2.1. □

Of course, we have an analogous statement for the Bergman projection, and in fact we can additionally provide a lower estimate in this case.

THEOREM 3.2. *Let* $\Omega \subsetneq \mathbb{C}$ *be a simply connected domain. Let* $\varphi : \bar{\mathbb{D}} \to \bar{\Omega}$ *be a conformal mapping. The following estimate holds for the Bergman projection* $\mathcal{B}$:

$$\left[|\varphi'|^{2-p}\right]_{A_p^+}^{\frac{1}{2p}} \lesssim \|\mathcal{B}\|_{L^p(\Omega)\to L^p(\Omega)} \lesssim C_p \left[|\varphi'|^{2-p}\right]_{A_p^+}^{\max\{1,\frac{1}{p-1}\}},$$

where $C_p = \max\{p, \frac{1}{p-1}\}$ *and the implicit constants are independent of* p *and* Ω.

PROOF. Lemma 2.4 gives that

$$\|\mathcal{B}\|_{L^p(\Omega)\to L^p(\Omega)} = \|\mathcal{B}_0\|_{L^p_\omega(\mathbb{D})},$$

where $\omega = |\varphi'|^{2-p}$. The upper bound then follows from [**24**, Theorem 2] or [**15**, Theorem 1.2] as before, while the lower bound can be found in [**15**, Theorem 1.3]. □

3.2. An example. We now provide an example of a domain in the plane where the upper bound of the L^p operator norm of the Szegő projection can be estimated directly, and in fact a lower bound may be computed as well. This example is inspired by a family of examples in [**19**]. Let $\mathbb{R}^2_+$ denote the upper half-plane in $\mathbb{C}$. Consider the following planar domains:

$$\Omega_1 = \Phi_1(\mathbb{R}^2_+), \Phi_1(z) = z^{3/2}e^{\frac{-i\pi}{4}} - \mathrm{i};$$

$$\Omega_2 = \Phi_2(\mathbb{R}^2_+), \Phi_2(z) = -z^{3/2}e^{\frac{-i\pi}{4}} + \mathrm{i};$$

$$\Omega_3 = \Phi_3(\mathbb{R}^2_+), \Phi_3(z) = z^{1/2}e^{\frac{-i\pi}{4}} - 2;$$

$$\Omega_4 = \Phi_4(\mathbb{R}^2_+), \Phi_4(z) = -z^{1/2}e^{\frac{-i\pi}{4}} + 2;$$

$$\Omega_5 = \Phi_5(\mathbb{R}^2_+), \Phi_5(z) = z^{1/2}e^{\frac{i\pi}{4}} - \left(\frac{1}{2} + \frac{3}{2}\mathrm{i}\right);$$

$$\Omega_6 = \Phi_6(\mathbb{R}^2_+), \Phi_6(z) = z^{1/2}e^{\frac{i\pi}{4}} + \left(\frac{1}{2} - \frac{3}{2}\mathrm{i}\right);$$

$$\Omega_7 = \Phi_7(\mathbb{R}^2_+), \Phi_7(z) = -z^{1/2}e^{\frac{i\pi}{4}} + \left(-\frac{1}{2} + \frac{3}{2}\mathrm{i}\right);$$

$$\Omega_8 = \Phi_8(\mathbb{R}^2_+), \Phi_8(z) = -z^{1/2}e^{\frac{i\pi}{4}} + \left(\frac{1}{2} + \frac{3}{2}\mathrm{i}\right).$$

Then, let $\Omega = \bigcap_{j=1}^8 \Omega_j$ (this is the same domain as $\Omega = \bigcap_{j=1}^4 \Omega_j$, but the eight domains are needed to describe the boundary of Ω locally as graphs of functions). It is clear Ω is a local graph domain with corresponding graph domains Ω_j, $1 \le j \le 8$. We provide a picture of this domain created using Mathematica below:

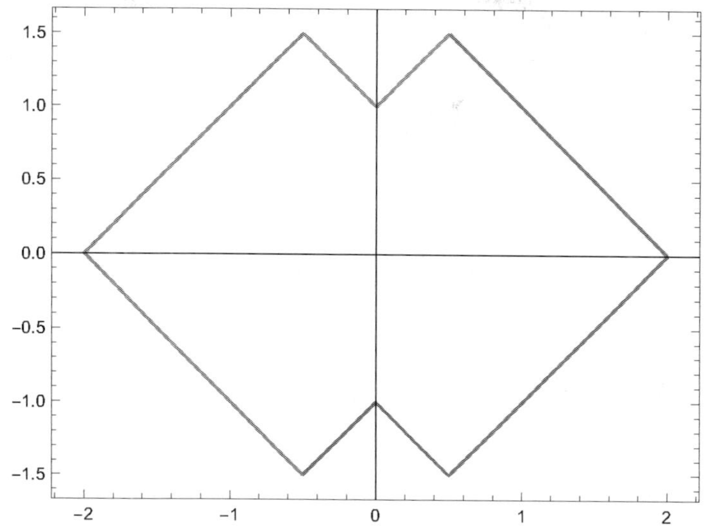

FIGURE 1. Lipschitz domain

It is obvious that Ω is a Lipschitz domain and it is easy to see that the Lipschitz constant is 1. We note that the points $(0, 1)$ and $(0, -1)$ in $\mathbb{R}^2 = \mathbb{C}$ are distinguished because the domain is not locally convex at those points. We know from previous work that since $M = 1$, $\mathcal{S}$ is bounded on $L^p(\Omega)$ for $\frac{6}{5} < p < 6$. The arguments in [**19**] establish that $\mathcal{S}$ is in fact bounded only in this p range, but the authors do not provide quantitative estimates on the $L^p(\Omega)$ operator norm of $\mathcal{S}$. Here we provide such estimates for the operator norm.

Let $\varphi : \bar{\mathbb{D}} \to \bar{\Omega}$ be a conformal map. We wish to work via the conformal maps on the upper half space defined above. In particular, we remark that a conformal map $\Psi : \mathbb{R}^2_+ \to \Omega$ which fixes the point at infinity has nontangential limits on the boundary of $\mathbb{R}^2_+$ almost everywhere. We now proceed to compute the $A_p(\partial\mathbb{D})$ characteristic of $|\varphi'|^{1-\frac{p}{2}}$ using the following proposition, which is a quantitative sharpening of [**19**, Proposition 2.4].

PROPOSITION 3.3. *Let Ω be a local graph domain with corresponding graph domains $\Omega_1, \ldots, \Omega_n$ and conformal maps $\Phi_j : \mathbb{R}^2_+ \to \Omega_j$ which fix the point at ∞. Let $1 < p < \infty$ and suppose $|\Phi_j'(x)|^r \in A_p(\mathbb{R})$ for some $r > 0$ and all $1 \le j \le n$. Let $\varphi : \bar{\mathbb{D}} \to \bar{\Omega}$ be a conformal map. Then $|\varphi'(e^{i\theta})|^r \in A_p(\partial\mathbb{D})$, and moreover:*

$$[|\varphi'|^r]_{A_p(\partial\mathbb{D})} \lesssim C_\Omega^p \max\{N_1, N_2\}$$

where

$$N_1 = \max_{1\le j\le n} \left[|\Phi_j'|^r\right]_{A_p(\mathbb{R})}$$

and

$$N_2 = \int_{\partial\mathbb{D}} |\varphi'|^r \, d\tilde{\theta} \left(\int_{\partial\mathbb{D}} |\varphi'|^{-\frac{r}{p-1}} \, d\tilde{\theta} \right)^{p-1}.$$

Here C_Ω is a constant that only depends on Ω.

PROOF. The proof is essentially contained in [**19**], so we omit some of the details. Let $\psi_j = \varphi^{-1} \circ \Phi_j$. The main gist of the argument is that it is possible to choose a finite covering $I_1, \ldots, I_n$ of $\partial\mathbb{D}$ together with intervals on $\mathbb{R}$ (identified as the boundary of $\mathbb{R}^2_+$), $J_j = \psi_j^{-1}(I_j)$, so that $|\psi_j'|$ is bounded above and below in a neighborhood of J_j and consequently $|\Phi_j'(z)| \approx |\varphi'(\psi_j(z))|$ for $z \in J_j$. Since there are finitely many such intervals I_j and $|\Phi_j'|^r \in A_p(\mathbb{R})$ for all j, it is clear that $|\varphi'|^r$ and $|\varphi'|^{\frac{-r}{p-1}}$ belong to $L^1(\partial\mathbb{D})$.

It is possible to choose $\delta > 0$ small enough so that if $|I| < \delta$, then $I \subseteq I_j$ for some $j \in \{1, \cdots, n\}$ (this is just the Lebesgue number lemma applied to $\{I_j\}$). The parameter δ will depend on the underlying domain Ω. We then can estimate using change of variables:

$$\begin{aligned}
&\frac{1}{|I|}\int_I |\varphi'|^r \, d\tilde{\theta} \left(\frac{1}{|I|}\int_I |\varphi'|^{-\frac{r}{p-1}} \, d\tilde{\theta}\right)^{p-1} \\
&\approx \frac{1}{|\psi_j^{-1}(I)|}\int_{\psi_j^{-1}(I)} |\varphi' \circ \psi_j|^r |\psi_j'| \, dx \left(\frac{1}{|\psi_j^{-1}(I)|}\int_{\psi_j^{-1}(I)} |\varphi' \circ \psi_j|^{-\frac{r}{p-1}} |\psi_j'| \, dx\right)^{p-1} \\
&\approx \frac{1}{|\psi_j^{-1}(I)|}\int_{\psi_j^{-1}(I)} |\Phi_j'|^r \, dx \left(\frac{1}{|\psi_j^{-1}(I)|}\int_{\psi_j^{-1}(I)} |\Phi_j'|^{-\frac{r}{p-1}} \, dx\right)^{p-1} \\
&\leq N_1.
\end{aligned}$$

All the implicit constants above evidently only depend on Ω.

If $|I| \geq \delta$, then we estimate as follows:

$$\begin{aligned}
\frac{1}{|I|}\int_I |\varphi'|^r \, d\tilde{\theta} \left(\frac{1}{|I|}\int_I |\varphi'|^{-\frac{r}{p-1}} \, d\tilde{\theta}\right)^{p-1} &\leq \frac{1}{\delta^p}\int_{\partial\mathbb{D}} |\varphi'|^r \, d\theta \left(\int_{\partial\mathbb{D}} |\varphi'|^{-\frac{r}{p-1}} \, d\theta\right)^{p-1} \\
&\leq \left(\frac{1}{\delta}\right)^p N_2.
\end{aligned}$$

This completes the proof. $\square$

We now compute the $A_p(\mathbb{R})$ characteristics of $|\Phi_1'(x)|^{1-\frac{p}{2}} = \frac{3}{2}|x|^{\frac{2-p}{4}}$ and $|\Phi_3'(x)| = \frac{1}{2}|x|^{\frac{p-2}{4}}$, which are the relevant derivatives for the domain Ω introduced at the beginning of the section. The computations for the other derivatives are obviously identical so we omit them.

First, we compute the following average for I an interval centered at 0 with length δ:

$$\begin{aligned}
\frac{1}{|I|}\int_I |x|^{\frac{2-p}{4}} \, dx &\approx \frac{1}{\delta}\int_0^\delta x^{\frac{2-p}{4}} \, dx \\
&= \delta^{\frac{2-p}{4}}\left(\frac{4}{6-p}\right)
\end{aligned}$$

and on the other hand

$$\begin{aligned}
\left(\frac{1}{|I|}\int_I |x|^{\frac{p-2}{4(p-1)}} \, dx\right)^{p-1} &\approx \left(\frac{1}{\delta}\int_0^\delta x^{\frac{p-2}{4(p-1)}} \, dx\right)^{p-1} \\
&= \delta^{\frac{p-2}{4}}\left(\frac{4(p-1)}{5p-6}\right)^{p-1}.
\end{aligned}$$

The product of these two factors gives

$$\left(\frac{4}{6-p}\right)\left(\frac{4(p-1)}{5p-6}\right)^{p-1}.$$

If I is not centered at the origin, we consider two cases. Let x_0 denote the center of I and R the radius. First suppose $|x_0| < 2R$. Then $I \subset 3I'$, where I' is the translate of I centered at 0. Note I and $3I'$ have comparable measure, so we can just apply the computation above to deduce

$$\left(\frac{1}{|I|}\int_I |x|^{\frac{2-p}{4}}\,dx\right)\left(\frac{1}{|I|}\int_I |x|^{\frac{p-2}{4(p-1)}}\,dx\right)^{p-1} \lesssim \left(\frac{4}{6-p}\right)\left(\frac{4(p-1)}{5p-6}\right)^{p-1}.$$

In the other case, I is such that $|x_0| \geq 2R$. In this case, if $x \in I$, $|x| \approx R$. It is then clear

$$\left(\frac{1}{|I|}\int_I |x|^{\frac{2-p}{4}}\,dx\right)\left(\frac{1}{|I|}\int_I |x|^{\frac{p-2}{4(p-1)}}\,dx\right)^{p-1} \approx 1.$$

It follows from these calculations that

$$\left[|\Phi_1'|^{1-\frac{p}{2}}\right]_{A_p(\mathbb{R})} \approx \left(\frac{1}{6-p}\right)\left(\frac{4(p-1)}{5p-6}\right)^{p-1}.$$

Similar calculations may be employed to deduce

$$\left[|\Phi_3'|^{1-\frac{p}{2}}\right]_{A_p(\mathbb{R})} \approx \left(\frac{1}{p+2}\right)\left(\frac{4(p-1)}{3p-2}\right)^{p-1}.$$

Note that the first characteristic blows up at the endpoints $p = \frac{6}{5}$ and $p = 6$ while the second one does not. In fact, it is easy to see that there exists $C > 0$ so that

$$\sup_I \fint |\Phi_3'|^{1-p/2}\,dx \leq C; \quad \sup_I \left(\fint |\Phi_3'|^{\frac{p/2-1}{p-1}}\,dx\right)^{p-1} \leq C. \tag{3.1}$$

The crucial behavior of the weight evidently occurs at the points $p_1 = \varphi^{-1}(\mathrm{i})$ and $p_2 = \varphi^{-1}(-\mathrm{i})$. Thus, we may conclude using (3.1) together with the arguments in Proposition 3.3 that

$$\left[|\varphi'|^{1-\frac{p}{2}}\right]_{A_p(\partial\mathbb{D})} \approx \left(\frac{1}{6-p}\right)\left(\frac{4(p-1)}{5p-6}\right)^{p-1}.$$

Note that is it not difficult to see the lower bound by testing on small intervals centered at p_1 or p_2.

With this information, we can obtain upper and lower bounds the norm of the Szegő projection in this case. First, we will need a preliminary lemma.

LEMMA 3.4. *Suppose $r < \frac{2}{5}$. If $I_1 = I_1(p_1, r)$ is an arc centered at $p_1 \in \partial\mathbb{D}$, then there exists a point $p_2 \in \partial\mathbb{D}$ and a disjoint arc $I_2 = I_2(p_2, r)$ centered at p_2 so that for for $j \in \{1,2\}$ and any function $f > 0$ supported in I_j and $z \in I_k$ with $j \neq k$ we have the estimate:*

$$|\mathcal{S}_0 f(z)| \geq \frac{1}{28\pi}\fint_{I_j} f\,d\tilde{\theta}.$$

PROOF. Let $S_0(z,w) = \frac{1}{1-\overline{w}z}$ denote the kernel of $\mathcal{S}_0$ with respect to the measure $d\tilde{\theta}$. An easy computation shows that S_0 satisfies the smoothness estimate

$$|S_0(z,w) - S_0(z,w')| \le \frac{3}{2}\frac{|w-w'|}{|w-z|^2} \tag{3.2}$$

provided $|w-w'| \le \frac{1}{3}|z-w|$. It suffices to prove the inequality for $j=1$. Now, let $w' = p_1$ as above. Select p_2 so that if $z \in I_2$ and $w \in I_1$, then $7r \ge |z-w| \ge 3r \ge 3|w-w'|$ (this is possible if $r < \frac{2}{5}$). Then we compute, using (3.2) and noting that $z \notin \operatorname{supp}(f)$:

$$\begin{aligned}
|\mathcal{S}_0 f(z)| &= \left|\int_{I_1} S_0(z,w) f(w)\, d\tilde{\theta}(w)\right| \\
&\ge \left|\int_{I_1} S_0(z,w') f(w)\, d\tilde{\theta}(w)\right| - \left|\int_{I_1} [S_0(z,w) - S_0(z,w')] f(w)\, d\tilde{\theta}(w)\right| \\
&\ge \frac{1}{|z-w'|}\int_{I_1} f(w)\, d\tilde{\theta}(w) - \int_{I_1} |S_0(z,w) - S_0(z,w')| f(w)\, d\tilde{\theta}(w) \\
&\ge \frac{3}{4|z-w|}\int_{I_1} f(w)\, d\tilde{\theta}(w) - \int_{I_1} \frac{3}{2}\frac{|w-w'|}{|w-z|^2} f(w)\, d\tilde{\theta}(w) \\
&\ge \frac{1}{4|z-w|}\int_{I_1} f(w)\, d\tilde{\theta}(w) \\
&\ge \frac{1}{28\pi} \fint_{I_1} f\, d\tilde{\theta},
\end{aligned}$$

as required. □

THEOREM 3.5. *The following estimate holds for the domain Ω defined at the beginning of Subsection* 3.2*:*

$$\begin{aligned}
\left(\frac{1}{6-p}\left(\frac{4(p-1)}{5p-6}\right)^{p-1}\right)^{\frac{1}{2p}} &\lesssim \|\mathcal{S}\|_{L^p(\Omega)\to L^p(\Omega)} \\
&\lesssim \left(\frac{1}{6-p}\left(\frac{4(p-1)}{5p-6}\right)^{p-1}\right)^{\max\{1,\frac{1}{p-1}\}}.
\end{aligned}$$

PROOF. The upper estimate follows from Theorem 3.1 and the above computation of $\left[|\varphi'|^{1-\frac{p}{2}}\right]_{A_p(\partial\mathbb{D})}$. The lower bound follows from a standard argument used for the Hilbert/Riesz transforms, see [**14**, Theorem 7.4.7]. We include this argument for completeness. Let $\omega = |\varphi'|^{1-\frac{p}{2}}$, let $I_1 = I_1(\zeta, r)$ be any boundary arc with $r < \frac{2}{5}$, and let I_2 be as in Lemma 3.4. First, note that by Lemma 3.4, if $f > 0$ is supported in I_1 we have the set containment

$$I_2 \subseteq \left\{ z \in \partial\mathbb{D} : |\mathcal{S}_0 f(z)| \ge C \fint_{I_1} f\, d\tilde{\theta} \right\}.$$

This then implies

$$\begin{aligned}\omega(I_2) &\leq \omega\left(\left\{z \in \partial\mathbb{D} : |\mathcal{S}_0 f(z)| \geq C \fint_{I_1} f\, d\tilde{\theta}\right\}\right)\\ &\leq \frac{\|\mathcal{S}_0\|^p_{L^p_\omega(\partial\mathbb{D}) \to L^p_\omega(\partial\mathbb{D})}}{C^p \left(\fint_{I_1} f\, d\tilde{\theta}\right)^p} \|f\|^p_{L^p_\omega(\partial\mathbb{D})}\\ &= \frac{\|\mathcal{S}\|^p_{L^p(\partial\Omega) \to L^p(\partial\Omega)}}{C^p \left(\fint_{I_1} f\, d\tilde{\theta}\right)^p} \|f\|^p_{L^p_\omega(\partial\mathbb{D})}.\end{aligned}$$

Assume without loss of generality that $\omega^{\frac{-1}{p-1}}$ is locally integrable (if not, replace ω by $\omega + \varepsilon$ and use a limiting argument). Then take $f = \omega^{\frac{-1}{p-1}} \chi_{I_1}$ and note that $\|f\|^p_{L^p_\omega(\partial\mathbb{D})} = \int_{I_1} \omega^{-\frac{1}{p-1}}\, d\theta$ and $|I_1| = |I_2|$. Using these facts and rearranging the above display, we obtain

$$\left(\fint_{I_2} \omega\, d\tilde{\theta}\right) \left(\fint_{I_1} \omega^{\frac{-1}{p-1}}\, d\tilde{\theta}\right)^{p-1} \leq \frac{\|\mathcal{S}\|^p_{L^p(\partial\Omega) \to L^p(\partial\Omega)}}{C^p}.$$

Interchanging the roles of I_1 and I_2 , we obtain the analogous inequality:

$$\left(\fint_{I_1} \omega\, d\tilde{\theta}\right) \left(\fint_{I_2} \omega^{\frac{-1}{p-1}}\, d\tilde{\theta}\right)^{p-1} \leq \frac{\|\mathcal{S}\|^p_{L^p(\partial\Omega) \to L^p(\partial\Omega)}}{C^p}.$$

Finally, multiplying these two inequalities together and noting that

$$\left(\fint_{I_2} \omega\, d\tilde{\theta}\right) \left(\fint_{I_2} \omega^{\frac{-1}{p-1}}\, d\tilde{\theta}\right)^{p-1} \geq 1$$

yields

$$\begin{aligned}&\left(\fint_{I_1} \omega\, d\tilde{\theta}\right) \left(\fint_{I_1} \omega^{\frac{-1}{p-1}}\, d\tilde{\theta}\right)^{p-1}\\ &\leq \left(\fint_{I_2} \omega\, d\tilde{\theta}\right) \left(\fint_{I_1} \omega^{\frac{-1}{p-1}}\, d\tilde{\theta}\right)^{p-1} \left(\fint_{I_1} \omega\, d\tilde{\theta}\right) \left(\fint_{I_2} \omega^{\frac{-1}{p-1}}\, d\tilde{\theta}\right)^{p-1}\\ &\leq \frac{\|\mathcal{S}\|^{2p}_{L^p(\partial\Omega) \to L^p(\partial\Omega)}}{C^{2p}}.\end{aligned}$$

Rearranging this inequality and taking a supremum over intervals I_1 with small arc length then yields the desired result. □

REMARK 3.6. Notice that the lower bound blows up at the endpoints $p = \frac{6}{5}$ and $p = 6$, giving a proof that the Szegő projection on this domain is only bounded for $\frac{6}{5} < p < 6$.

OPEN QUESTION 3.7. As the reader may have noted, the dependences on p in the upper and lower bounds appearing in Theorem 3.5 are slightly different. It would be quite interesting to determine $\|\mathcal{S}\|_{L^p(\Omega) \to L^p(\Omega)}$ exactly or asymptotically in this specific case, or else come up with a different example to show that the upper bound in Theorem 2.1 is in fact sharp.

3.3. Weak-type estimates. We close the discussion of L^p behavior with a couple of open questions. In particular, one could ask about the weak-type endpoint behavior of the operators $\mathcal{S}$ and $\mathcal{B}$, even in the unweighted case, when the projections are bounded in a limited range. For instance, it would be quite interesting to see if endpoint weak-type estimates hold for the domains in Corollaries 2.7, 2.8, 2.9, and 2.10 (for example, does $\mathcal{S}$ satisfy a weak-type $(\frac{4}{3}, \frac{4}{3})$ estimate if Ω is a local graph domain?). The direct approach via conformal mapping seems to fail here, so the problem requires novel ideas. It is possible, though perhaps more difficult, that provided one solves the unweighted problem, one could use similar machinery to obtain weighted weak-type estimates.

In contrast, for certain classes of nicer domains which admit full range L^p bounds, including convex domains and domains with sufficient smoothness (for example, Dini-smooth domains), weak-type estimates have already been established (see [**5**, Theorem 2.1]). However it would also be interesting to investigate the case $p = 1$ when Ω is nonsmooth, but nevertheless the Bergman and Szegő are bounded in the full reflexive range, such as the vanishing chord-arc domains. Recall that Ω has the vanishing chord-arc property if the ratio $\frac{\sigma(pq)}{|p-q|}$ for $p, q \in \partial\Omega$ tends to 1 as p approaches q, where $\sigma(pq)$ denotes the length of the shortest arc joining p and q.

Open Question 3.8. Investigate unweighted/weighted weak-type estimates for the Bergman/Szegő projections at the endpoints of the L^p boundedness ranges for local graph domains and Lipschitz domains.

Open Question 3.9. Investigate unweighted/weighted weak-type estimates for the Bergman/Szegő projections at the endpoint $p = 1$ for vanishing chord-arc domains.

4. Bergman vs. Szegő

It is of interest to compare the Bergman and Szegő projections via the Poisson extension operator $\mathcal{P}$ (see, for example [**18**]). Assuming that the boundary of Ω is sufficiently smooth, recall that the Poisson extension of a continuous function f on $\partial\Omega$ is given by the (harmonic in Ω) function

$$\mathcal{P}f(z) = \int_{\partial\Omega} P(z, w) f(w)\, d\sigma(w), z \in \Omega,$$

where $P(z, w)$ denotes the Poisson kernel for Ω. Notice that the two operators $\mathcal{BP}$ and $\mathcal{PS}$ produce holomorphic functions in the interior of Ω from boundary data, and if the domain Ω is sufficiently regular, both will be bounded maps from $L^p(\partial\Omega)$ to $L^p(\Omega)$. In fact, in the case that Ω has sufficently smooth boundary, it can be shown that both $\mathcal{BP}$ and $\mathcal{PS}$ are bounded operators from $L^2(\partial\Omega)$ to the Sobolev space $L^2_{1/2}(\partial\Omega)$ [**18**] (there is also an L^p version of the Sobolev result, but we will not concern ourselves with it here).

Motivated by this Sobolev improvement, one might wonder about the compactness of the operators $\mathcal{BP}$ and $\mathcal{PS}$ as maps from $L^p(\partial\Omega)$ to $L^p(\Omega)$ (this may follow from some version of Sobolev embedding, but the author has not checked the details). However, this can be seen directly with rather minimal effort. We consider the case when Ω is Dini smooth, which means there is a parametrization $r(t)$ of the boundary Jordan curve C so that $r'(t)$ is Dini continuous. Recall a function

$f : \mathbb{R} \to \mathbb{C}$ is Dini continuous if

$$\int_0^1 \frac{\omega(f,\delta)}{\delta}\, d\delta < \infty; \quad \omega(f,\delta) := \sup_{|x-y|\le\delta} |f(x) - f(y)|.$$

Notice that Dini continuity is slightly weaker than Lipschitz/Hölder continuity and all Dini smooth domans are C^1, though the converse does not hold. We will use the key fact that if Ω is Dini-smooth and $\varphi : \mathbb{D} \to \Omega$ is a conformal map, then φ' extends continuously to $\bar{\mathbb{D}}$ and is nonvanishing on $\bar{\mathbb{D}}$ (see [**22**, Theorem 3.5]). We have the following result for this category of domains:

PROPOSITION 4.1. *Let $\Omega \subset \mathbb{C}$ be a Dini-smooth bounded domain. Then the operators $\mathcal{BP}$ and $\mathcal{PS}$ are compact maps from $L^p(\partial\Omega)$ to $L^p(\Omega)$ for $1 < p < \infty$.*

PROOF. We first note that the Bergman and Szegő projections are bounded on L^p for $1 < p < \infty$ in this situation by Theorems 3.1 and 3.2, for instance. Therefore, it suffices to show that the Poisson extension operator $\mathcal{P}$ is compact from $L^p(\partial\Omega)$ to $L^p(\Omega)$.

Indeed, this can be seen directly using the formula $\mathcal{P} = C_{\varphi^{-1}}\mathcal{P}_0 C_\varphi$ given in [**18**, pp. 972], where C_φ denotes the composition operator with symbol φ (which denotes the conformal map as before). This formula apriori holds for continuous functions on $\partial\Omega$, but it is easy to show it meaningfully extends to $L^p(\partial\Omega)$. To begin with, since $|\varphi'|$ is bounded above and below on $\bar{\mathbb{D}}$, the maps C_φ and C_φ^{-1} are bounded from $L^p(\partial\Omega)$ to $L^p(\partial\mathbb{D})$ and $L^p(\mathbb{D})$ to $L^p(\Omega)$, respectively. We have therefore reduced the problem to showing that $\mathcal{P}_0$ is compact from $L^p(\partial\mathbb{D})$ to $L^p(\mathbb{D})$ for $1 < p < \infty$ (actually, the argument will show that one can include the endpoint $p = 1$ for this statement).

To this end, take f_n converging weakly to zero in $L^p(\partial\mathbb{D})$. To prove the operator $\mathcal{P}$ is compact, it suffices to show that for all $\varepsilon > 0$, there exists an $N \in \mathbb{N}$ so that for all $n > N$, we have

$$\|\mathcal{P} f_n\|^p_{L^p(\mathbb{D})} < \varepsilon. \tag{4.1}$$

Since the f_n converge weakly, we know that there exists a $B > 0$ such

$$\sup_n \|f_n\|_{L^p(\partial\mathbb{D})} \le B.$$

We first claim that $\mathcal{P} f_n$ converges to 0 pointwise on $\mathbb{D}$. To see this, write

$$|\mathcal{P} f_n(re^{i\theta})| = \left| \frac{1}{2\pi} \int_0^{2\pi} \frac{(1-r^2) f_n(e^{it})}{|e^{it} - re^{i\theta}|^2}\, dt \right|. \tag{4.2}$$

For fixed r, θ, the function $g_{r,\theta}(e^{it}) = \frac{(1-r^2)}{|e^{it}-re^{i\theta}|^2}$ clearly belongs to $L^{p'}(\partial\mathbb{D})$ (in fact, it belongs to $L^\infty(\partial\mathbb{D})$), so then by weak convergence it is clear that (4.2) converges to 0 as $n \to \infty$. We have thus established pointwise convergence.

We now claim that for all $\varepsilon > 0$ we can choose r_0 sufficiently close to 1 so that

$$\sup_n \frac{1}{\pi} \int_0^{2\pi} \int_{r_0}^1 |\mathcal{P} f_n(re^{i\theta})|^p r\, dr\, d\theta < \frac{\varepsilon}{2}. \tag{4.3}$$

Assuming claim (4.3), we may write, for all $n \in \mathbb{N}$:

$$\begin{aligned}\|\mathcal{P}f_n\|^p_{L^p(\mathbb{D})} &= \frac{1}{\pi}\int_0^{2\pi}\int_0^1 |\mathcal{P}f_n(re^{i\theta})|^p r\,dr\,d\theta \\ &= \frac{1}{\pi}\int_0^{2\pi}\int_0^{r_0} |\mathcal{P}f_n(re^{i\theta})|^p r\,dr\,d\theta + \frac{1}{\pi}\int_0^{2\pi}\int_{r_0}^1 |\mathcal{P}f_n(re^{i\theta})|^p r\,dr\,d\theta \\ &< \frac{1}{\pi}\int_0^{2\pi}\int_0^{r_0} |\mathcal{P}f_n(re^{i\theta})|^p r\,dr\,d\theta + \frac{\varepsilon}{2}.\end{aligned}$$

Notice that if $0 < r < r_0$, we have the uniform estimate

$$|\mathcal{P}f_n(re^{i\theta})|^p r \le \frac{r_0}{(1-r_0)^{2p}}\left|\frac{1}{2\pi}\int_0^{2\pi} f_n(e^{it})\,dt\right|^p \le \frac{r_0 B^p}{(1-r_0)^{2p}}.$$

Now, using for example the Dominated Convergence Theorem with constant dominating function $F(r,\theta) = \frac{r_0 B^p}{(1-r_0)^{2p}}$ we may choose N sufficiently large so that for all $n > N$, we have

$$\frac{1}{\pi}\int_0^{2\pi}\int_0^{r_0} |\mathcal{P}f_n(re^{i\theta})|^p r\,dr\,d\theta < \frac{\varepsilon}{2}.$$

Assuming claim (4.3), we have established (4.1).

It remains to prove (4.3). This is a straightforward consequence of Minkowski's inequality, which yields

$$\sup_{0<r<1}\frac{1}{2\pi}\int_0^{2\pi} |\mathcal{P}f_n(re^{i\theta})|^p\,d\theta \le \|f_n\|^p_{L^p(\partial\mathbb{D})} \le B^p.$$

(see, for example, [**13**, pp. 14]). □

REMARK 4.2. The author does not know of a reference for this fact concerning the compactness of the Poisson extension on the disk, but it is probably well known to the experts. The result appears to be mentioned in passing in [**11**] in the context of a bounded domain $\Omega \subset \mathbb{R}^n$ with C^∞ smooth boundary and for $p = 2$.

REMARK 4.3. The reason for the Dini-smooth hypothesis for this result is that this guarantees the boundedness of the Bergman and Szegő projections for $1 < p < \infty$, and also the composition maps C_φ and $C_{\varphi^{-1}}$, as φ' is bounded above and below on $\bar{\mathbb{D}}$. However, it is very plausible that this result extends to domains of class C^1. In particular, it is known that $\mathcal{B}$ and $\mathcal{S}$ are bounded on L^p for $1 < p < \infty$ for C^1 domains (see [**19**]), and in fact as we mentioned this can be weakened to vanishing chord-arc domains. One would have to check the compactness of $\mathcal{P}$ in this situation, which is not immediate as C_φ and $C_{\varphi^{-1}}$ may not be bounded maps.

A more interesting question, perhaps, is to study how the boundary regularity improves the mapping properties of the difference operator $\mathcal{BP} - \mathcal{PS}$. For domains with some degree of smoothness, one would expect the difference operator $\mathcal{BP} - \mathcal{PS}$ to be better behaved than each of the individual operators in an appropriate sense due to cancellation. Indeed, in the special case of the unit disk $\mathbb{D}$, one actually shows that $\mathcal{B}_0\mathcal{P}_0 = \mathcal{P}_0\mathcal{S}_0$, but clearly this complete cancellation cannot be expected to hold for general domains.

Given a positive integer k and $0 < \alpha < 1$, we say a domain Ω has $C^{k,\alpha}$ boundary if its boundary can be locally represented as the graph of a $C^{k,\alpha}$ function. It is known that if φ is a conformal map associated to a domain Ω with $C^{k,\alpha}$ boundary, then $\varphi \in C^{k,\alpha}(\bar{\mathbb{D}})$ and φ is nonvanishing on $\bar{\mathbb{D}}$. The authors in [**18**] analyzed the mapping properties of the difference operator for $C^{k,\alpha}$ domains with $k \geq 2$ in terms of a gain in the Sobolev or Lipschitz space scale. In particular, they proved that the difference operator $\mathcal{BP} - \mathcal{PS}$ maps $L^2(\partial\Omega)$ to the Sobolev space $L^2_{3/2}(\Omega)$, representing a gain of a full derivative in the Sobolev scale compared to the operators individually. The authors also showed that $\mathcal{BP} - \mathcal{PS}$ is a bounded map from $L^\infty(\partial\Omega)$ to $\Lambda_1(\Omega)$, where $\Lambda_1(\Omega)$ is a Hölder-Zygmund space [**18**, Theorem 1.2]. We will not concern ourselves with the precise definition of $\Lambda_1(\Omega)$ here because this result is not our focus.

Motivated by Proposition 4.1, we might wonder about the compactness of the individual operators $\mathcal{BP}$ and $\mathcal{PS}$ at the endpoints $p = 1$ and $p = \infty$. However, in contrast to the other values of p, it is clear that the operator $\mathcal{P}_0$ is not compact from $L^\infty(\partial\mathbb{D})$ to $L^\infty(\mathbb{D})$ (it is in fact an isometry), and the projections $\mathcal{B}$ and $\mathcal{S}$ are in general not bounded at either of the endpoints. Thus, the compactness (even the boundedness) of $\mathcal{BP}$ and $\mathcal{PS}$ cannot be expected to hold at the endpoints.

In a similar spirit as [**18**], we can apply a characterization of compact integral operators on L^1 and L^∞ given in [**12**] in a simple way to show that the difference operator $\mathcal{BP} - \mathcal{PS}$ has the improved property of compactness on L^1 and L^∞ rather than just L^p, if we assume the boundary is just slightly more regular. We remark that our assumptions on regularity are less than those in [**18**]; this is heuristically because we are proving compactness rather than quantitative Sobolev results.

THEOREM 4.4. *Suppose Ω has $C^{1,\alpha}$ boundary and $0 < \alpha < 1$. Then $\mathcal{BP} - \mathcal{PS}$ is a compact operator from $L^1(\partial\Omega)$ to $L^1(\Omega)$ and $L^\infty(\partial\Omega)$ to $L^\infty(\Omega)$.*

PROOF. First, using the formula given in [**18**, pp. 974], we may write

$$\mathcal{BP} - \mathcal{PS} = \tau^{-1} \circ \left([\mathcal{B}_0, \mathcal{M}_{\varphi'}]\mathcal{P}_0\mathcal{M}_{(\varphi')^{-1/2}} + \mathcal{M}_{\varphi'}\mathcal{P}_0[\mathcal{S}_0, \mathcal{M}_{(\varphi')^{-1/2}}]\right) \circ \tau_{1/2},$$

where the notation $[A, B]$ means the commutator $AB - BA$ and $\mathcal{M}_f$ denotes the multiplication operator with symbol f.

Notice that, on account of the fact that Ω is $C^{1,\alpha}$, the function φ' is bounded above and below on $\bar{\mathbb{D}}$. It is then clear from Lemmas 2.3 and 2.4 that $\tau_{1/2}$ is a linear homeomorphism from $L^p(\partial\Omega)$ to $L^p(\partial\mathbb{D})$, and similarly τ^{-1} is a linear homeomorphism from $L^p(\mathbb{D})$ to $L^p(\Omega)$. Moreover, the operators $\mathcal{M}_{\varphi'}$ and $\mathcal{M}_{(\varphi')^{-1/2}}$ are evidently bounded on $L^p(\mathbb{D})$ and $L^p(\partial\mathbb{D})$ respectively. We also know the Poisson extension $\mathcal{P}_0$ is a bounded operator from $L^p(\partial\mathbb{D})$ to $L^p(\mathbb{D})$ for $1 \leq p \leq \infty$.

Therefore, it clearly suffices to prove the compactness of the two commutators $[\mathcal{B}_0, \mathcal{M}_{\varphi'}]$ and $[\mathcal{S}_0, \mathcal{M}_{(\varphi')^{-1/2}}]$. We only prove the compactness of $[\mathcal{B}_0, \mathcal{M}_{\varphi'}]$, as the other commutator is similar. Moreover, we only prove compactness for $p = \infty$; the result for $p = 1$ will follow using the same argument.

Let $k_z(w) = K(z, w) = \frac{(\varphi'(z) - \varphi'(w))}{(1-\overline{w}z)^2}$ denote the kernel of the operator $[\mathcal{B}_0, \mathcal{M}_{\varphi'}]$. It is easy to check that $K(z, w)$ is uniformly continuous on the set $\{(z, w) \in \bar{\mathbb{D}} \times \bar{\mathbb{D}} :$

$|z-w|\geq r\}$ for each $r>0$. We have, using our hypothesis that Ω is $C^{1,\alpha}$:

$$\begin{aligned} |k_z(w)| &= \frac{|\varphi'(z)-\varphi'(w)|}{|1-\overline{w}z|^2} \\ &\leq C_\varphi \frac{|z-w|^\alpha}{|1-\overline{w}z|^2} \\ &\leq \frac{C_\varphi}{|z-w|^{2-\alpha}}. \end{aligned}$$

It is thus obvious

$$\sup_{z\in\mathbb{D}} \int_{\mathbb{D}} |k_z(w)|\, dA(w) < \infty.$$

Moreover, an easy computation shows

$$\int_{D_\delta(z)} \frac{1}{|z-w|^{2-\alpha}}\, dA(w) = \frac{2}{\alpha}\delta^\alpha. \tag{4.4}$$

By a characterization given in [**12**, Corollary 5.1], it is sufficient to show that the set of functions $\{k_z\}_{z\in\mathbb{D}}$ is relatively compact in $L^1(\mathbb{D})$. This can be checked using a version of the Riesz-Kolmogorov theorem given in [**17**, Lemma 1]:

$$\lim_{r\to 0} \sup_{z\in\mathbb{D}} \int_{\mathbb{D}} \left| k_z(w) - \fint_{D_r(w)} k_z\, dA \right| dA(w) = 0.$$

Fix an arbitrary $\varepsilon>0$. Let $\delta^\alpha=\varepsilon$ and let $r<\delta$. We will show that we can make a sufficiently small choice of r so

$$\sup_{z\in\mathbb{D}} \int_{\mathbb{D}} \left| k_z(w) - \fint_{D_r(w)} k_z\, dA \right| dA(w) \lesssim \delta^\alpha.$$

First, we show

$$\sup_{z\in\mathbb{D}} \int_{D_\delta(z)} \left| k_z(w) - \fint_{D_r(w)} k_z\, dA \right| dA(w) \lesssim \delta^\alpha.$$

To see this, note that by (4.4) and the triangle inequality it is clearly enough to show

$$\sup_{z\in\mathbb{D}} \int_{D_\delta(z)} \left| \fint_{D_r(w)} k_z\, dA \right| dA(w) \lesssim \delta^\alpha.$$

For fixed $z\in\mathbb{D}$, let $G_z=\{w : |z-w|\geq 2r\}$. Note that if $w\in G_z$ and $\zeta\in D_r(w)$, then $|z-w|\leq 2|z-\zeta|$. We further split the region of integration and estimate for fixed z:

$$\begin{aligned}
&\int_{D_\delta(z)\cap G_z}\left|\fint_{D_r(w)} k_z\,dA\right| dA(w)\\
&\leq \int_{D_\delta(z)\cap G_z}\left(\frac{1}{|D_r(w)|}\int_{D_r(w)}|k_z(\zeta)|\,dA(\zeta)\right) dA(w)\\
&\leq 2^{2-\alpha}C_\varphi\int_{D_\delta(z)\cap G_z}\left(\frac{1}{|D_r(w)|}\int_{D_r(w)}\frac{1}{|z-w|^{2-\alpha}}\,dA(\zeta)\right) dA(w)\\
&\leq 2^{2-\alpha}C_\varphi\int_{D_\delta(z)}\frac{1}{|z-w|^{2-\alpha}}\,dA(w)\\
&\leq \frac{8}{\alpha}C_\varphi\delta^\alpha.
\end{aligned}$$

On the other hand,

$$\begin{aligned}
\int_{D_\delta(z)\cap G_z^c}\left|\fint_{D_r(w)} k_z\,dA\right| dA(w) &\leq \int_{D_{2r}(z)}\left(\frac{1}{|D_r(w)|}\int_{D_{3r}(z)}|k_z(\zeta)|\,dA(\zeta)\right) dA(w)\\
&\leq \frac{24}{\alpha}C_\varphi r^\alpha\\
&\lesssim \delta^\alpha,
\end{aligned}$$

establishing the first claim.

For the other region of integration, by uniform continuity we can choose r small enough so

$$\sup_{z\in\mathbb{D}}\int_{\mathbb{D}\setminus B(z,\delta)}\left|k_z(w)-\fint_{D_r(w)} k_z\,dA\right| dA(w) < \delta^\alpha,$$

establishing the result. □

The proof above makes use of fairly strong hypotheses on the boundary smoothness of Ω. It is natural to ask whether these hypothesis can be weakened, especially since $\mathcal{B}$ and $\mathcal{S}$ are both bounded operators on L^p for all p on domains that are less regular.

Open Question 4.5. Determine if there is a class larger than the $C^{k,\alpha}$ domains ($k \geq 1$)) so that $\mathcal{BP} - \mathcal{PS}$ is compact from $L^p(\partial\Omega)$ to $L^p(\Omega)$ for $1 \leq p \leq \infty$. In particular, does the result hold Ω is Dini-smooth, or even better when Ω is class C^1 (or less smooth)?

References

[1] Pascal Auscher and José María Martell, *Weighted norm inequalities, off-diagonal estimates and elliptic operators. I. General operator theory and weights*, Adv. Math. **212** (2007), no. 1, 225–276, DOI 10.1016/j.aim.2006.10.002. MR2319768

[2] Alexandru Aleman, Sandra Pott, and María Carmen Reguera, *Characterizations of a limiting class B_∞ of Békollé-Bonami weights*, Rev. Mat. Iberoam. **35** (2019), no. 6, 1677–1692, DOI 10.4171/rmi/1097. MR4029779

[3] Sheldon Axler, *The Bergman space, the Bloch space, and commutators of multiplication operators*, Duke Math. J. **53** (1986), no. 2, 315–332, DOI 10.1215/S0012-7094-86-05320-2. MR850538

[4] Theresa C. Anderson and Armen Vagharshakyan, *A simple proof of the sharp weighted estimate for Calderón-Zygmund operators on homogeneous spaces*, J. Geom. Anal. **24** (2014), no. 3, 1276–1297, DOI 10.1007/s12220-012-9372-7. MR3223553

[5] David Bekollé, *Inégalité à poids pour le projecteur de Bergman dans la boule unité de* $\mathbf{C}^n$ (French), Studia Math. **71** (1981/82), no. 3, 305–323, DOI 10.4064/sm-71-3-305-323. MR667319

[6] David Békollé, *Projections sur des espaces de fonctions holomorphes dans des domaines plans* (French), Canad. J. Math. **38** (1986), no. 1, 127–157, DOI 10.4153/CJM-1986-007-1. MR835039

[7] Frédéric Bernicot, Dorothee Frey, and Stefanie Petermichl, *Sharp weighted norm estimates beyond Calderón-Zygmund theory*, Anal. PDE **9** (2016), no. 5, 1079–1113, DOI 10.2140/apde.2016.9.1079. MR3531367

[8] David V. Cruz-Uribe, José Maria Martell, and Carlos Pérez, *Weights, extrapolation and the theory of Rubio de Francia*, Operator Theory: Advances and Applications, vol. 215, Birkhäuser/Springer Basel AG, Basel, 2011, DOI 10.1007/978-3-0348-0072-3. MR2797562

[9] Yaohua Deng, Li Huang, Tao Zhao, and Dechao Zheng, *Bergman projection and Bergman spaces*, J. Operator Theory **46** (2001), no. 1, 3–24. MR1862176

[10] Oliver Dragičević, Loukas Grafakos, María Cristina Pereyra, and Stefanie Petermichl, *Extrapolation and sharp norm estimates for classical operators on weighted Lebesgue spaces*, Publ. Mat. **49** (2005), no. 1, 73–91, DOI 10.5565/PUBLMAT_49105_03. MR2140200

[11] Miroslav Engliš, *Boundary singularity of Poisson and harmonic Bergman kernels*, J. Math. Anal. Appl. **429** (2015), no. 1, 233–272, DOI 10.1016/j.jmaa.2015.03.081. MR3339073

[12] Simon P. Eveson, *Compactness criteria for integral operators in* L^∞ *and* L^1 *spaces*, Proc. Amer. Math. Soc. **123** (1995), no. 12, 3709–3716, DOI 10.2307/2161898. MR1291766

[13] John B. Garnett, *Bounded analytic functions*, Pure and Applied Mathematics, vol. 96, Academic Press, Inc. [Harcourt Brace Jovanovich, Publishers], New York-London, 1981. MR628971

[14] Loukas Grafakos, *Classical Fourier analysis*, 3rd ed., Graduate Texts in Mathematics, vol. 249, Springer, New York, 2014, DOI 10.1007/978-1-4939-1194-3. MR3243734

[15] Zhenghui Huo, Nathan A. Wagner, and Brett D. Wick, *Bekollé–Bonami estimates on some pseudoconvex domains*, Bull. Sci. Math. **170** (2021), Paper No. 102993, 36, DOI 10.1016/j.bulsci.2021.102993. MR4263006

[16] Tuomas Hytönen and Anna Kairema, *Systems of dyadic cubes in a doubling metric space*, Colloq. Math. **126** (2012), no. 1, 1–33, DOI 10.4064/cm126-1-1. MR2901199

[17] Iya A. Ivanishko and Veniamin G. Krotov, *Compactness of embeddings of Sobolev type on metric measure spaces* (Russian, with Russian summary), Mat. Zametki **86** (2009), no. 6, 829–844, DOI 10.1134/S0001434609110200; English transl., Math. Notes **86** (2009), no. 5-6, 775–788. MR2643451

[18] Kenneth D. Koenig and Loredana Lanzani, *Bergman versus Szegő via conformal mapping*, Indiana Univ. Math. J. **58** (2009), no. 2, 969–997, DOI 10.1512/iumj.2009.58.3841. MR2514395

[19] Loredana Lanzani and Elias M. Stein, *Szegö and Bergman projections on nonsmooth planar domains*, J. Geom. Anal. **14** (2004), no. 1, 63–86, DOI 10.1007/BF02921866. MR2030575

[20] Steven G. Krantz and Song-Ying Li, *Boundedness and compactness of integral operators on spaces of homogeneous type and applications. II*, J. Math. Anal. Appl. **258** (2001), no. 2, 642–657, DOI 10.1006/jmaa.2000.7403. MR1835564

[21] Nikolai K. Nikolski, *Operators, functions, and systems: an easy reading. Vol. 1*, Mathematical Surveys and Monographs, vol. 92, American Mathematical Society, Providence, RI, 2002. Hardy, Hankel, and Toeplitz; Translated from the French by Andreas Hartmann. MR1864396

[22] Christian Pommerenke, *Boundary behaviour of conformal maps*, Grundlehren der mathematischen Wissenschaften [Fundamental Principles of Mathematical Sciences], vol. 299, Springer-Verlag, Berlin, 1992, DOI 10.1007/978-3-662-02770-7. MR1217706

[23] Sandra Pott and Maria Carmen Reguera, *Sharp Békollé estimates for the Bergman projection*, J. Funct. Anal. **265** (2013), no. 12, 3233–3244, DOI 10.1016/j.jfa.2013.08.018. MR3110501

[24] Rob Rahm, Edgar Tchoundja, and Brett D. Wick, *Weighted estimates for the Berezin transform and Bergman projection on the unit ball*, Math. Z. **286** (2017), no. 3-4, 1465–1478, DOI 10.1007/s00209-016-1809-4. MR3671584

[25] Aleksandr A. Solov′ev, *Estimates in L^p of the integral operators that are connected with spaces of analytic and harmonic functions* (Russian), Dokl. Akad. Nauk SSSR **240** (1978), no. 6, 1301–1304. MR499758

[26] Aleksandr A. Solov′ëv, *Continuity of the harmonic projection in L^p-spaces* (Russian, with English summary), Zap. Nauchn. Sem. Leningrad. Otdel. Mat. Inst. Steklov. (LOMI) **126** (1983), 191–195. Investigations on linear operators and the theory of functions, XII. MR697438

[27] Ke He Zhu, *BMO and Hankel operators on Bergman spaces*, Pacific J. Math. **155** (1992), no. 2, 377–395. MR1178032

Department of Mathematics, Brown University, 151 Thayer Street, Providence, Rhode Island, 02912

Email address: nathan_wagner@brown.edu

Contemporary Mathematics
Volume **792**, 2024
https://doi.org/10.1090/conm/792/15897

Bounds on 10th moments of (x, x^3) for ellipsephic sets

Theresa C. Anderson, Bingyang Hu, Yu-Ru Liu, and Alan Talmage

ABSTRACT. Let $\mathcal{A}$ be an ellipsephic set which satisfies digital restrictions in a given base. Using the method developed by Hughes and Wooley, we bound the number of integer solutions to the system of equations

$$\sum_{i=1}^{2}(x_i^3 - y_i^3) = \sum_{i=3}^{5}(x_i^3 - y_i^3)$$
$$\sum_{i=1}^{2}(x_i - y_i) = \sum_{i=3}^{5}(x_i - y_i),$$

with $\mathbf{x}, \mathbf{y} \in \mathcal{A}^5$. The fact that ellipsephic sets with small digit sumsets have fewer solutions of linear equations allows us to improve the general bounds obtained by Hughes and Wooley and also the corresponding efficient congruencing estimates. We also generalize our result from the curve (x, x^3) to $(x, \phi(x))$, where ϕ is a polynomial with integer coefficients and $\deg(\phi) \geq 3$.

1. introduction

The discrete restriction conjecture has recently been of wide interest for researchers in both harmonic analysis and number theory (for example, see [**8**], [**9**] and [**10**]). To recall the conjecture for the cuve (x, x^3), let $\mathbf{a} = \{a(n)\}_{n\in\mathbb{Z}}$ and

$$E\mathbf{a}\,(\alpha, \beta) := \sum_{|n|\leq N} a(n)e\left(\alpha n^3 + \beta n\right) \quad \text{with} \quad \alpha, \beta \in \mathbb{R}.$$

A weaker version of the conjecture can be stated as follows.

CONJECTURE 1.1. *For each $p \in [1, \infty]$ and $\epsilon > 0$, there exists a constant $C_{p,\epsilon} > 0$, such that for all $N \in \mathbb{N}$ and all sequence $\mathbf{a} = \{a(n)\}_{n\in\mathbb{Z}} \in \ell^2(\mathbb{Z})$, one has*

$$\|E\mathbf{a}\|_{L^p(\mathbb{T}^2)} \leq C_{p,\epsilon} N^\epsilon \left(1 + N^{\frac{1}{2}-\frac{4}{p}}\right) \|\mathbf{a}\|_{\ell^2(\mathbb{Z})}.$$

By proving the bound for $p = 6$, Bourgain [**5**] established Conjecture 1.1 for the cases $1 \leq p \leq 6$. Little further progress was made before Hu and Li [**8**] proved the case $p = 14$. Lai and Ding [**10**] extended the range to $p \geq 12$. Recently, Hughes and

2020 *Mathematics Subject Classification.* Primary 11L07, 42B05, 35Q53.
Key words and phrases. Mean value estimates, ellipsephic sets, KdV-like equations.
The first author was supported by an NSF DMS Grant in analysis and number theory.
The third author was supported by an NSERC Discovery Grant.
The authors would like to thank K. Biggs and T. Wooley for helpful discussions related to the project. They also thank the referee for providing helpful comments.

Wooley [**9**] proved the bound for $p = 10$ and hence established the conjecture for $p \geq 10$. For more details about the discrete restriction conjecture and its relation to KdV equations, we refer interested readers to [**8**], [**9**] and [**10**].

Given a set $\mathcal{S} \subset \mathbb{Z}$, let $\mathcal{S}(N) = \mathcal{S} \cap [-N, N]$ and $S = |\mathcal{S}(N)|$. For $s \in \mathbb{N} = \{1, 2, \cdots\}$, we denote by $J_s(\mathcal{S}(N))$ the number of solutions to the system of equations

$$\sum_{i=1}^{s} (x_i^3 - y_i^3) = 0$$
$$\sum_{i=1}^{s} (x_i - y_i) = 0,$$

with $\mathbf{x}, \mathbf{y} \in \mathcal{S}(N)^s$. In order to prove Conjecture 1.1 with $p = 10$, Hughes and Wooley studied in [**9**] the quantity $J_5(\mathcal{S}(N))$ for any set $\mathcal{S}$. In particular, they proved that there exists a positive constant κ such that

$$J_5(\mathcal{S}(N)) \ll N \exp\left(\kappa \frac{\log N}{\log \log N}\right) S^5.$$

The bound for $J_5(\mathcal{S}(N))$ can be improved if additional structures are employed on the set $\mathcal{S}$. In this paper, we consider the cases of ellipsephic sets.

The term *ellipsephic set* was introduced by Biggs in [**3**] and [**4**]. She used the terminology to mimic the word *ellipséphique* used in the French mathematical literature to denote integers with missing digits (for example, see [**1**] and [**2**]). Let ℓ be a positive integer and $\mathcal{A}_\ell \subset \{0, 1, \cdots, \ell - 1\}$ with $2 \leq |\mathcal{A}_\ell| \leq \ell - 1$. In other words, $\mathcal{A}_\ell$ contains at least two elements of $\{0, 1, \cdots, \ell - 1\}$, but at most $(\ell - 1)$ elements from the set. We say $\mathcal{A}$ is an *ellipsephic set in the base* ℓ if for any $n \in \mathcal{A}$,

$$n = \sum_i a_i \ell^i \quad \text{with} \quad a_i \in \mathcal{A}_\ell.$$

An ellipsephic set $\mathcal{A}$ is said to be *with small digit sumsets* if there exists some constant $K > 0$, such that $|\mathcal{A}_\ell + \mathcal{A}_\ell| \leq K|\mathcal{A}_\ell|$. For example, if $\ell \geq 3$ and $\mathcal{A}_\ell = \{0, 1\}$, then $\mathcal{A}_\ell + \mathcal{A}_\ell \in \{0, 1, 2\}$ and we have $K = 3/2$. In this paper, we will show that if $\mathcal{A}$ is an ellipsephic set with small digit sumsets, then we can obtain the following bound for $J_5(\mathcal{A}(N))$.

THEOREM 1.2. *Let $\mathcal{A}$ be an ellipsephic set with $\mathcal{A}_\ell = \mathcal{A} \cap [0, \ell - 1]$. Write $\mathcal{A}(N) = \mathcal{A} \cap [-N, N]$ with $A = |\mathcal{A}(N)|$. Suppose that $|\mathcal{A}_\ell + \mathcal{A}_\ell| \leq K|\mathcal{A}_\ell|$ for some constant $K > 0$. Then there exists a positive constant κ such that*

$$J_5(\mathcal{A}(N)) \ll K^4 N^{4 \log_\ell K} \exp\left(\kappa \frac{\log N}{\log \log N}\right) A^6.$$

We recall that for a general set $\mathcal{S}(N)$ with $S = |\mathcal{S}(N)|$, Hughes and Wooley proved in [**9**, Theorem 2.1] that

$$J_5(\mathcal{S}(N)) \ll N \exp\left(\kappa \frac{\log N}{\log \log N}\right) S^5.$$

To compare the above result with Theorem 1.2, we first notice that the set $\mathcal{S}$ ($\mathcal{S}_\ell$ and S, respectively) in [**9**] plays the role of $\mathcal{A}$ ($\mathcal{A}_\ell$ and A, respectively) in our setting.

Write r for both $|\mathcal{S}_\ell|$ and $|\mathcal{A}_\ell|$ with $2 \leq r \leq \ell - 1$. Then $A \leq r^{\log_\ell N + 1} = rN^{\log_\ell r}$. It follows that if $4\log_\ell K + \log_\ell r < 1$, i.e., $K^4 r < \ell$, then

$$K^4 N^{4\log_\ell K} A \leq rK^4 N^{4\log_\ell K + \log_\ell r} = o_{r,K}(N).$$

Hence the bound for ellipsephic sets in Theorem 1.2 is sharper in this case. The condition $K^4 r < \ell$ is often satisfied. For example, if $\ell \geq 3$ and $\mathcal{A}_\ell \in \{0, 1\}$, then $r = 2$ and $K = 3/2$. In this case, $K^4 r < \ell$ provided that $\ell > 2 \cdot (3/2)^4$, i.e., $\ell \geq 11$.

Using his efficient congruencing method, Wooley proved in [**14**] that for a general set $\mathcal{S}$,

$$J_3(\mathcal{S}(N)) \ll N^\epsilon S^3$$

for any $\epsilon > 0$. By trivially bounding the additional four variables by S^4, this gives us

$$J_5(\mathcal{S}(N)) \ll N^\epsilon S^7.$$

Let $r = |\mathcal{S}_\ell|$. Then $S \leq rN^{\log_\ell r}$. We notice that the above bound is stronger than the $N^{1+\epsilon}S^5$ bound in [**9**, Theorem 2.1] provided that $S^2 = o(N)$, which holds if $2\log_\ell r < 1$, i.e., $r^2 < \ell$. By noticing the set $\mathcal{S}$ ($\mathcal{S}_\ell$ and S, respectively) here plays the role of $\mathcal{A}$ ($\mathcal{A}_\ell$ and A, respectively) in Theorem 1.2, we can also compare the nested efficient congruencing bound $N^\epsilon S^7$ with the $K^4 N^{4\log_\ell K + \epsilon} A^6$ bound in Theorem 1.2. We see that if $4\log_\ell K < \log_\ell r$, i.e., $K^4 < r$, then

$$K^4 N^{4\log_\ell K} = o(rN^{\log_\ell r}),$$

so $K^4 N^{4\log_\ell K} = o(S)$. Hence the bound in Theorem 1.2 is sharper in this case. The conditions $K^4 < r$ and $r^2 < \ell$ are often satisfied. For example, if $\mathcal{A} = \{0, 1, \cdots, 13\}$, then $\mathcal{A} + \mathcal{A} \in \{0, 1, \cdots, 26\}$. We have $K = 27/14$ and $r = 14$. By taking $\ell > 14^2$, we have $K^4 < r < \ell^{1/2}$.

Let $\phi(x)$ be a polynomial with integer coefficients and $\deg(\phi) \geq 3$. We can generalize Theorem 1.2 from the curve (x, x^3) to the curve $(x, \phi(x))$. Let $J_{5,\phi}(\mathcal{A}(N))$ denote the number of solutions to the system of equations

$$\sum_{i=1}^{5} (\phi(x_i) - \phi(y_i)) = 0$$

$$\sum_{i=1}^{5} (x_i - y_i) = 0,$$

with $\mathbf{x}, \mathbf{y} \in \mathcal{A}(N)^5$. We will prove the following result.

THEOREM 1.3. *Let $\mathcal{A}$ be an ellipsephic set with $\mathcal{A}_\ell = \mathcal{A} \cap [0, \ell - 1]$. Write $\mathcal{A}(N) = \mathcal{A} \cap [-N, N]$ with $A = |\mathcal{A}(N)|$. Suppose that $|\mathcal{A}_\ell + \mathcal{A}_\ell| \leq K|\mathcal{A}_\ell|$. Then for any $\epsilon > 0$, we have*

$$J_{5,\phi}(\mathcal{A}(N)) \ll K^4 N^{4\log_\ell K + \epsilon} A^6.$$

We assume here and throughout that the implicit constant in the symbol $\ll$ may depend on ε, s, k, and the coefficients of ϕ.

Let $|\mathcal{A}_\ell| = r$ with $2 \leq r \leq \ell - 1$. Similar to the remarks after Theorem 1.2, the result in Theorem 1.3 is sharper than the general bounds in [2, Theorem 3.4] and the corresponding nested efficient congruencing estimates, provided that $K^4 < r < \ell^{1/2}$.

We now restrict our attention to the case when ℓ is a prime. A subset $\mathcal{R} \subset \mathbb{N} \cup \{0\}$ is called a E_2^*-set if for any $n \in \mathbb{N}$, we have

$$\#\left\{(a_1, a_2) \in \mathcal{R}^2 : a_1 + a_2 = n\right\} \ll n^\epsilon \tag{1.1}$$

for any $\epsilon > 0$. Let $\mathcal{R}_\ell = \mathcal{R} \cap [0, \ell - 1]$ and suppose that $2 \leq |\mathcal{R}_\ell| \leq \ell - 1$. Given a prime ℓ and a E_2^*-set $\mathcal{R}$, a set $\mathcal{E} = \mathcal{E}_\ell^\mathcal{R}$ is called a $(\ell, 2)^*$-ellipsephic set if

$$\mathcal{E} = \left\{ n = \sum_i a_i \ell^i : a_i \in \mathcal{R}_\ell \text{ for all } i \right\}.$$

In this setting, Biggs [**4**, Theorem 1.2] proved that

$$J_{5,\phi}(\mathcal{E}(N)) \ll N^\epsilon E^5,$$

where $E = |\mathcal{E}(N)|$. Her bound is essentially optimal as we get $J_{5,\phi}(\mathcal{E}(N)) \gg E^5$ from the diagonal solutions. She also obtained similar bounds for general E_t^*-sets with $t \geq 2$.

The optimal result of Biggs and Theorem 1.3 are applied to sets of different nature. To illustrate their difference, we first notice that the set $\mathcal{E}$ ($\mathcal{E}_\ell$ and E, respectively) in [**4**] plays the role of $\mathcal{A}$ ($\mathcal{A}_\ell$ and A, respectively) in our setting. In [**11**], Landau proved that the set of squares satisfies the condition (1.1). Write r for both $|\mathcal{E}_\ell|$ and $|\mathcal{A}_\ell|$. Since the set of squares is sparse, the set $\mathcal{E}_\ell + \mathcal{E}_\ell$ could be of size r^2 for sufficiently large ℓ. On the other hand, if an ellipsephic set $\mathcal{A}$ satisfies $|\mathcal{A}_\ell + \mathcal{A}_\ell| < Kr$, then Theorem 1.3 provides meaningful improvement only if $K^4 < r$. This condition $K < r^{1/4}$ is not always satisfied for large r if we take an ellipsephic set $\mathcal{E}$ with square digits since $\mathcal{E}_\ell + \mathcal{E}_\ell$ could be of size r^2. Thus one can say the result of Biggs provides useful estimates for "large K," while Theorem 1.3 is meaningful for "small K."

We will prove Theorem 1.2 in Section 2 and Theorem 1.3 in Section 3. The key idea of our paper is to make use of the fact that ellipsephic sets with small digit sumsets have fewer solutions of linear equations. More precisely, we can bound elements of the form $2\mathcal{A} - 2\mathcal{A}$ for ellipsephic sets $\mathcal{A}$ with small digit sumsets more efficiently than general sets (see Lemma 2.1). We will highlight this idea with variations of Theorem 1.3 at the end of the paper.

2. Proof of Theorem 1.2

To prove Theorem 1.2, we need the following lemma.

LEMMA 2.1. *Let $\mathcal{A}$ be an ellipsephic set with $\mathcal{A}_\ell = \mathcal{A} \cap [0, \ell - 1]$. Write $\mathcal{A}(N) = \mathcal{A} \cap [-N, N]$ with $A = |\mathcal{A}(N)|$. Suppose that $|\mathcal{A}_\ell + \mathcal{A}_\ell| \leq K|\mathcal{A}_\ell|$. Then*

$$|2\mathcal{A} - 2\mathcal{A}| \ll K^4 N^{4 \log_\ell K} A.$$

PROOF. Since $\mathcal{A}_\ell \subset [0, \ell - 1]$, by viewing $\mathcal{A}_\ell$ as a subset of the abelian group $\mathbb{Z}_{4\ell-3} = \{2 - 2\ell, \cdots, -1, 0, 1, \cdots, 2\ell - 2\}$, we have $\mathcal{A}_\ell + \mathcal{A}_\ell \subset \mathbb{Z}_{4\ell-3}$. By the Plünnecke-Ruzsa inequality [**12**], [**13**], we have $|2\mathcal{A}_\ell - 2\mathcal{A}_\ell| \leq K^4|\mathcal{A}_\ell|$. Since $\mathcal{A} \subset [-N, N]$, each element of $\mathcal{A}$ is formed of $\leq \log_\ell N + 1$ digits, each of which is in $\mathcal{A}_\ell$. Hence, each element of $2\mathcal{A} - 2\mathcal{A}$ corresponds to at least one element of

$(2\mathcal{A}_\ell - 2\mathcal{A}_\ell)^{\log_\ell N+1}$. Thus we have

$$\begin{aligned}|2\mathcal{A} - 2\mathcal{A}| &\le |2\mathcal{A}_\ell - 2\mathcal{A}_\ell|^{\log_\ell N+1}\\ &\le K^{4(\log_\ell N+1)}|\mathcal{A}_\ell|^{\log_\ell N+1}\\ &= K^4 N^{4\log_\ell K} A.\end{aligned}$$

□

REMARK. Let $\mathcal{A}$ be an ellipsephic set satisfying the conditions in Lemma 2.1. For $m, n \in \mathbb{N}$, using the same argument as the above proof, we can show that

$$|m\mathcal{A} - n\mathcal{A}| \ll K^{m+n} N^{(m+n)\log_\ell K} A.$$

Let

$$E\mathbb{1}_{\mathcal{A}(N)}(\alpha, \beta) := \sum_{n\in\mathcal{A}(N)} e\left(\alpha n^3 + \beta n\right).$$

By the orthogonal relation of the exponential function, we see that

$$J_5(\mathcal{A}(N)) = \int_{\mathbb{T}^2} \left|E\mathbb{1}_{\mathcal{A}(N)}(\alpha, \beta)\right|^{10} d\alpha d\beta,$$

where $\mathbb{T} = [0, 1)$. Hence to prove Theorem 1, it is equivalent to show that there exists a positive constant κ such that

$$\int_{\mathbb{T}^2} \left|E\mathbb{1}_{\mathcal{A}(N)}(\alpha, \beta)\right|^{10} d\alpha d\beta \ll K^4 N^{4\log_\ell K} \exp\left(\kappa \frac{\log N}{\log\log N}\right) A^6.$$

PROOF OF THEOREM 1.2. Write $a = \mathbb{1}_{\mathcal{A}(N)}$. The tenth moment $\|Ea\|_{10}^{10}$ counts the number of solutions to the system of equations

$$\begin{aligned}\sum_{i=1}^{2}\left(x_i^3 - y_i^3\right) &= \sum_{i=3}^{5}\left(x_i^3 - y_i^3\right)\\ \sum_{i=1}^{2}(x_i - y_i) &= \sum_{i=3}^{5}(x_i - y_i),\end{aligned}$$

with $\mathbf{x}, \mathbf{y} \in \mathcal{A}(N)^5$. By the second equation above, we let

$$h := x_1 - y_1 + x_2 - y_2 = x_3 - y_3 + x_4 - y_4 + x_5 - y_5 \in 2\mathcal{A} - 2\mathcal{A}.$$

We now write $\|Ea\|_{10}^{10}$ as

$$\sum_{h\in 2\mathcal{A}-2\mathcal{A}} \int_{\mathbb{T}}\int_{\mathbb{T}} |Ea(\alpha_1, \alpha_2)|^4 e(-\alpha_2 h) d\alpha_2 \int_{\mathbb{T}} |Ea(\alpha_1, \alpha_3)|^6 e(-\alpha_3 h) d\alpha_3 d\alpha_1.$$

By the triangle inequality and Lemma 2.1, this gives

$$\|Ea\|_{10}^{10} \le K^4 N^{4\log_\ell K} A \int_{\mathbb{T}^3} |Ea(\alpha_1, \alpha_2)|^4 |Ea(\alpha_1, \alpha_3)|^6 d\alpha_1 d\alpha_2 d\alpha_3. \tag{2.1}$$

Let $c_t(k)$ denote the number of solutions of the simultaneous equations

$$\sum_{i=1}^{t}(x_i^3 - y_i^3) = k \qquad \text{and} \qquad \sum_{i=1}^{t}(x_i - y_i) = 0.$$

In the proof of [**9**, Theorem 2.1], Hughes and Wooley proved that

$$\int_{\mathbb{T}^3} |Ea(\alpha_1,\alpha_2)|^4 |Ea(\alpha_1,\alpha_3)|^6 \, d\alpha_1 d\alpha_2 d\alpha_3 \ll \sum_{|k|\le 4N^3} c_2(k)c_3(k)$$
$$\ll \exp\left(\kappa \frac{\log N}{\log\log N}\right) A^5.$$

It follows that

$$\|Ea\|_{10}^{10} \ll K^4 N^{4\log_\ell K} A \exp\left(\kappa \frac{\log N}{\log\log N}\right) A^5$$
$$\ll K^4 N^{4\log_\ell K} \exp\left(\kappa \frac{\log N}{\log\log N}\right) A^6.$$

□

REMARK. We see in Section 1 a remark after Theorem 1.2 that if $K^4 < r = |\mathcal{A}_\ell|$, then the bound of Theorem 1.2 is sharper than the bound derived from the efficient congruencing method. One can find examples to satisfy $K^m < r$ for all $m \in \mathbb{N}$, provided that r is sufficiently large. For example, for a large ℓ, if $\mathcal{A} = \{0, 1, \cdots, q\}$ with $2q < \ell$, then $\mathcal{A} + \mathcal{A} \in \{0, 1, \cdots, 2q\}$. We have

$$K = (2q+1)/(q+1) = 2 - 1/(q+1) < 2.$$

Hence by taking $q = 2^m - 1$, we get $K^m < r = q + 1$.

3. Proof of Theorem 1.3

Let $\phi(x)$ be a polynomial with integer coefficients and $\deg(\phi) \ge 3$. Let

$$F\mathbb{1}_{\mathcal{A}(N)}(\alpha,\beta) := \sum_{n\in\mathcal{A}(N)} e\left(\alpha\phi(n) + \beta n\right).$$

By the orthogonal relation of the exponential function, we see that

$$J_{5,\phi}(\mathcal{A}(N)) = \int_{\mathbb{T}^2} \left|F\mathbb{1}_{\mathcal{A}(N)}(\alpha,\beta)\right|^{10} d\alpha d\beta.$$

Hence to prove Theorem 1.3, it is equivalent to show that for any $\epsilon > 0$, we have

$$\int_{\mathbb{T}^2} \left|F\mathbb{1}_{\mathcal{A}(N)}(\alpha,\beta)\right|^{10} d\alpha d\beta \ll K^4 N^{4\log_\ell K + \epsilon} A^6.$$

Proof of Theorem 1.3 Write $a = \mathbb{1}_{\mathcal{A}(N)}$. The tenth moment $\|Fa\|_{10}^{10}$ counts the number of solutions to the system of equations

$$\sum_{i=1}^{2} (\phi(x_i) - \phi(y_i)) = \sum_{i=3}^{5} (\phi(x_i) - \phi(y_i))$$
$$\sum_{i=1}^{2} (x_i - y_i) = \sum_{i=3}^{5} (x_i - y_i),$$

with $\mathbf{x}, \mathbf{y} \in \mathcal{A}(N)^5$. Using the same argument as the one in Theorem 1.2, we get

$$\|Fa\|_{10}^{10} \le K^4 N^{4\log_\ell K} A \int_{\mathbb{T}^3} |Fa(\alpha_1,\alpha_2)|^4 |Fa(\alpha_1,\alpha_3)|^6 \, d\alpha_1 d\alpha_2 d\alpha_3. \tag{3.1}$$

Let $c_t(k)$ counts the number of solutions of the simultaneous equations

$$\sum_{i=1}^{t}(\phi(x_i)-\phi(y_i))=k \qquad \text{and} \qquad \sum_{i=1}^{t}(x_i-y_i)=0,$$

with $\mathbf{x},\mathbf{y}\in\mathcal{A}^t$. In the proof of [**9**, Theorem 3.4], Hughes and Wooley proved that there exists a constant C, depending on ϕ, such that

$$\int_{\mathbb{T}^3}|Fa(\alpha_1,\alpha_2)|^4\,|Fa(\alpha_1,\alpha_3)|^6\,d\alpha_1 d\alpha_2 d\alpha_3 \leq \sum_{|k|\leq CN^k} c_2(k)c_3(k)$$
$$\ll N^{\epsilon}A^5.$$

It follows that

$$\|Fa\|_{10}^{10}\ll K^4N^{4\log_\ell K}AN^{\epsilon}A^5\ll K^4N^{4\log_\ell K+\epsilon}A^6.$$

□

The improved bounds in Theorem 1.2 and Theorem 1.3 come from the fact that we can bound elements of the form $2\mathcal{A}-2\mathcal{A}$ for ellipsephic sets $\mathcal{A}$ with small digit sumsets more efficiently than general sets. To highlight the idea, we consider a variation of Theorem 1.3.

Given a set $\mathcal{S}\subset\mathbb{Z}$, let $\mathcal{S}(N)=\mathcal{S}\cap[-N,N]$ and $S=|\mathcal{S}(N)|$. Suppose that $|2\mathcal{S}(N)-2\mathcal{S}(N)|\leq P(S)$ for some function P of S. Let $\phi(x)$ be a polynomial with integer coefficients and $\deg(\phi)\geq 3$. Following the proof of Theorem 1.3, we have

THEOREM 3.1. *Let $\mathcal{S}(N)\subset\mathbb{Z}\cap[-N,N]$ with $|\mathcal{S}(N)|=S$. Suppose that $|2\mathcal{S}(N)-2\mathcal{S}(N)|\leq P(S)$ for some function P of S. Then for any $\epsilon>0$, we have*

$$J_{5,\phi}(\mathcal{S}(N))\ll P(S)N^{\epsilon}S^5.$$

For a set $\mathcal{S}(N)\subset\mathbb{Z}\cap[-N,N]$, suppose that $|\mathcal{S}(N)+\mathcal{S}(N)|\leq KS$ for some constant K (by Freiman's theorem [**6**] [**7**], sets satisfying this condition are contained in a generalized arithmetic progression). By the Plünnecke-Ruzsa inequality [**12**], [**13**], we have $|2\mathcal{S}(N)-2\mathcal{S}(N)|\leq K^4S$. Hence as a direct consequence of Theorem 3.1, we have

COROLLARY 3.2. *Let $\mathcal{S}(N)\subset\mathbb{Z}\cap[-N,N]$ with $|\mathcal{S}(N)|=S$. Suppose that $|\mathcal{S}(N)+\mathcal{S}(N)|\leq KS$ for some constant K. Then for any $\epsilon>0$, we have*

$$J_{5,\phi}(\mathcal{S}(N))\ll K^4N^{\epsilon}S^6.$$

References

[1] Karam Aloui, *Sur les entiers ellipséphiques: somme des chiffres et répartition dans les classes de congruence* (French, with English summary), Period. Math. Hungar. **70** (2015), no. 2, 171–208, DOI 10.1007/s10998-014-0066-8. MR3343999

[2] Karam Aloui, Christian Mauduit, and Mohamed Mkaouar, *Somme des chiffres et répartition dans les classes de congruence pour les palindromes ellipséphiques* (French, with English and French summaries), Acta Math. Hungar. **151** (2017), no. 2, 409–455, DOI 10.1007/s10474-017-0688-4. MR3620841

[3] Kirsti D. Biggs, *Efficient congruencing in ellipsephic sets: the quadratic case*, Acta Arith. **200** (2021), no. 4, 331–348, DOI 10.4064/aa191216-8-2. MR4331907

[4] Kirsti D. Biggs, *Efficient congruencing in ellipsephic sets: the quadratic case*, Acta Arith. **200** (2021), no. 4, 331–348, DOI 10.4064/aa191216-8-2. MR4331907

[5] Jean Bourgain, *Fourier transform restriction phenomena for certain lattice subsets and applications to nonlinear evolution equations. II. The KdV-equation*, Geom. Funct. Anal. **3** (1993), no. 3, 209–262, DOI 10.1007/BF01895688. MR1215780

[6] G. A. Fridman,*Addition of finite sets*, Soviet Math. Dokl., **5**: 1366–1370.

[7] Gregory A. Freĭman, *Nachala strukturnoĭ teorii slozheniya mnozhestv* (Russian), Kazan. Gosudarstv. Ped. Inst.; Elabuž. Gosudarstv. Ped. Inst., Kazan, 1966. MR0360495

[8] Yi Hu and Xiaochun Li, *Discrete Fourier restriction associated with KdV equations*, Anal. PDE **6** (2013), no. 4, 859–892, DOI 10.2140/apde.2013.6.859. MR3092732

[9] Kevin Hughes and Trevor D. Wooley, *Discrete restriction for (x, x^3) and related topics*, Int. Math. Res. Not. IMRN **20** (2022), 15612–15631, DOI 10.1093/imrn/rnab113. MR4609773

[10] Xudong Lai and Yong Ding, *A note on the discrete Fourier restriction problem*, Proc. Amer. Math. Soc. **146** (2018), no. 9, 3839–3846, DOI 10.1090/proc/13975. MR3825838

[11] Edmund Landau. *Über die Anzahl der Gitterpunkte in geweissen Bereichen*, Nachr. Ges. Wiss. Goettingen, Math.-Phys. Kl. **1912** (1912), 687–770.

[12] Helmut Plünnecke, *Eine zahlentheoretische Anwendung der Graphentheorie* (German), J. Reine Angew. Math. **243** (1970), 171–183, DOI 10.1515/crll.1970.243.171. MR266892

[13] Imre Z. Ruzsa, *An application of graph theory to additive number theory*, Sci. Ser. A Math. Sci. (N.S.) **3** (1989), 97–109. MR2314377

[14] Trevor D. Wooley, *Nested efficient congruencing and relatives of Vinogradov's mean value theorem*, Proc. Lond. Math. Soc. (3) **118** (2019), no. 4, 942–1016, DOI 10.1112/plms.12204. MR3938716

Department of Mathematical Sciences, Carnegie Mellon University, Pittsburgh, PA 15213

Email address: tanders2@andrew.cmu.edu

Department of Mathematics, Purdue University, 150 N. University St., West Lafayette, IN, 47907, U.S.A

Email address: hu776@purdue.edu

Department of Pure Mathematics, University of Waterloo, 200 University Ave. West, Waterloo, ON, N2L 3G1, Canada

Email address: yrliu@uwaterloo.ca

Department of Pure Mathematics, University of Waterloo, 200 University Ave. West, Waterloo, ON, N2L 3G1, Canada

Email address: alan.talmage@uwaterloo.ca

Contemporary Mathematics
Volume **792**, 2024
https://doi.org/10.1090/conm/792/15898

Oscillation inequalities for Radon averages

Bartosz Langowski

ABSTRACT. We prove one-parameter oscillation estimates for the family of Radon means. The result generalizes the result of Jones, Kaufman, Rosenblatt and Wierdl.

1. Introduction

1.1. Motivation and the main result. The seminal paper [**1**] of Bourgain initiated an extensive study of oscillation and variational inequalities in harmonic analysis and ergodic theory. The importance of such inequalities is related to their applicability to pointwise convergence problems, in particular in the settings where more traditional methods fail. In the present manuscript we prove L^p oscillation inequalities for the family of Radon averaging operators along polynomial mappings. Our result extends the assertions in [**4**] and also complements the results of [**5**].

For a totally ordered infinite set $\mathbb{I} \subseteq \mathbb{R}$ we let $\mathfrak{S}(\mathbb{I})$ be a family of all increasing sequences with values in $\mathbb{I}$. Then for a sequence $I = (I_j : j \in \mathbb{N}) \in \mathfrak{S}(\mathbb{I})$, the truncated oscillation semi–norm is defined for any function $\mathfrak{a} : \mathbb{I} \to \mathbb{C}$ by setting

$$O_I(\mathfrak{a}_t : t \in \mathbb{I}) = \Big(\sum_{k=1}^{\infty} \sup_{t \in [I_k, I_{k+1})} |\mathfrak{a}_t - \mathfrak{a}_{I_k}|^2 \Big)^{1/2}.$$

Another object frequently used to obtain quantitative information on rates of convergence is the r-variational semi–norm V^r of a function $\mathfrak{a} : \mathbb{I} \to \mathbb{C}$ acting on an ordered set $\mathbb{I}$ defined by setting

$$V^r(\mathfrak{a}_t : t \in \mathbb{I}) = \sup_{J \in \mathbb{N}} \sup_{\substack{t_0 < \ldots < t_J \\ t_j \in \mathbb{I}}} \Big(\sum_{j=1}^{J} |\mathfrak{a}_{t_j} - \mathfrak{a}_{t_{j-1}}|^r \Big)^{1/r},$$

where the supremum is taken over all finite increasing sequences in $\mathbb{I} \subseteq (0, \infty)$. The semi–norm V^r proved to be applicable mostly for $r > 2$, see e.g. [**1**].

In [**4**] it was shown that the family of Hardy–Littlewood averages

$$\mathcal{A}_t f(x) = \frac{1}{|B_t(x)|} \int_{B_t(x)} f(y)\, dy, \qquad t \in \mathbb{R}_+,$$

2020 *Mathematics Subject Classification.* Primary 42B25; Secondary 42B10.
Key words and phrases. Oscillation inequality, Radon average.

where $B_t(x)$ is an Euclidean ball in $\mathbb{R}$ of radius t centered at x, satisfies for $p \in (1, \infty)$ and every $f \in L^p(\mathbb{R})$ the following uniform estimate

$$\sup_{I \in \mathfrak{S}(\mathbb{N})} \left\| O_I(\mathcal{A}_t f : t \in \mathbb{R}_+) \right\|_{L^p(\mathbb{R})} \leq C_p \|f\|_{L^p(\mathbb{R})}, \tag{1.1}$$

with some $C_p > 0$. Strictly speaking, the main result of [**4**] concerns a discrete variant of the above averages, but the proof presented there works also in the continuous setting. The oscillation and also variational estimates in the context of Hardy–Littlewood means received considerable attention in the recent years, see for example [**6, 7, 16**].

The motivation behind the present paper is to generalize the result of [**4**] by proving inequality (1.1) with Hardy–Littlewood averages replaced by a large class of averaging operators of Radon type modelled on polynomial mappings. To introduce the setting of our interest let $P(s) = P(s_1, \dots, s_m) = \big(P_1(s_1, \dots, s_m), \dots, P_d(s_1, \dots, s_m)\big) : \mathbb{R}^m \to \mathbb{R}^d$ be a polynomial mapping such that each P_j, $j = 1, \dots, d$, is of the form $P_j(s) = \prod_{i=1}^m s_i{}^{\alpha_{i,j}}$ with some $\alpha_{1,j}, \dots, \alpha_{m,j} \in \mathbb{N} \cup \{0\}$ (we allow the case $\alpha_{1,j} + \dots + \alpha_{m,j} = 0$). We fix a convex bounded open set $\Omega \subset \mathbb{R}^m$ containing the origin and for every $x \in \mathbb{R}^d$ and $t \in \mathbb{R}_+$ we define the Radon averaging operators by setting

$$M_t^P f(x) := \frac{1}{|\Omega_t|} \int_{\Omega_t} f(x - P(y)) \mathrm{d}y,$$

where $\Omega_t = \{x \in \mathbb{R}^m : t^{-1}x \in \Omega\}$.

Jones, Seeger and Wright [**5**] proved the estimate

$$\|V^r(M_t^P f : t \in \mathbb{R}_+)\|_{L^p(\mathbb{R}^d)} \leq C_p \|f\|_{L^p(\mathbb{R}^d)} \tag{1.2}$$

with any $r > 2$. On the other hand, it is well known that the above bound does not hold for $r = 2$, due to the failure of the Lepingle's inequality for martingales in the case $r = 2$, see [**8**].

In the following main result of the present manuscript we prove oscillation inequallity for Radon averages $M_t^P f$. It provides a natural substitute for the variational estimate in the case $r = 2$.

THEOREM 1.3. *For every $p \in (1, \infty)$ there exists a constant $C_p > 0$ such that for every $f \in L^p(\mathbb{R}^d)$ we have*

$$\sup_{I \in \mathfrak{S}(\mathbb{R}_+)} \|O_I(M_t^P f : t \in \mathbb{R}_+)\|_{L^p(\mathbb{R}^d)} \leq C_p \|f\|_{L^p(\mathbb{R}^d)}. \tag{1.4}$$

The above theorem extends the results of [**4**] by allowing a substantially larger class of families of operators. As we mentioned before, it can be also thought of as a completion of the results from [**5**]. To prove (1.4) we follow the standard approach proposed in [**4**] which relies on splitting the oscillatory semi–norm into long and short parts: the first measuring oscillations along a certain lacunary subsequence and the latter corresponding to oscillations within intervals with endpoints of comparable magnitude. The treatment of the short part differs from the approach in [**4**] and [**5**]. In particular, we employ the method of handling oscillating objects developed recently in [**10**], which allows for a substantial simplification of the short part analysis, compared to [**5**]. Noteworthily, the approach proposed in the present manuscript gives an alternative, simpler proof of the estimate (1.2) from [**5**].

1.2. Overview of the paper and methods. The paper is organized as follows. In the remainder of this section we gather notation that will be frequently used in the sequel. In Section 2 we prove the oscillatory inequality for the family of Radon averages. To this end we treat separately long and short oscillations. For the long part we use a relevant estimate for martingale averages from [**4**], see Theorem 2.4, and then we argue as in [**5**] constructing a suitable martingale which approximates well the lacunary Radon averages. To transition from the estimate for martingales to the desired bound for Radon means, we need to pass through the smooth approximations of Radon means. The result for them is contained in Theorem 2.9. The difference between the martingale and smoothed Radon averages is measured by means of a square function for which we use an L^p bound from [**5**]. For the short oscillations part our approach differs substantially from the method used in [**5**]. We present a much simpler argument relying on a robust technique of obtaining maximal and variational estimates for fairly general families of operators established in the recent paper [**10**] (see also [**3**] for an earlier result in the similar spirit). At its heart lies an elementary numerical inequality, which has lately proved to be of remarkable importance in the study of variational and oscillatory estimates, see e.g. [**9, 10, 12, 13**].

1.3. Notation. Throughout the whole paper $C > 0$ is an absolute constant which may change from line to line. For two nonnegative quantities A, B we will write $A \lesssim_\delta B$ ($A \gtrsim_\delta B$) to say that there is an absolute constant $C_\delta > 0$ (which possibly depends on $\delta > 0$) such that $A \le C_\delta B$ ($A \ge C_\delta B$). We will write $A \simeq_\delta B$ when $A \lesssim_\delta B$ and $A \gtrsim_\delta B$ hold simultaneously.

Let $\mathbb{N} = \{1, 2, \dots\}$ be the set of positive integers. We also denote by $\mathbb{D} = \{\tau^n : n \in \mathbb{Z}\}$ the set of powers of a fixed constant $\tau > 1$. Furthermore, we set $\mathbb{R}_+ = (0, \infty)$. Given a totally ordered set $\mathbb{I} \subseteq \mathbb{R}$ we also let $\mathfrak{S}(\mathbb{I})$ be the set of all strictly increasing sequences taking values in $\mathbb{I}$.

By $\ell^p(\mathbb{Z}^d)$ for $1 \le p < \infty$, we mean the Lebesgue spaces corresponding to $\mathbb{Z}^d$ with the norm

$$\|f\|_{\ell^p(\mathbb{Z}^d)} = \Big(\sum_{n\in\mathbb{Z}^d} |f(n)|^p\Big)^{1/p}.$$

By $L^p(\mathbb{R}^d)$ for $1 \le p < \infty$, we mean the Lebesgue spaces corresponding to $\mathbb{R}^d$ with the norm

$$\|f\|_{L^p(\mathbb{R}^d)} = \Big(\int_{\mathbb{R}^d} |f(x)|^p dx\Big)^{1/p}.$$

Moreover, by $\mathcal{S}(\mathbb{R}^d)$ we denote the Schwartz class of functions.

For a finite measure μ on $\mathbb{R}^d$ we define its Fourier transform by setting

$$\widehat{\mu}(\xi) = \int_{\mathbb{R}^d} \exp(-2\pi i \langle x, \xi\rangle) \mathrm{d}\mu(x),$$

where $\langle \cdot, \cdot \rangle$ is the standard inner product on $\mathbb{R}^d$. In particular for $f \in L^1(\mathbb{R}^d)$ we let

$$\widehat{f}(\xi) = \int_{\mathbb{R}^d} f(x) \exp(-2\pi i \langle x, \xi\rangle) \mathrm{d}x,$$

We also write $*$ for the convolution on $\mathbb{R}^d$.

2. The oscillation inequality

2.0.1. *Oscillation inequalities for martingales.* We begin with fixing useful notation and terminology. Let (X, ρ) be a quasi–metric space with a symmetric quasi–metric $\rho : X \times X \to [0, \infty)$ such that for all $x, y, z \in X$:

$$\rho(x, y) \leq A_0\big(\rho(x, z) + \rho(z, y)\big) \tag{2.1}$$

where $1 \leq A_0 < \infty$ is an absolute constant independent of x, y, z. We denote by $B(x, r) = \{y \in X : \rho(y, x) < r\}$ a quasi–metric ball with center $x \in X$ and radius $r > 0$.

A measure μ on a quasi–metric space (X, ρ) is called doubling if there exists $0 < A_1 < \infty$ such that for all $x \in X$ and $r > 0$ the doubling condition holds

$$\mu(B(x, 2r)) \leq A_1 \mu(B(x, r)). \tag{2.2}$$

Finally, a quadruple $(X, \mathcal{B}(X), \rho, \mu)$ is called a space of homogeneous type if a nonempty set X is endowed with a quasi–metric ρ for which all quasi–metric balls are open and a nonnegative Borel measure μ, which is finite for all quasi–metric balls and satisfies the doubling condition.

As it was shown by Christ [**2**], see Proposition 2.3 below, for any space $(X, \mathcal{B}(X), \rho, \mu)$ of homogeneous type there exists a system of dyadic cubes.

PROPOSITION 2.3. *Let $(X, \mathcal{B}(X), \rho, \mu)$ be a space of homogeneous type with a quasi–triangle inequality constant $1 \leq A_0 < \infty$ and a doubling constant $0 < A_1 < \infty$, see (2.1) and (2.2). Then there are constants $\kappa > 1$, $0 < C_1, C_2, a_0, \eta < \infty$, and $N_0 \in \mathbb{N}$ depending only on A_0 and A_1, and a family of open sets*

$$\mathcal{D} = \bigcup_{k \in \mathbb{Z}} \mathcal{D}_k,$$

where $\mathcal{D}_k = (Q^k_\alpha : \alpha \in I_k)$ is a collection of open sets (we will call them the kth generation) with the following properties:

(1) *For every $k \in \mathbb{Z}$ we have*

$$\mu\Big(X \setminus \bigcup_{Q \in \mathcal{D}_k} Q\Big) = 0.$$

(2) *If $l \leq k$ and $Q_1 \in \mathcal{D}_k$ and $Q_2 \in \mathcal{D}_l$ then either $Q_2 \subseteq Q_1$ or $Q_1 \cap Q_2 = \emptyset$.*
(3) *For every $Q \in \mathcal{D}_l$ and $l \leq k$ there exists a unique $Q_0 \in \mathcal{D}_k$ such that $Q \subseteq Q_0$.*
(4) *For every $Q_0 \in \mathcal{D}_k$ we have*

$$|\{Q \in \mathcal{D}_{k-1} : Q \subseteq Q_0\}| \leq N_0.$$

(5) *For every $k \in \mathbb{Z}$ and for every $Q \in \mathcal{D}_k$ there is $c_Q \in Q$ such that $B(c_Q, a_0\kappa^k) \subseteq Q \subseteq B(c_Q, C_1\kappa^k)$. We shall call c_Q the center of Q.*
(6) *For every $k \in \mathbb{Z}$ and for every $Q \in \mathcal{D}_k$ we have*

$$\mu\big(\{x \in Q : \operatorname{dist}(x, X \setminus Q) \leq t\kappa^k\}\big) \leq C_2 t^\eta \mu(Q) \quad \textit{for all} \quad t \in \mathbb{R}_+.$$

Now, given the homogeneous space $(X, \mathcal{B}(X), \rho, \mu)$ we fix a system of sets $\mathcal{D}$ on $(X, \mathcal{B}(X), \rho, \mu)$ with all above-mentioned properties. For simplicity we shall call the elements of $\mathcal{D}$ dyadic cubes. We also define a natural filtration corresponding to $\mathcal{D}$ by setting $\mathcal{F}_k = \sigma(\mathcal{D}_k)$ for all $k \in \mathbb{Z}$. Then the corresponding conditional expectation operator is defined for $f \in L^1_{\rm loc}(X)$ by

$$\mathbb{E}_k f(x) = \mathbb{E}[f|\mathcal{F}_k](x) = \sum_{Q\in\mathcal{D}_k} \mathbf{1}_Q(x)\frac{1}{|Q|}\int_Q f(y)\mathrm{d}\mu(y).$$

The martingale difference operator is denoted by $\mathbb{D}_k f = \mathbb{E}_k f - \mathbb{E}_{k-1} f$. We see that $\mathbb{D}_k f$ is constant on each cube $Q \in \mathcal{D}_{k-1}$ and has integral 0 on each cube $Q \in \mathcal{D}_k$. The following result was proved in [**4**].

THEOREM 2.4. *For every $p \in (1,\infty)$ there exists a constant $C_p > 0$ such that for every function $f \in L^p(X)$ we have*

$$\sup_{I\in\mathfrak{S}(\mathbb{Z})} \Big\| O_I(\mathbb{E}_n f : n \in \mathbb{Z}) \Big\|_{L^p(X)} \le C_p \|f\|_{L^p(X)}.$$

2.0.2. *Oscillation inequalities for smooth bump functions.* We now use Theorem 2.4 to obtain the oscillatory inequality for operators given by convolution with smooth bump functions transformed by a suitable group of nonisotropic dilations. We will work with the Euclidean space $X = \mathbb{R}^d$ equipped with the Lebesgue measure $d\mu(x) = dx$, and with a quasi–metric ρ induced by a suitable quasi–norm. To be more precise, for a polynomial mapping $P(s) = P(s_1,\ldots,s_m) : \mathbb{R}^m \to \mathbb{R}^d$ such that each P_j, $j = 1,\ldots,d$, is of the form $P_j(s) = \prod_{i=1}^m s_i{}^{\alpha_{i,j}}$ as in Section 1, we let A be a diagonal $d \times d$ matrix with the diagonal entries $|\alpha_1|,\ldots,|\alpha_d|$, where $|\alpha_j| = \sum_{i=1}^m \alpha_{i,j}$, for $j = 1,\ldots,d$. In what follows we will assume that $|\alpha_j| > 0$ for each j. The argument in the opposite case requires only minor modifications and will be omitted. We define the group of dilations on $\mathbb{R}^d$ given by a multiplication by the matrix

$$t^A := \exp(A\log t) = \begin{pmatrix} t^{|\alpha_1|} & 0 & \cdots & 0 \\ 0 & t^{|\alpha_2|} & \cdots & 0 \\ \vdots & \vdots & \ddots & \vdots \\ 0 & 0 & \cdots & t^{|\alpha_d|} \end{pmatrix}, \qquad \text{for} \quad t \in \mathbb{R}_+. \tag{2.5}$$

We will consider a smooth A-homogeneous quasi–norm

$$\mathfrak{q}(\xi) = \max_{j\in\{1,\ldots,d\}} |\xi_j|^{1/|\alpha_j|}, \qquad \xi \in \mathbb{R}^d.$$

It means that $\mathfrak{q} : \mathbb{R}^d \to [0,\infty)$ is a continuous function, smooth on $\mathbb{R}^d \setminus \{0\}$, and satisfies the following conditions:

(1) $\mathfrak{q}(x) = 0 \iff x = 0$;
(2) there is $C \ge 1$ such that for all $x, y \in \mathbb{R}^d$ we have $\mathfrak{q}(x+y) \le C(\mathfrak{q}(x)+\mathfrak{q}(y))$;
(3) $\mathfrak{q}(t^A x) = t\mathfrak{q}(x)$ for all $t \in \mathbb{R}_+$ and $x \in \mathbb{R}^d$.

It is not difficult to check, using (2.5), that there exist $\gamma, \delta > 0$ (depending on eigenvalues of A) such that for all $x \in \mathbb{R}^d$

$$\mathfrak{q}(x)^\gamma \lesssim |x| \lesssim \mathfrak{q}(x)^\delta, \qquad \text{if} \quad |x| \le 1,$$
$$\mathfrak{q}(x)^\delta \lesssim |x| \lesssim \mathfrak{q}(x)^\gamma, \qquad \text{if} \quad |x| \ge 1.$$

The size of the constant $\kappa > 1$ in Proposition 2.3 will determine the suitable scale in the decomposition of the oscillation semi–norm into the short and long parts, see (2.10) and the following definitions. The main point here is that in the long part, which measures oscillations over a lacunary subsequence of indices, we are going to approximate Radon means with the martingale sequence arising from the system of 'dyadic cubes' described in Proposition 2.3. In order for this approximation to be possible we need to know that the sequence $\{\kappa^n\}_{n\in\mathbb{Z}}$ of (essentially) diameters of these cubes has at least as high density as the lacunary sequence of scales in the long part of the oscillation semi–norm. It motivates one to define for a given $\kappa > 1$

$$\tau = \lfloor \kappa \rfloor + 1.$$

That leads to the lacunary sequence $\mathbb{D} = (\tau^n : n \in \mathbb{Z})$. Note that if $\kappa \in (1, 2)$, then $\tau = 2$, so in this case $\mathbb{D}$ is simply the set of classical dyadic numbers. We shall abuse the terminology by referring to $\mathbb{D}$ as the dyadic system.

We now find a sequence of Littlewood–Paley projections corresponding to the quasi–norm $\mathfrak{q}$. For this purpose, we take a smooth function $\phi_0 : [0, \infty) \to [0, \infty)$ such that $0 \le \phi_0 \le \mathbf{1}_{[1/\tau,\tau]}$ and its dilates $\phi_k(x) = \phi_0(\tau^k x)$ satisfying

$$\sum_{k\in\mathbb{Z}} \phi_k^2 = \mathbf{1}_{(0,\infty)}. \tag{2.6}$$

For each $k \in \mathbb{Z}$ we define the Littlewood–Paley operator $\tilde{S}_k$ by setting $\widehat{\tilde{S}_k f} = \psi_k \widehat{f}$, with $\psi_k(\xi) = \phi_k(\mathfrak{q}(\xi))$ being a smooth function on $\mathbb{R}^d$. Due to (2.6) we see that the family $(S_k \colon k \in \mathbb{Z})$, where $S_k = \tilde{S}_k^2$, is a resolution of the identity on $L^2(\mathbb{R}^d)$, i.e. the identity

$$\sum_{k\in\mathbb{Z}} S_k = \mathrm{Id} \tag{2.7}$$

holds in the strong operator topology on $L^2(\mathbb{R}^d)$. Moreover, [**14**, Theorem II.1.5] ensures that for every $p \in (1, \infty)$ the Littlewood–Paley inequality

$$\Big\| \Big(\sum_{k\in\mathbb{Z}} |S_k f|^2 \Big)^{1/2} \Big\|_{L^p(\mathbb{R}^d)} + \Big\| \Big(\sum_{k\in\mathbb{Z}} |\tilde{S}_k f|^2 \Big)^{1/2} \Big\|_{L^p(\mathbb{R}^d)} \lesssim \|f\|_{L^p(\mathbb{R}^d)} \tag{2.8}$$

holds for every $f \in L^p(\mathbb{R}^d)$.

Let $(\Phi_t : t \in \mathbb{R}_+)$ be a family of Schwartz functions such that

$$\Phi_t(x) = t^{-\operatorname{tr} A} \Phi(t^{-A} x),$$

where Φ is a nonnegative Schwartz function on $\mathbb{R}^d$ with integral one and t^{-A} is the inverse matrix to t^A. Note that on the frequency side $\widehat{\Phi_t}(\xi) = \widehat{\Phi}(t^A \xi)$, so the scaling is consistent with the scale of the Littlewood–Paley projections introduced before.

Our main result for bump functions is the following theorem.

THEOREM 2.9. *For every $p \in (1, \infty)$ there exists a constant $C_p > 0$ such that for every function $f \in L^p(\mathbb{R})$ we have*

$$\sup_{I \in \mathfrak{S}(\mathbb{R}_+)} \|O_I(\Phi_t * f : t \in \mathbb{R}_+)\|_{L^p(\mathbb{R}^d)} \le C_p \|f\|_{L^p(\mathbb{R}^d)}.$$

Before presenting the proof we need some preparatory results. First we shall suitably decompose the oscillatory semi-norm. It will allow us to separate the dyadic contribution in the oscillations. Recall that $\mathbb{D} = (\tau^n : n \in \mathbb{Z})$ and for a sequence $I = (I_k : k \in \mathbb{Z}) \in \mathfrak{S}(\mathbb{R}_+)$ let

$$\mathbb{A} = \{k \in \mathbb{Z} : [I_k, I_{k+1}) \cap \mathbb{D} \ne \emptyset\}. \tag{2.10}$$

Then define

$$O_{I,\text{long}}\left(\mathfrak{a}_t\colon t\in\mathbb{R}_+\right)=\Big(\sum_{k\in\mathbb{A}}\sup_{I_k\le t<I_{k+1}}|\mathfrak{a}_t-\mathfrak{a}_{I_k}|^2\Big)^{1/2},$$

$$O_{I,\text{short}}\left(\mathfrak{a}_t\colon t\in\mathbb{R}_+\right)=\Big(\sum_{k\in\mathbb{Z}\setminus\mathbb{A}}\sup_{I_k\le t<I_{k+1}}|\mathfrak{a}_t-\mathfrak{a}_{I_k}|^2\Big)^{1/2}$$

and observe that

$$O_I\left(\mathfrak{a}_t\colon t\in\mathbb{R}_+\right)\le O_{I,\text{long}}\left(\mathfrak{a}_t\colon t\in\mathbb{R}_+\right)+O_{I,\text{short}}\left(\mathfrak{a}_t\colon t\in\mathbb{R}_+\right). \tag{2.11}$$

Motivated by the developments in [**5**] we introduce the variational semi-norm of a function $(\mathfrak{a}_t : t\in\mathbb{R}_+)$ by setting

$$\begin{aligned}V_{2,m}\left(\mathfrak{a}_t\colon t\in\mathbb{R}_+\right)&:=V^2\big(\mathfrak{a}_t\colon t\in[\tau^m,\tau^{m+1})\big)\\&=\sup_{\substack{\tau^m\le t_1<\cdots<t_J<\tau^{m+1}\\ J\in\mathbb{Z}_+}}\Big(\sum_{k=1}^{J-1}\big|\mathfrak{a}_{t_{k+1}}-\mathfrak{a}_{t_k}\big|^2\Big)^{1/2},\qquad m\in\mathbb{Z},\end{aligned}$$

and the short 2-variation square function

$$\mathcal{S}_2\left(\mathfrak{a}_t\colon t\in\mathbb{R}_+\right)=\left(\sum_{m\in\mathbb{Z}}[V_{2,m}\left(\mathfrak{a}_t\colon t\in\mathbb{R}_+\right)]^2\right)^{1/2}.$$

Let $d:\mathbb{R}_+\to\mathbb{D}$ be given by

$$d(x)=\sup\{y\in\mathbb{D}: y\le x\}.$$

For $I=(I_k : k\in\mathbb{Z})\in\mathfrak{S}(\mathbb{R}_+)$ define

$$I_{\text{dyad}}=(d(I_k): k\in\mathbb{A}),$$

where $\mathbb{A}$ is as in (2.10). Note that due to the definition of $\mathbb{A}$, the sequence I_{dyad} is also increasing.

We shall need the following estimate, which corresponds to [**5**, Lemma 1.3].

LEMMA 2.12. *For any function $(\mathfrak{a}_t : t\in\mathbb{R}^+)$ and for any $I=(I_k : k\in\mathbb{N})\in\mathfrak{S}(\mathbb{R}_+)$ the following relations hold*

(1) $O_{I,\text{short}}\left(\mathfrak{a}_t\colon t\in\mathbb{R}_+\right)\le\mathcal{S}_2\left(\mathfrak{a}_t\colon t\in\mathbb{R}_+\right)$,
(2) $O_{I,\text{long}}\left(\mathfrak{a}_t\colon t\in\mathbb{R}_+\right)\le 3\mathcal{S}_2\left(\mathfrak{a}_t\colon t\in\mathbb{R}_+\right)+\sqrt{3}O_{I_{\text{dyad}}}\left(\mathfrak{a}_{\tau^n}\colon n\in\mathbb{Z}\right)$,
(3) $O_I\left(\mathfrak{a}_t\colon t\in\mathbb{R}_+\right)\le 4\mathcal{S}_2\left(\mathfrak{a}_t\colon t\in\mathbb{R}_+\right)+\sqrt{3}O_{I_{\text{dyad}}}\left(\mathfrak{a}_{\tau^n}\colon n\in\mathbb{Z}\right)$.

PROOF. Clearly (3) follows from (1) and (2) by (2.11).

Part (1) is straightforward, since if $k\in\mathbb{Z}/\mathbb{A}$, then the corresponding interval $[I_k,I_{k+1})$ is contained inside a certain dyadic interval.

It remains to prove (2). We have

$$\begin{aligned}&\left(O_{I,\text{long}}\left(\mathfrak{a}_t\colon t\in\mathbb{R}_+\right)\right)^2=\sum_{k\in\mathbb{A}}\sup_{I_k\le t<I_{k+1}}|\mathfrak{a}_t-\mathfrak{a}_{I_k}|^2\\&\le 3\left(\sum_{k\in\mathbb{A}}\sup_{I_k\le t<I_{k+1}}|\mathfrak{a}_t-\mathfrak{a}_{d(t)}|^2+\sum_{k\in\mathbb{A}}\sup_{I_k\le t<I_{k+1}}|\mathfrak{a}_{d(t)}-\mathfrak{a}_{d(I_k)}|^2+\sum_{k\in\mathbb{A}}|\mathfrak{a}_{d(I_k)}-\mathfrak{a}_{I_k}|^2\right)\\&\le 3\Big(2\left(\mathcal{S}_2\left(\mathfrak{a}_t\colon t\in\mathbb{R}_+\right)\right)^2+\left(O_{I_{\text{dyad}}}\left(\mathfrak{a}_{\tau^n}\colon n\in\mathbb{Z}\right)\right)^2+\left(\mathcal{S}_2\left(\mathfrak{a}_t\colon t\in\mathbb{R}_+\right)\right)^2\Big).\end{aligned}$$

That concludes the proof. □

In view of Lemma 2.12 the proof of Theorem 2.9 reduces to showing two estimates

$$\sup_{I\in\mathfrak{S}(\mathbb{R}_+)}\Big\|O_{I_{\mathrm{dyad}}}(\Phi_{\tau^n}*f:n\in\mathbb{Z})\Big\|_{L^p(\mathbb{R}^d)}\leq C_p\|f\|_{L^p(\mathbb{R}^d)}$$

and

$$\|\mathcal{S}_2(\Phi_t*f:t\in\mathbb{R}_+)\|_{L^p(\mathbb{R}^d)}\leq C_p\|f\|_{L^p(\mathbb{R}^d)}.$$

The first inequality is treated in Proposition 2.13. Before formulating this result we need to introduce a martingale associated with the family of nonisotropic dilations $(t^A : t\in\mathbb{R}_+)$. To this end observe that Proposition 2.3 assures the existence of a family of "dyadic cubes" $\mathcal{D}=\bigcup_{k\in\mathbb{Z}}\mathcal{D}_k$ satisfying conditions stated in that proposition with the underlying metric-measure space $(\mathbb{R}^d,\mathcal{B}(\mathbb{R}^d),\rho,dx)$, where ρ is a quasi–metric induced by a t^A-homogeneous quasi–norm $\mathfrak{q}$, see the beginning of this subsection. Next, given a function f consider the martingale consisting of conditional expectations $(\mathbb{E}_k f : k\in\mathbb{Z})$ adapted to the filtration generated by families $(\mathcal{D}_k : k\in\mathbb{Z})$. In the proof of the next proposition we shall use this martingale to approximate smooth Radon means on dyadic scales.

PROPOSITION 2.13. *For every $p\in(1,\infty)$ there exists a constant $C_p>0$ such that for every function $f\in L^p(\mathbb{R}^d)$ we have*

$$\sup_{I\in\mathfrak{S}(\mathbb{R}_+)}\Big\|O_{I_{\mathrm{dyad}}}(\Phi_{\tau^n}*f:n\in\mathbb{Z})\Big\|_{L^p(\mathbb{R}^d)}\leq C_p\|f\|_{L^p(\mathbb{R}^d)}.$$

PROOF. For every $n\in\mathbb{Z}$ let $l(n)$ be the integer such that $\tau^{l(n)}\leq\kappa^n<\tau^{l(n)+1}$, with $\kappa>1$ being the constant from Proposition 2.3. Note that since $\tau>\kappa$, thus the map $l:\mathbb{Z}\to\mathbb{Z}$ is surjective. Therefore, modifying a sequence $\{\Phi_{\tau^n}*f\}_{n\in\mathbb{Z}}$ by possibly duplicating some of its elements (which does not affect the oscillation semi–norm) it suffices to estimate the L^p norm of

$$O_{I_{\mathrm{dyad}}}(\Phi_{\tau^{l(n)}}*f:n\in\mathbb{Z}).$$

Now we can estimate pointwise

$$\begin{aligned}O_{I_{\mathrm{dyad}}}(\Phi_{\tau^{l(n)}}*f:n\in\mathbb{Z})&\leq O_{I_{\mathrm{dyad}}}(\mathbb{E}_nf:n\in\mathbb{Z})+O_{I_{\mathrm{dyad}}}(\Phi_{\tau^{l(n)}}*f-\mathbb{E}_nf:n\in\mathbb{Z})\\&\leq O_{I_{\mathrm{dyad}}}(\mathbb{E}_nf:n\in\mathbb{Z})+\mathfrak{D}f,\end{aligned}$$

where for a function $f:\mathbb{R}^d\to\mathbb{R}$ we let $\mathfrak{D}f(x)=\big(\sum_{n\in\mathbb{Z}}|\Phi_{\tau^{l(n)}}*f(x)-\mathbb{E}_nf(x)|^2\big)^{1/2}$. The first term on the right-hand side above is bounded by $\|f\|_{L^p(\mathbb{R}^d)}$ for any $p\in(1,\infty)$ due to Theorem 2.4. It remains to show the estimate

$$\|\mathfrak{D}f\|_{L^p(\mathbb{R}^d)}\lesssim\|f\|_{L^p(\mathbb{R}^d)},\quad 1<p<\infty.$$

This however was done in [**5**] for $\tau=2$ and the proof for general $\tau\geq2$ does not require any substantial modifications. □

To finish the proof of Theorem 2.9 it remains to deal with short variations. To this end we need the following key result, which appeared in [**10**] in the case $\tau=2$.

THEOREM 2.14. *Let $\tau\geq2$ be an integer. Assume that $(X,\mathcal{B},\mathfrak{m})$ is a σ-finite measure space endowed with a sequence of linear operators $(\mathbb{S}_j)_{j\in\mathbb{Z}}$ satisfying* (2.7) *and* (2.8). *Let* $(A_t)_{t\in\mathbb{U}}$ *be a family of positive convolution operators such that the square function estimate*

$$\Big\|\Big(\sum_{k\in\mathbb{Z}}\sum_{i=0}^{\tau^l-1}|(A_{\tau^k+\tau^{k-l}(i+1)}-A_{\tau^k+\tau^{k-l}i})\mathbb{S}_{j+k}f|^2\Big)^{1/2}\Big\|_{L^2}\le\tau^{-\frac{l}{2}}a_{j,l}\|f\|_{L^2} \tag{2.15}$$

holds for all $j\in\mathbb{Z}$ *and* $l\in\mathbb{N}$ *with some numbers* $a_{j,l}\ge 0$ *such that for every* $0<\varepsilon<\rho$ *we have*

$$\sum_{l\ge 0}\sum_{j\in\mathbb{Z}}\tau^{-\varepsilon l}a_{j,l}^{\rho}<\infty.$$

Let $q_0\in[1,2)$ *and* $\alpha\in[0,1]$ *be such that* $\alpha q_0\le 1$. *Suppose that we have the operator norm Hölder type condition*

$$\|A_{t+h}-A_t\|_{L^{q_0}\to L^{q_0}}\lesssim\Big(\frac{h}{t}\Big)^{\alpha},\qquad t\in\mathbb{R}_+\ \text{and}\ h\in(0,1] \tag{2.16}$$

and the exponent q_1 *satisfies*

$$q_0\le 2-\frac{2-q_0}{2-\alpha q_0}<q_1\le 2.$$

Then, if

$$\|\sup_{n\in\mathbb{Z}}|A_{\tau^n}f|\|_{L^{q_1}}\lesssim\|f\|_{L^{q_1}} \tag{2.17}$$

holds for all $f\in L^{q_1}(X)$, *then the estimate*

$$\Big\|\mathcal{S}_2(A_tf:t\in\mathbb{R}_+)\Big\|_{L^{q_1}}\lesssim\|f\|_{L^{q_1}}, \tag{2.18}$$

holds with the implicit constant which is a constant multiple of

$$\mathbf{a}:=\sum_{l\ge 0}\sum_{j\in\mathbb{Z}}\tau^{-\left(\alpha\frac{2-q_1}{2-q_0}\frac{q_0}{2}+\frac{1}{2}\frac{q_1-q_0}{2-q_0}-\frac{2-q_1}{2-q_0}\frac{1}{2}\right)l}a_{j,l}^{\frac{q_1-q_0}{2-q_0}}<\infty.$$

Moreover, the implication from (2.17) *to* (2.18) *also holds with* q_1 *replaced by* q_1'.

PROOF. In view of [**10**, Section 2] it suffices to prove that the following numerical identity

$$\big(V^2(\mathfrak{a}_j:j\in[0,\tau^s))\big)\le\sqrt{2(\tau-1)}\sum_{i=0}^{s}\Big(\sum_{j=0}^{\tau^{s-i}-1}|\mathfrak{a}_{(j+1)\tau^i}-\mathfrak{a}_{j\tau^i}|^2\Big)^{1/2},\qquad s\in\mathbb{N}, \tag{2.19}$$

holds for any $\tau=2,3,\ldots$

Fix $0\le m<n\le\tau^s$. We will first prove that $[m,n)$ is a finite disjoint union of intervals from collections $\mathcal{I}_i=\{[j\tau^i,(j+1)\tau^i):0\le j\le\tau^{s-i}-1\}$, $i=0,1,\ldots,s$, such that for each i there are at most $2(\tau-1)$ intervals of length τ^i in this decomposition. To prove it we begin with setting $m_0=m$. Then for a chosen m_p we select m_{p+1} in such a way that $[m_p,m_{p+1})$ is the largest interval of length τ^i for some $i\in\mathbb{N}$ contained inside $[m_p,n)$. We note that by maximality there can occur at most $\tau-1$ consecutive intervals of the same length in this construction.

Now we consider two cases. If the sequence of lengths of the constructed intervals is nondecreasing then intervals of a given length can appear at most $\tau-1$ and we are done.

In the opposite case there exists p such that $m_{p+1}-m_p > m_{p+2}-m_{p+1}$. Notice that then the sequence $\{m_{p+k+1} - m_{p+k}\}_{k\geq 1}$ is nonincreasing. Therefore in this case a given length can appear at most $2(\tau - 1)$ times.

The rest of the proof follows the steps of the proof of [**9**, Lemma 2.1]. Let $k_0 < k_1 < \ldots < k_J \leq 2^s$ be any increasing sequence. For each $j \in \{0, \ldots, J-1\}$ we apply the above construction getting a decomposition

$$[k_j, k_{j+1}) = \bigcup_{p=0}^{P_j} [u_p^j, u_{p+1}^j)$$

for some $P_j \geq 1$, where intervals $[u_p^j, u_{p+1}^j)$ belong to families $\mathcal{I}_i$, $i = 0, 1, \ldots, s$, and for each i there are at most $2(\tau - 1)$ intervals from $\mathcal{I}_i$. Then

$$|a_{k_{j+1}} - a_{k_j}| \leq \sum_{p=0}^{P_j} |a_{u_{p+1}^j} - a_{u_p^j}| = \sum_{i=0}^{s} \sum_{p:\, [u_p^j, u_{p+1}^j) \in \mathcal{I}_i} |a_{u_{p+1}^j} - a_{u_p^j}|.$$

Hence, by Minkowski's inequality

$$\begin{aligned}\Big(\sum_{j=0}^{J-1} |a_{k_{j+1}} - a_{k_j}|^2\Big)^{1/2} &\leq \Big(\sum_{j=0}^{J-1} \Big(\sum_{i=0}^{s} \sum_{p:\, [u_p^j, u_{p+1}^j) \in \mathcal{I}_i} |a_{u_p^j} - a_{u_{p+1}^j}|\Big)^2\Big)^{1/2} \\ &\leq \sum_{i=0}^{s} \Big(\sum_{j=0}^{J-1} \Big(\sum_{p:\, [u_p^j, u_{p+1}^j) \in \mathcal{I}_i} |a_{u_p^j} - a_{u_{p+1}^j}|\Big)^2\Big)^{1/2}.\end{aligned}$$

Since for a given $i \in \{0, 1, \ldots, 2^s\}$ and $j \in \{0, 1, \ldots, J-1\}$ the inner sums contain at most $2(\tau - 1)$ elements, we obtain by Cauchy–Schwarz inequality

$$\Big(\sum_{j=0}^{J-1} |a_{k_{j+1}} - a_{k_j}|^2\Big)^{1/2} \leq \sqrt{2(\tau - 1)} \sum_{i=0}^{s} \Big(\sum_{j=0}^{J-1} \sum_{p:\, [u_p^j, u_{p+1}^j) \in \mathcal{I}_i} |a_{u_p^j} - a_{u_{p+1}^j}|^2\Big)^{1/2}$$

which is bounded by the right-hand side of (2.19). □

Now we are ready to state and prove the main estimate for short variations.

PROPOSITION 2.20. *For every $p \in (1, \infty)$ there exists a constant $C_p > 0$ such that for every function $f \in L^p(\mathbb{R}^d)$ we have*

$$\|\mathcal{S}_2(\Phi_t * f : t \in \mathbb{R}_+)\|_{L^p(\mathbb{R}^d)} \leq C_p \|f\|_{L^p(\mathbb{R}^d)}. \tag{2.21}$$

PROOF. It suffices to verify that $A_t f = \Phi_t * f$ satisfies the assumptions of Theorem 2.14 with $\mathbb{S}_k = \tilde{S}_k$, and $a_{j,l} = \min\{1, \tau^l \tau^{-\epsilon|j|}\}$.

We begin with noting that (2.16) holds for all $q_0 \geq 1$ and $\alpha = 1$. Next, we have

$$|\widehat{\Phi}_t(\xi) - \widehat{\Phi}_t(0)| \lesssim (t\mathfrak{q}(\xi))^{\epsilon}, \quad t\mathfrak{q}(\xi) \leq 1, \tag{2.22}$$

$$|\widehat{\Phi}_t(\xi)| \lesssim (t\mathfrak{q}(\xi))^{-\epsilon}, \quad t\mathfrak{q}(\xi) \geq 1. \tag{2.23}$$

which implies

$$\sup_{\mathfrak{q}(\xi) \simeq \tau^{-j-k}} |\widehat{\Phi}_{\tau^k + \tau^{k-l}(i+1)}(\xi) - \widehat{\Phi}_{\tau^k + \tau^{k-l} i}(\xi)| \lesssim \tau^{-\epsilon|j|}.$$

Moreover, since (2.16) holds in particular for $q_0 = 2$ and $\alpha = 1$, thus Plancherel's identity gives

$$\sup_{\xi\in\mathbb{R}^d} |\widehat{\Phi}_{\tau^k+\tau^{k-l}(i+1)}(\xi) - \widehat{\Phi}_{\tau^k+\tau^{k-l}i}(\xi)| \lesssim \tau^{-l}.$$

Combining the above relations leads to

$$\sup_{\mathfrak{q}(\xi)\simeq\tau^{-j-k}} |\widehat{\Phi}_{\tau^k+\tau^{k-l}(i+1)}(\xi) - \widehat{\Phi}_{\tau^k+\tau^{k-l}i}(\xi)| \lesssim \min\{\tau^{-l}, \tau^{-\epsilon|j|}\}.$$

Now using Plancherel's theorem, the above bound and then (2.8) we get

$$\begin{aligned}
&\Big\|\Big(\sum_{k\in\mathbb{Z}}\sum_{i=0}^{\tau^l-1}|(A_{\tau^k+\tau^{k-l}(i+1)} - A_{\tau^k+\tau^{k-l}i})\tilde{S}_{j+k}f|^2\Big)^{1/2}\Big\|_{L^2(\mathbb{R}^d)}\\
&= \Big\|\Big(\sum_{k\in\mathbb{Z}}\sum_{i=0}^{\tau^l-1}|(\widehat{\Phi}_{\tau^k+\tau^{k-l}(i+1)} - \widehat{\Phi}_{\tau^k+\tau^{k-l}i})\widehat{\tilde{S}_{j+k}f}|^2\Big)^{1/2}\Big\|_{L^2(\mathbb{R}^d)}\\
&\leq \tau^{l/2}\min\{\tau^{-l}, \tau^{-\epsilon|j|}\}\Big(\sum_{k\in\mathbb{Z}}\|\widehat{\tilde{S}_{j+k}f}\|^2_{L^2(\mathbb{R}^d)}\Big)^{1/2}\\
&\lesssim \tau^{-l/2}\min\{1, \tau^l\,\tau^{-\epsilon|j|}\}\|f\|_{L^2(\mathbb{R}^d)},
\end{aligned}$$

which gives (2.15) with $a_{j,l} = \min\{1, \tau^l\,\tau^{-\epsilon|j|}\}$.

Finally, for any $p \in (1, \infty)$ we have

$$\|\sup_{n\in\mathbb{N}} |\Phi_{\tau^n} * f|\|_{L^p(\mathbb{R}^d)} \lesssim \|f\|_{L^p(\mathbb{R}^d)},$$

by [**3**, Theorem A] (it follows also from Proposition 2.13, since oscillation estimates imply maximal estimates), so (2.17) holds with any $q_1 > 1$. □

2.0.3. *Oscillation inequalities for Radon means.* We conclude with showing how Theorem 2.9 implies the main result of the paper, Theorem 1.3.

Proof of Theorem 1.3. In view of (2.11) and Lemma 2.12 to prove Theorem 1.3 it suffices to show

$$\sup_{I\in\mathfrak{S}(\mathbb{R}_+)}\Big\|O_{I_{\mathrm{dyad}}}(M^P_{\tau^n}f : n \in \mathbb{Z})\Big\|_{L^p(\mathbb{R}^d)} \leq C_p\|f\|_{L^p(\mathbb{R}^d)} \tag{2.24}$$

and

$$\|\mathcal{S}_2(M^P_t f : t \in \mathbb{R}_+)\|_{L^p(\mathbb{R}^d)} \leq C_p\|f\|_{L^p(\mathbb{R}^d)}. \tag{2.25}$$

We begin with observing that for any $f : \mathbb{R}^d \to \mathbb{R}$ and $t \in \mathbb{R}_+$ we have

$$M^P_t f(x) = \mu_t * f(x), \qquad x \in \mathbb{R}^d,$$

where the measures $(\mu_t : t \in \mathbb{R}^+)$ on $\mathbb{R}^d$ are given by

$$\widehat{\mu}_t(\xi) = \frac{1}{|\Omega_t|}\int_{\Omega_t}\exp\big(-2\pi i\langle\xi, P(u)\rangle\big)\,du, \qquad \xi \in \mathbb{R}^d.$$

Note that all the measures μ_t are dilates of the measure $\mu := \mu_1$. By the stationary phase estimate, see [**10**, Proposition B.2], there exists $\epsilon > 0$ such that

$$\begin{aligned}
|\widehat{\mu}_t(\xi) - \widehat{\mu}_t(0)| &\lesssim (t\mathfrak{q}(\xi))^{\epsilon}, \quad t\mathfrak{q}(\xi) \leq 1,\\
|\widehat{\mu}_t(\xi)| &\lesssim (t\mathfrak{q}(\xi))^{-\epsilon}, \quad t\mathfrak{q}(\xi) \geq 1.
\end{aligned}$$

Furthermore, for any $p \in [1,\infty)$ the following smoothness condition holds

$$\|\mu_{t+\epsilon} - \mu_t\|_{L^p(\mathbb{R}^d)\to L^p(\mathbb{R}^d)} \lesssim \frac{\epsilon}{t},$$

uniformly in $0 < 2\epsilon < t$ and $t \in \mathbb{R}_+$. Consequently, a repetition of the argument from the proof of Proposition 2.20 gives (2.21).

It remains to show (2.24). To this end let $\phi \in \mathcal{S}(\mathbb{R}^d)$ be a nonnegative, compactly supported function with $\int_{\mathbb{R}^d} \phi(x)\,dx = 1$. We decompose

$$\begin{aligned}\|O_{I_{\text{dyad}}}(M^P_{\tau^n} f : n \in \mathbb{Z})\|_{L^p(\mathbb{R}^d)} \\ \le \Big\|O_{I_{\text{dyad}}}((\mu*\phi)_{\tau^n} * f : n\in\mathbb{Z})\Big\|_{L^p(\mathbb{R}^d)} + \Big\|\Big(\sum_{n\in\mathbb{Z}} |(\mu - \mu*\phi)_{\tau^n} * f|^2\Big)^{1/2}\Big\|_{L^p(\mathbb{R}^d)}.\end{aligned}$$

Since $\mu*\phi \in \mathcal{S}(\mathbb{R}^d)$ and $\int \mu*\phi = 1$, the first term on the right-hand side satisfies the desired estimate by Propostion 2.20 (applied with $\Phi_t(x) = t^{-\operatorname{tr} A}\mu*\phi(t^{-A}x)$). The relevant bound for the second term is an immediate consequence of [**3**, Theorem B], in view of (2.22) and (2.23), the condition $\int(\mu - \mu*\phi) = 0$ and maximal estimates

$$\|\sup_n |\mu_{\tau^n} * f|\|_{L^p(\mathbb{R}^d)} + \|\sup_n |(\mu*\phi)_{\tau^n} * f|\|_{L^p(\mathbb{R}^d)} \lesssim \|f\|_{L^p(\mathbb{R}^d)},$$

which follow from [**3**, Theorem A]. □

Acknowledgments

The author would like to kindly thank the anonymous referee for reading carefully the manuscript and sharing useful remarks that led to the improvement of the article.

References

[1] Jean Bourgain, *Pointwise ergodic theorems for arithmetic sets*, Inst. Hautes Études Sci. Publ. Math. **69** (1989), 5–45. With an appendix by the author, Harry Furstenberg, Yitzhak Katznelson and Donald S. Ornstein. MR1019960

[2] Michael Christ, *A $T(b)$ theorem with remarks on analytic capacity and the Cauchy integral*, Colloq. Math. **60/61** (1990), no. 2, 601–628, DOI 10.4064/cm-60-61-2-601-628. MR1096400

[3] Javier Duoandikoetxea and José L. Rubio de Francia, *Maximal and singular integral operators via Fourier transform estimates*, Invent. Math. **84** (1986), no. 3, 541–561, DOI 10.1007/BF01388746. MR837527

[4] Roger L. Jones, Robert Kaufman, Joseph M. Rosenblatt, and Máté Wierdl, *Oscillation in ergodic theory*, Ergodic Theory Dynam. Systems **18** (1998), no. 4, 889–935, DOI 10.1017/S0143385798108349. MR1645330

[5] Roger L. Jones, Andreas Seeger, and James Wright, *Strong variational and jump inequalities in harmonic analysis*, Trans. Amer. Math. Soc. **360** (2008), no. 12, 6711–6742, DOI 10.1090/S0002-9947-08-04538-8. MR2434308

[6] Ben Krause and Pavel Zorin-Kranich, *Weighted and vector-valued variational estimates for ergodic averages*, Ergodic Theory Dynam. Systems **38** (2018), no. 1, 244–256, DOI 10.1017/etds.2016.27. MR3742545

[7] Ben Krause. *On higher-dimensional oscillation in ergodic theory*, Preprint, arXiv:1309.2336, (2013).

[8] Dominique Lépingle, *La variation d'ordre p des semi-martingales* (French), Z. Wahrscheinlichkeitstheorie und Verw. Gebiete **36** (1976), no. 4, 295–316, DOI 10.1007/BF00532696. MR420837

[9] Mariusz Mirek, Elias M. Stein, and Bartosz Trojan, *$\ell^p(\mathbb{Z}^d)$-estimates for discrete operators of Radon type: Variational estimates*, Invent. Math. **209** (2017), no. 3, 665–748, DOI 10.1007/s00222-017-0718-4. MR3681393

[10] Mariusz Mirek, Elias M. Stein, and Pavel Zorin-Kranich, *A bootstrapping approach to jump inequalities and their applications*, Anal. PDE **13** (2020), no. 2, 527–558, DOI 10.2140/apde.2020.13.527. MR4078235

[11] Mariusz Mirek, Elias M. Stein, and Pavel Zorin-Kranich, *Jump inequalities for translation-invariant operators of Radon type on* $\mathbb{Z}^d$, Adv. Math. **365** (2020), 107065, 57, DOI 10.1016/j.aim.2020.107065. MR4067360

[12] Mariusz Mirek, Tomasz Z. Szarek, and James Wright, *Oscillation inequalities in ergodic theory and analysis: one-parameter and multi-parameter perspectives*, Rev. Mat. Iberoam. **38** (2022), no. 7, 2249–2284, DOI 10.4171/rmi/1383. MR4525137

[13] Mariusz Mirek and Bartosz Trojan, *Discrete maximal functions in higher dimensions and applications to ergodic theory*, Amer. J. Math. **138** (2016), no. 6, 1495–1532, DOI 10.1353/ajm.2016.0045. MR3595493

[14] N. M. Rivière, *Singular integrals and multiplier operators*, Ark. Mat. **9** (1971), 243–278, DOI 10.1007/BF02383650. MR440268

[15] José L. Rubio de Francia, *Maximal functions and Fourier transforms*, Duke Math. J. **53** (1986), no. 2, 395–404, DOI 10.1215/S0012-7094-86-05324-X. MR850542

[16] Pavel Zorin-Kranich, *Variational truncations of singular integrals on spaces of homogeneous type*, Preprint, `arXiv:2009.04541`, (2020).

Department of Mathematics, Indiana University, 831 East 3rd St., Bloomington, Indiana 47405; and Faculty of Pure and Applied Mathematics, Wrocław University of Science and Technology, Wyb. Wyspiańskiego 27, 50–370 Wrocław, Poland

Email address: `balango@iu.edu`

Contemporary Mathematics
Volume **792**, 2024
https://doi.org/10.1090/conm/792/15899

Refinements of the 2-dimensional Strichartz estimate on the maximum wave packet

Hong Wang and Lingfu Zhang

ABSTRACT. The Strichartz estimates for Schrödinger equations can be improved when the data is spread out in either physical or frequency space. In this paper we give refinements of the 2-dimensional homogeneous Strichartz estimate on the maximum size of a single wave packet. Different approaches are used in the proofs, including arithmetic approaches, polynomial partitioning, and the l^2 Decoupling Theorem, for different cases. We also give examples to show that the refinements we obtain cannot be further improved when $2 \leq p \leq 4$ and $p = 6$.

1. Introduction

We start with the *initial value problem* (IVP) of the $n+1$ dimensional *linear Schördinger equation*:

$$\begin{cases} \mathbf{i}u_t - \Delta u = 0, \\ u(x,0) = u_0(x) \end{cases} \tag{1.1}$$

where $u : \mathbb{R}^n \times \mathbb{R} \to \mathbb{C}$ is the unknown, $u_0 : \mathbb{R}^n \to \mathbb{C}$ is the given initial data, and $\Delta = \sum_{i=1}^n \frac{\partial^2}{\partial x_i^2}$ is the Laplacian operator. Assuming that u_0 is Schwartz, denote $f = \widehat{u}$ to be the Fourier transform. As a classical result, the solution to (1.1) is given by

$$u(x,t) = \int_{\mathbb{R}^n} \mathrm{e}^{\mathbf{i}(|\omega|^2 t + \omega \cdot x)} f(\omega)\, \mathrm{d}\omega, \tag{1.2}$$

see, e.g. [**Eva10**, Section 4.3] [**Tao06**, Section 2.2], in slightly different notations. The well-known Strichartz estimate bounds the ($n+1$ dimensional) L^p norm of the solution u by the (n-dimensional) L^2 norm of f.

THEOREM 1.1. [**Str77**, Theorem 1] *For* $p = \frac{2(n+2)}{n}$, *there is*

$$\|u\|_{L^p(\mathbb{R}^n \times \mathbb{R})} \lesssim \|f\|_{L^2(\mathbb{R}^n)}, \tag{1.3}$$

and the constant behind $\lesssim$ *only relies on* p *and* n.

2020 *Mathematics Subject Classification.* Primary 42B10, 42B20; Secondary 42B37.

Key words and phrases. Strichartz estimate, Schrödinger equation, wave packet, polynomial partitioning, decoupling.

The first author was supported in part by NSF Grant #2141426.

The second author was supported in part by the Miller Institute for Basic Research in Science at UC Berkeley.

This result can also be derived by the Tomas-Stein theorem in the restriction problem (see e.g., [**Ste86**] [**Tom75**], and see [**Tao04**] for a survey of the restriction problem). Let $E : L^2(\mathbb{R}^n) \to L^p(\mathbb{R}^{n+1})$ denote the operator sending f to u, and it is also referred to as the extension operator in some literature.

In this text we focus on dimension $n = 1$ (thus 2 time-space dimensions), where $p = 6$ in Theorem 1.1. As Fourier transform is unitary, and the space L^2 norm of the solution to (1.1) is conserved, there is

$$\|Ef\|_{L^2(B_R)} \leq \|Ef\|_{L^2_x L^2_t([-R,R])} = (2R)^{\frac{1}{2}}\|f\|_{L^2}, \tag{1.4}$$

where B_R is the ball centered at the origin with radius R. Applying Hölder's inequality to (1.3) and (1.4), there is

$$\|Ef\|_{L^p(B_R)} \lesssim R^{\frac{3}{2p}-\frac{1}{4}}\|f\|_{L^2}, \tag{1.5}$$

for any $2 \leq p \leq 6$ and $R > 0$.

The estimate (1.5) is sharp for all $2 \leq p \leq 6$, but if we make additional assumptions about the function f, then we can prove better estimates. Specifically, we break f into wave packets

$$f = \sum_{\theta,v} f_{\theta,v}, \tag{1.6}$$

where each $f_{\theta,v}$ is supported in an interval of length $O(R^{-\frac{1}{2}})$, and its Fourier transform is essentially supported in an interval of length $O(R^{\frac{1}{2}})$. In $\mathbb{R}^2$ each $Ef_{\theta,v}$ is essentially supported in a $R^{\frac{1}{2}} \times R$ tube inside B_R. We denote

$$S = \frac{\max_{\theta,v}\|f_{\theta,v}\|_{L^2}}{\|f\|_{L^2}}, \tag{1.7}$$

which measures the size of a single wave packet compared with the total size. Then key results of this paper, which are special cases for Theorem 3.1, are as following.

THEOREM 1.2 (Key Points). *For arbitrarily small $\epsilon > 0$, there is*

$$\|Ef\|_{L^4(B_R)} \lesssim_\epsilon R^{\frac{1}{8}+\epsilon}S^{\frac{1}{4}-\epsilon}\|f\|_{L^2}, \tag{1.8}$$

$$\|Ef\|_{L^5(B_R)} \lesssim_\epsilon R^{\frac{1}{20}+\epsilon}S^{\frac{1}{5}-\epsilon}\|f\|_{L^2}, \tag{1.9}$$

$$\|Ef\|_{L^6(B_R)} \lesssim_\epsilon R^{\frac{1}{6}+\epsilon}S^{\frac{2}{3}-\epsilon}\|f\|_{L^2}. \tag{1.10}$$

Compare with (1.5): for $p = 4, 5$, we have an extra power of S in the right hand side of (1.8) and (1.9), respectively; for $p = 6$, (1.10) is a refinement only when $S \lesssim R^{-\frac{1}{4}}$.

Note that essentially there is $R^{-\frac{1}{2}} \lesssim S \lesssim 1$. Using Hölder's inequality, we can obtain refinements of (1.5) for all $2 < p \leq 6$. We also give examples, which imply that our results for $2 < p \leq 4$ cannot be further improved, and for $5 \leq p \leq 6$ they can not be improved either if we assume that $S \lesssim R^{-\frac{1}{4}}$.

The motivation of these refinements is the observation that (1.5) is not sharp when f is spread out in either physical or frequency space. To be more precise, let's look at two examples. First consider $f_0 : \mathbb{R} \to \mathbb{C}$, which is a smooth bump function supported in $[-2, 2]$, and equals 1 in $[-1, 1]$. Through a typical argument of stationary phase, we can bound $|Ef_0(x,t)|$ by $|t|^{-\frac{1}{2}}$ when $|t| \gtrsim |x|$, and $|x|^{-1}$ when $|x| \lesssim |t|$. Then there is

$$\|Ef_0\|_{L^p(B_R)} \lesssim \begin{cases} R^{\frac{2}{p}-\frac{1}{2}} \sim R^{\frac{2}{p}-\frac{1}{2}}\|f_0\|_{L^2}, & 2 \le p \le 4 \\ 1 \sim \|f_0\|_{L^2}, & p > 4, \end{cases} \tag{1.11}$$

which is strictly better than (1.5) for $2 < p < 6$. We also have a sharp example for (1.5): let f_1 be a smooth bump function supported in $\left[-2R^{-\frac{1}{2}}, 2R^{-\frac{1}{2}}\right]$, and equals 1 in $\left[-R^{-\frac{1}{2}}, R^{-\frac{1}{2}}\right]$. When $|x| < R^{\frac{1}{2}}$ and $|t| < R$ there is $|Ef_1(x,t)| \sim 1$. This implies that

$$\|Ef_1\|_{L^p(B_R)} \gtrsim R^{\frac{3}{2p}-\frac{1}{2}} \sim R^{\frac{3}{2p}-\frac{1}{4}}\|f_1\|_{L^2}, \quad 2 \le p \le 6, \tag{1.12}$$

which saturates the estimate (1.5) for all $2 \le p \le 6$.

When comparing these two examples, we note that in a wave packet decomposition (of size R), f_0 has $\sim R^{\frac{1}{2}}$ nonzero wave packets of (almost) equal size, while f_1 has ~ 1 nonzero wave packet. Then for f_0, there is $S \sim R^{-\frac{1}{4}}$; and for f_1, there is $S \sim 1$. This motivates us to refine (1.5) when S is small, by inserting a power of S to its right hand side.

For the proofs, we use different approaches for different cases of this problem. For $p = 4$, which is an even integer, we use arithmetic approaches to fully exploit the structure of the wave packet decomposition. For $p = 5$ we adapt polynomial partitioning, a method from incidental geometry and introduced by Larry Guth to the restriction estimates [**Gut16a**] [**Gut16b**]. For $p = 6$, we apply the recent proved l^2 Decoupling Theorem [**BD15**]. Let us point out that the methods of polynomial partitioning and l^2 Decoupling Theorem capture different aspects in the refinements: the polynomial method deals with the intersecting of "tubes" in B_R, while l^2 Decoupling Theorem deals with cancellations in frequencies.

At the end we make a conjecture about the sharp estimates, for the remaining cases. We expect that a combination of the approaches used in this text will eventually settle this problem.

2. Problem setup

2.1. Notations and standard results. In this section we introduce notations and some some basic results under this context. These results will also be widely referred to in the proofs in subsequent sections.

There is a standard result from the method of stationary phase, and we state it here for future reference. The proof of this can be easily found in most harmonic analysis textbooks (see, e.g. [**SM93**, Chapter VIII] [**Wol03**, Chapter 6]).

THEOREM 2.1 (Principle of nonstationary phase). *Let both $\phi : \mathbb{R}^n \to \mathbb{R}$ and $\varphi : \mathbb{R}^n \to \mathbb{C}$ be smooth, and $|\nabla\phi|$ is bounded away from zero in the support of φ. Then for any positive integer N, we have*

$$\left|\int_{\mathbb{R}^n} e^{\mathrm{i}\lambda\phi(x)}\varphi(x)\,\mathrm{d}x\right| \lesssim_N \lambda^{-N}. \tag{2.1}$$

In other words, we say that the integral fast decays as λ grows.

We formally give the definition of the extension operator E. By rescaling it suffices to consider the case where f is supported in $[-1,1]$.

DEFINITION 2.2. For any smooth function $f : [-1,1] \to \mathbb{C}$, let

$$Ef(x,t) := \int_{[-1,1]} e^{\mathrm{i}(\omega^2 t + \omega x)} f(\omega)\, \mathrm{d}\omega. \tag{2.2}$$

We also identify E with the inverse Fourier transform in $\mathbb{R}^2$, in the following sense:

DEFINITION 2.3. Denote P to be the *truncated parabola* in $\mathbb{R}^2$: $P := \{(\omega, \omega^2) : \omega \in [-1,1]\}$. Let $\mathrm{d}\sigma$ be the distribution supported on P, such that its projection to the first coordinate is the uniform measure on $[-1,1]$. For each function $f : [-1,1] \to \mathbb{C}$, denote $f^* : P \to \mathbb{C}$ to be the function with $f^*(\omega, \omega^2) = f(\omega)$. By direct computation, the $\mathbb{R}^2$ Fourier transform of Ef is $\widehat{Ef} = f^*\, \mathrm{d}\sigma$.

We will use a function to "cut out" the L^p norm inside B_R.

DEFINITION 2.4. Let $\eta : \mathbb{R}^2 \to \mathbb{C}$ be a Schwartz function, such that $\widehat{\eta}$ is supported in B_1, nonnegative, and is constant 1 inside $B_{0.99}$. By direct computation, $\eta \sim 1$ inside B_1; and by Theorem 2.1,

$$|\eta(x,t)| \lesssim_N (1 + |x| + |t|)^{-N}, \tag{2.3}$$

for any positive integer N.

Let $\eta_R : \mathbb{R}^2 \to \mathbb{C}$ such that $\eta_R(x,t) = \eta(x/R, t/R)$. Then $\eta_R \sim 1$ inside B_R and

$$|\eta_R(x,t)| \lesssim_N \left(1 + (|x| + |t|)R^{-1}\right)^{-N}, \tag{2.4}$$

for any positive integer N.

DEFINITION 2.5. For any measurable set $A \subset \mathbb{R}^2$, use $\mathrm{Vol}(A)$ to denote the Lebesgue measure (*area*) of A.

2.2. Wave packet decomposition. Now we state the wave packet decomposition that will be used in this text. The version we adapt here is from [**Tao03**], and the sketch proof given below also follows the proof there. We point out that there are some other versions of the wave packet decomposition in literature, and interested readers can find some discussions in [**Bou91**] and [**TVV98**].

First let us introduce the notation of a *tube*.

DEFINITION 2.6. For any $R > 0$, $\theta \in [-1,1]$ and $v \in \mathbb{R}$, denote $RT_{\theta,v}$ to be the *tube*:

$$RT_{\theta,v} = \left\{(t,x) \,\middle|\, |x - v - \theta t| \le R^{\frac{1}{2}} \right\}. \tag{2.5}$$

For any smooth function $f : [-1,1] \to \mathbb{C}$, we can decompose it into wave packets, each supported in an interval of length $\sim R^{-\frac{1}{2}}$, and the corresponding Ef is decomposed such that inside B_R, each wave packet is essentially supported in a tube.

More precisely, we have the following *wave-packet decomposition of size R*.

PROPOSITION 2.7. [**Tao03**, Lemma 4.1] *For any $R>0$ and smooth $f : [-1,1] \to \mathbb{C}$, there exists a decomposition*

$$f = \sum_{\substack{\theta \in R^{-\frac{1}{2}}\mathbb{Z} \bigcap [-1,1], \\ v \in R^{\frac{1}{2}}\mathbb{Z}}} f_{\theta,v} = \sum_{\substack{\theta \in R^{-\frac{1}{2}}\mathbb{Z} \bigcap [-1,1], \\ v \in R^{\frac{1}{2}}\mathbb{Z}}} c_{\theta,v}\phi_{\theta,v}, \tag{2.6}$$

where each $f_{\theta,v}$ *(and* $\phi_{\theta,v}$*) is supported in* $[\theta - 3R^{-\frac{1}{2}}, \theta + 3R^{-\frac{1}{2}}]$*. Each coefficient* $c_{\theta,v} \in \mathbb{C}$ *satisfies that*

$$\sum_{\theta,v} |c_{\theta,v}|^2 \sim \|f\|_{L^2}^2, \tag{2.7}$$

and for each $\theta_o \in R^{-\frac{1}{2}}\mathbb{Z} \bigcap [-1,1], v_o \in R^{\frac{1}{2}}\mathbb{Z}$*,*

$$|c_{\theta_o,v_o}| \lesssim \max_{\theta,v} \|f_{\theta,v}\|_{L^2}. \tag{2.8}$$

Each $\phi_{\theta,v}$ *also satisfies*

$$|E\phi_{\theta,v}(x,t)| \lesssim_N R^{-\frac{1}{4}} \left(1 + |x - v - \theta t| R^{-\frac{1}{2}}\right)^{-N} \tag{2.9}$$

for any positive integer N*.*

Sketch proof. First write $f = \sum_{\theta \in R^{-\frac{1}{2}}\mathbb{Z} \bigcap [-1,1]} f_\theta$, where each f_θ is supported in the $R^{-\frac{1}{2}}$-neighborhood of θ. Since the support for each f_θ overlaps with at most 2 others, there is

$$\sum_\theta \|f_\theta\|_{L^2}^2 \sim \|f\|_{L^2}^2. \tag{2.10}$$

By the Poisson summation formula we find a Schwartz function $\gamma : \mathbb{R} \to \mathbb{C}$, such that $\widehat{\gamma}$ is supported in $[-1,1]$, and $\sum_{k\in\mathbb{Z}} \gamma(x-k) = 1$ for any $x \in \mathbb{R}$. Now denote

$$f_{\theta,v} = \left(\widehat{f_\theta} \cdot \gamma\left(R^{-\frac{1}{2}}(x-v)\right)\right)^\vee, \tag{2.11}$$

then each $f_{\theta,v} = f_\theta * \gamma\left(R^{-\frac{1}{2}}(x-v)\right)^\vee$ is supported in the $3R^{-\frac{1}{2}}$-neighborhood of θ. Take

$$c_{\theta,v} := R^{\frac{1}{4}} MEf_\theta(v,0), \tag{2.12}$$

where

$$MEf_\theta(x,0) := \sup_{r>0} \frac{1}{2r} \int_{[x-r,x+r]} |Ef_\theta(x',0)| \, dx' \tag{2.13}$$

is the Hardy–Littlewood maximal function.

Take Schwartz function $\varphi : \mathbb{R} \to \mathbb{C}$, such that $\widehat{\varphi} = 1$ inside the $3R^{-\frac{1}{2}}$-neighborhood of θ and vanishes outside the $4R^{-\frac{1}{2}}$-neighborhood of θ. And φ also satisfies

$$|\varphi(x)| \lesssim_N R^{-\frac{1}{2}} \left(1 + |x| R^{-\frac{1}{2}}\right)^{-N} \tag{2.14}$$

for any positive integer N.

Since f_θ is supported in the $3R^{-\frac{1}{2}}$-neighborhood of θ, we have that $Ef_\theta(\cdot,0) = Ef_\theta(\cdot,0) * \varphi$. This implies that

$$MEf_\theta(x_1,0) \sim MEf_\theta(x_2,0), \tag{2.15}$$

for any $|x_1 - x_2| \le R^{\frac{1}{2}}$. Then we conclude that

$$\sum_v |c_{\theta,v}|^2 \sim \sum_v \int_{[v-R^{\frac{1}{2}}, v+R^{\frac{1}{2}}]} |MEf_\theta(x,0)|^2 \, dx \lesssim \int |Ef_\theta(x,0)|^2 \, dx \le \|f_\theta\|_{L^2}^2. \tag{2.16}$$

Let's consider (2.8). For any θ_o and v_o, if for each $x \in \left[v_o - R^{\frac{1}{2}}, v_o + R^{\frac{1}{2}}\right]$, there is a neighborhood $\mathcal{A}_x$ of x, such that $|\mathcal{A}_x| \le R^{\frac{1}{2}}$, and

$$MEf_{\theta_o}(x,0) = \frac{1}{|\mathcal{A}_x|} \int_{\mathcal{A}_x} |Ef_{\theta_o}(x',0)| \, dx', \tag{2.17}$$

then $MEf_{\theta_o} = M(\chi_o Ef_{\theta_o})$ inside $\left[v_o - R^{\frac{1}{2}}, v_o + R^{\frac{1}{2}}\right]$, where χ_o is the indicator function of the interval $\left[v_o - 2R^{\frac{1}{2}}, v_o + 2R^{\frac{1}{2}}\right]$, and

$$\begin{aligned} |c_{\theta_o,v_o}|^2 &\sim \int_{\left[v_o - R^{\frac{1}{2}}, v_o + R^{\frac{1}{2}}\right]} |M(\chi_o Ef_{\theta_o})|^2 \, dx \\ &\lesssim \int_{\left[v_o - 2R^{\frac{1}{2}}, v_o + 2R^{\frac{1}{2}}\right]} |Ef_{\theta_o}(x,0)|^2 \, dx \\ &\lesssim \int \sum_{k=-2}^{2} \left|Ef_{\theta_o, v_o + kR^{\frac{1}{2}}}(x,0)\right|^2 dx \\ &\lesssim \max_{\theta,v} \|f_{\theta,v}\|_{L^2}^2. \end{aligned} \tag{2.18}$$

Otherwise, if (2.8) does not hold for each $x \in \left[v_o - R^{\frac{1}{2}}, v_o + R^{\frac{1}{2}}\right]$, meaning that there is $x_o \in [v_o - R^{\frac{1}{2}}, v_o + R^{\frac{1}{2}}]$, and a neighborhood $\mathcal{A}_o$ of x_o, such that $|\mathcal{A}_o| > R^{\frac{1}{2}}$, and

$$MEf_{\theta_o}(x_o,0) = \frac{1}{|\mathcal{A}_o|} \int_{\mathcal{A}_o} |Ef_{\theta_o}(x',0)| \, dx', \tag{2.19}$$

then there is a v', covered by $\mathcal{A}_o$, such that

$$|c_{\theta_o,v_o}|^2 \lesssim \int_{\left[v' - R^{\frac{1}{2}}, v' + R^{\frac{1}{2}}\right]} |Ef_{\theta_o}(x',0)|^2 \, dx' \lesssim \|f_{\theta_o,v'}\|_{L^2}^2 \lesssim \max_{\theta,v} \|f_{\theta,v}\|_{L^2}^2. \tag{2.20}$$

Now we show (2.9). It suffices to prove the following estimate:

$$|Ef_{\theta,v}(x,t)| \lesssim_N R^{-\frac{1}{4}} \left(1 + |x - v - \theta t| R^{-\frac{1}{2}}\right)^{-N} MEf_\theta(v,0) \tag{2.21}$$

for any positive integer N. This follows direct stationary phase computation, and details can be found in the proof of [**Tao03**, Lemma 4.1].

From (2.9), direct computation implies that

$$\int |Ef_{\theta,v}(x,0)|^2 \, dx \lesssim c_{\theta,v}^2, \quad \int |Ef_\theta(x,0)|^2 \, dx \lesssim \sum_v c_{\theta,v}^2, \tag{2.22}$$

which, with (2.16) implies (2.7). □

The bound (2.9) has the following implication.

COROLLARY 2.8. *Let $f = \sum_{\theta,v} f_{\theta,v}$ be a wave packet decomposition of size R. Denote $M = \max_{\theta,v} \|c_{\theta,v}\|$. Then there is*

$$\sum_{\theta,v : q \notin R^{1+\delta} T_{\theta,v}} |Ef_{\theta,v}(x,t)| \lesssim_{\delta,N,\lambda} R^{-N} \|f\|_{L^2}^{\lambda} M^{1-\lambda} \tag{2.23}$$

for any point $(x,t) \in \mathbb{R}^2$, and $\lambda \in (0,1]$.

PROOF. By (2.9), for each $\theta \in \mathbb{R}^{-\frac{1}{2}}\mathbb{Z}\bigcap[-1,1]$, there is

(2.24)
$$\begin{aligned}\sum_{v:(x,t)\notin R^{1+\delta}T_{\theta,v}} &|Ef_{\theta,v}(x,t)| \\ &\lesssim_L \sum_{v:(x,t)\notin R^{1+\delta}T_{\theta,v}} |c_{\theta,v}|R^{-\frac{1}{4}}\left(1+|x-v-\theta t|R^{-\frac{1}{2}}\right)^L \\ &\lesssim M^{1-\lambda}\left(\sum_v |c_{\theta,v}|^2\right)^{\frac{\lambda}{2}} \\ &\quad\cdot\left(\sum_{v:(x,t)\notin R^{1+\delta}T_{\theta,v}} R^{-\frac{1}{4}\left(1-\frac{\lambda}{2}\right)^{-1}}\left(1+|x-v-\theta t|R^{-\frac{1}{2}}\right)^{L\left(1-\frac{\lambda}{2}\right)^{-1}}\right)^{1-\frac{\lambda}{2}}.\end{aligned}$$

We have

$$\sum_{v:(x,t)\notin R^{1+\delta}T_{\theta,v}}\left(1+|x-v-\theta t|R^{-\frac{1}{2}}\right)^{L\left(1-\frac{\lambda}{2}\right)^{-1}} \lesssim \sum_{n>R^{\frac{\delta}{2}}} n^{-2L} \sim R^{\frac{\delta}{2}\cdot(1-2L)}, \tag{2.25}$$

since $v \in R^{\frac{1}{2}}\mathbb{Z}$, and $|x-v-\theta t|R^{-\frac{1}{2}} \gtrsim R^{\frac{\delta}{2}}$. Summing over θ, by Cauchy–Schwarz inequality there is

$$\begin{aligned}\sum_{\theta,v:q\notin R^{1+\delta}T_{\theta,v}} |Ef_{\theta,v}(x,t)| &\lesssim_L M^{1-\lambda}\sum_\theta\left(\sum_v |c_{\theta,v}|^2\right)^{\frac{\lambda}{2}} R^{\frac{\delta}{2}\cdot(1-2L)} \\ &\lesssim R^{\frac{1}{2}-\frac{\lambda}{4}+\frac{\delta}{2}\cdot(1-2L)}\|f\|_{L^2}^{\lambda}M^{1-\lambda}.\end{aligned} \tag{2.26}$$

Taking L large enough finishes the proof. □

Let's consider wave packets in different scales. There is a relation when transiting from the wave packet decomposition of one size to another.

LEMMA 2.9. *Let* $0 < R_1 < R_2$, *and* $\left(\frac{R_2}{R_1}\right)^{\frac{1}{2}}$ *an integer. For smooth* $f : [-1,1] \to \mathbb{C}$, *with wave packet decomposition of size* R_2*:*

$$f = \sum_{\tau,w} c_{\tau,w}\phi_{\tau,w} = \sum_{\tau,w} f_{\tau,w} = \sum_\tau f_\tau, \tag{2.27}$$

then there exists a wave packet decomposition for size R_1*:*

$$f = \sum_{\theta,v} c_{\theta,v}\phi_{\theta,v} = \sum_{\theta,v} f_{\theta,v} = \sum_\theta f_\theta, \tag{2.28}$$

such that for each θ, v, *there is*

$$|c_{\theta,v}| \lesssim \left(\frac{R_2}{R_1}\right)^{\frac{1}{4}}\max_{\tau,w}|c_{\tau,w}|. \tag{2.29}$$

PROOF. We can do the wave packet decomposition of size R_1 in the following way: we can write $R_2^{-\frac{1}{2}}\mathbb{Z}\bigcap[-1,1]$ as a disjoint union $\bigcup_{\theta\in R_1^{-\frac{1}{2}}\mathbb{Z}\bigcap[-1,1]}\mathcal{K}_\theta$. where each $\mathcal{K}_\theta \subset \left[\theta - R_2^{-\frac{1}{2}}, \theta + R_2^{-\frac{1}{2}}\right]$. Then we let

$$f_\theta = \sum_{\tau \in \mathcal{K}_\theta} f_\tau. \tag{2.30}$$

And the the remaining of the decomposition follows the proof in Proposition 2.7.

Note that $|\mathcal{K}_\theta| \lesssim (R_2/R_1)^{\frac{1}{2}}$. For any θ, v, from the decomposition, there is a neighborhood $\mathcal{A}$ of v, such that

$$|c_{\theta,v}|^2 \leq R_1^{\frac{1}{2}} \left(\frac{1}{|\mathcal{A}|}\int_{\mathcal{A}} |Ef_\theta|\right)^2 \leq \frac{R_1^{\frac{1}{2}}}{|\mathcal{A}|}\int_{\mathcal{A}} |Ef_\theta|^2 \lesssim \sum_{\tau\in\mathcal{K}_\theta} \frac{R_2^{\frac{1}{2}}}{|\mathcal{A}|}\int_{\mathcal{A}} |Ef_\tau|^2. \tag{2.31}$$

At each point $|Ef_\tau|^2 \sim \sum_w |c_{\tau,w}|^2$, thus is bounded by $R_2^{-\frac{1}{2}} \max_w |c_{\tau,w}|^2$. Then each $\frac{R_2^{\frac{1}{2}}}{|\mathcal{A}|}\int_{\mathcal{A}} |Ef_\tau|^2$ is bounded by $\max_w |c_{\tau,w}|^2$. Summing over τ leads to (2.29). □

3. Main results

3.1. The problem and obtained results. For any wave packet decomposition (of size R) of smooth function f , denote

$$S = \frac{\max_{\theta,v} \|f_{\theta,v}\|_{L^2}}{\|f\|_{L^2}} \sim \frac{\max_{\theta,v} |c_{\theta,v}|}{\|f\|_{L^2}}. \tag{3.1}$$

We would like to find all $\alpha, \beta \geq 0$, such that the following inequality always holds for any $\epsilon > 0$, sufficiently small (relying on p, α, β):

$$\|Ef\|_{L^p(B_R)} \leq C(\alpha,\beta,p,\epsilon) R^{\alpha+\epsilon} S^{\beta-\epsilon} \|f\|_{L^2}, \tag{3.2}$$

where $C(\alpha,\beta,p,\epsilon)$ is a constant. We also denote $M = \max_{\theta,v} \|f_{\theta,v}\|_{L^2} = S\|f\|_{L^2}$ to be the size of largest wave packet. We will use both notations (S and M) in our proofs.

Let us emphasize that the decomposition is *not unique*, and can lead to different S. The constant $C(\alpha,\beta,p,\epsilon)$ is independent of the function f and the choice of the decomposition.

Also, if for some p, the pair α_1, β_1, and α_2, β_2 make (3.2) hold, respectively, then for any $\lambda \in [0,1]$, so does the pair $\lambda\alpha_1 + (1-\lambda)\alpha_2$, $\lambda\beta_1 + (1-\lambda)\beta_2$.

The results to be presented in this text are as following:

THEOREM 3.1. *For $p \in [2,6]$, the inequality* (3.2) *holds for any smooth function $f : [-1,1] \to \mathbb{C}$, and any of its wave packet decomposition, if p, α, β satisfies*

$$\begin{cases} \beta \leq 1 \\ 4p\alpha + p \geq 6 \\ 2p\alpha - p\beta + p \geq 4 \\ 4\alpha - \beta \geq 0 \\ 12p\alpha - 4p\beta + 3p \geq 14 \end{cases}, \tag{3.3}$$

and only if p, α, β satisfies

$$\begin{cases} \beta \leq 1 \\ 4p\alpha + p \geq 6 \\ 2p\alpha - p\beta + p \geq 4 \\ 4\alpha - \beta \geq 0 \end{cases}. \tag{3.4}$$

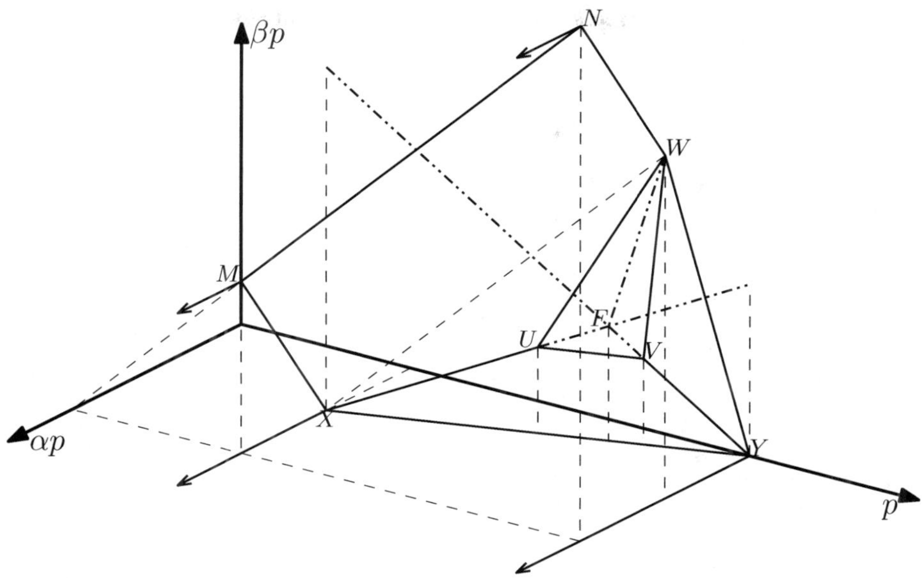

FIGURE 1. The parameter space

The results are better illustrated via a diagram of the parameter space (see Figure 1): we prove that (3.2) holds in the convex hull supported by M, N, X, Y, U, V, W, and its extension in the αp direction; whether (3.2) holds inside the tetrahedron $UVWF$ is still open; and for the remaining areas between $p = 2$ and $p = 6$ we've shown that (3.2) cannot be true.

It's worth noting that the above result means that we've found all possible pairs α, β for $2 \le p \le 4$ and $p = 6$; and for $5 \le p < 6$, the bounds we have cannot be improved when $S \ge R^{-\frac{1}{4}}$.

3.2. Initial steps and the structure of proof. There are a few observations which make the proof much clearer.

First, it is obvious that if (3.2) holds for p, α, β, then it also holds for p, α', β, for $\alpha' > \alpha$.

Besides, if (3.2) holds for some p_1, α_1, β_1, and p_2, α_2, β_2, then for any $\lambda \in [0, 1]$, (3.2) holds for $p_1\lambda + p_2(1 - \lambda), \frac{\alpha_1\lambda + \alpha_2(1-\lambda)}{p_1\lambda + p_2(1-\lambda)}, \frac{\beta_1\lambda + \beta_2(1-\lambda)}{p_1\lambda + p_2(1-\lambda)}$. This directly follows Hölder's inequality.

Also, note that inside B_R, there are *essentially* $\lesssim R$ wave packets, and it suffices to consider the cases where $R^{-\frac{1}{2}} \lesssim S \lesssim 1$. In other words, we have the following Lemma:

LEMMA 3.2. *If* (3.2) *holds for some* p, α, β, *then for any* $\lambda > 0$, *it also holds for* $p, \alpha + \lambda, \beta + 2\lambda$, *with* $\beta + 2\lambda \le 1$.

PROOF. Take $\delta = \frac{\epsilon}{2}$. Let $f' = \sum_{\theta, v: R^{1+\delta} T_{\theta,v} \bigcap B_R \neq \emptyset} f_{\theta,v}$, and this gives a wave packet decomposition of f'. Then $\|f'\|_{L^2}^2 \sim \sum_{\theta, v: R^{1+\delta} T_{\theta,v} \bigcap B_R \neq \emptyset} |c_{\theta,v}|^2 \lesssim R^{1+\frac{\delta}{2}} M^2$. By Corollary 2.8, there is

$$\|Ef - Ef'\|_{L^p B_R} \lesssim_{\delta, N, \lambda} R^{-N} S^{\beta + 2\lambda - \delta} \|f - f'\|_{L^2} \lesssim R^{-N} S^{\beta + 2\lambda - \delta} \|f\|_{L^2} \tag{3.5}$$

for any $N>0$, since $\beta+2\lambda-\delta\in[0,1)$ (when δ sufficiently small). Using the fact that (3.2) holds for p,α,β, we conclude

$$\begin{aligned}\|Ef\|_{L^p(B_R)} &\lesssim_\delta \|Ef'\|_{L^p(B_R)} + R^{\alpha+\delta}S^{\beta+2\lambda-\delta}\|f\|_{L^2}\\ &\lesssim_\delta R^{\alpha+\delta}M^{\beta-\delta}\|f'\|_{L^2}^{1+\delta-\beta} + R^{\alpha+\delta}S^{\beta+2\lambda}\|f\|_{L^2}\\ &\lesssim_\delta R^{\alpha+\delta+\lambda(1+\frac{\delta}{2})}M^{\beta-\delta+2\lambda}\|f'\|_{L^2}^{1+\delta-\beta-2\lambda} + R^{\alpha+\lambda+\delta}S^{\beta+2\lambda}\|f\|_{L^2}\\ &\lesssim R^{\alpha+\lambda+\epsilon}S^{\beta+2\lambda-\epsilon}\|f\|_{L^2}. \end{aligned}\tag{3.6}$$

□

Then it suffices to prove (3.2) for the end points:

- $p=2$, $\alpha=\frac{1}{2}$, $\beta=0$. This corresponds to point $X=(2,1,0)$ in Figure 1, and is part of (1.5).
- $p=4$, $\alpha=\frac{1}{8}$, $\beta=\frac{1}{4}$. This corresponds to point $U=\left(4,\frac{1}{2},1\right)$ in Figure 1, and is proved in Section 4, with arithmetic approaches.
- $p=5$, $\alpha=\frac{1}{20}$, $\beta=\frac{1}{5}$. This corresponds to point $V=\left(5,\frac{1}{4},1\right)$ in Figure 1, and is proved by polynomial partitioning in Section 5.
- $p=6$, $\alpha=0$, $\beta=0$. This corresponds to point $Y=(6,0,0)$ in Figure 1, and is also part of (1.5).
- $p=6$, $\alpha=\frac{1}{6}$, $\beta=\frac{2}{3}$. This corresponds to point $W=(6,1,4)$ in Figure 1, and is proved using l^2 Decoupling Theorem in Section 6.

In Section 7 we give examples, showing that (3.4) is a necessary condition for (3.2) to hold. In Section 8 we make the conjecture, which implies that (3.4) is also sufficient.

4. Explicit computation: Proof for $p=4$

In this section we prove the bound for $p=4$ by direct computation, writing the integral in L^4 norm as a sum of integrals of the wave packets. By doing this we can exploit the fact that different wave packets are far away in either physical or frequency space.

We point out that bounds given in this approach are always sharp; but it can be used only when p is an even integer.

PROPOSITION 4.1. *When $p=4$, (3.2) holds for $\alpha=\frac{1}{8}$ and $\beta=\frac{1}{4}$.*

PROOF. By Plancherel Theorem there is

$$\begin{aligned}\|Ef\|_{L^4(B_R)} &= \int_{B_R}\left|\sum_\theta Ef_\theta\right|^4\\ &\lesssim \int\left|\eta_R\sum_\theta Ef_\theta\right|^4\\ &= \int\left(\eta_R\sum_\theta Ef_\theta\right)^2\left(\overline{\eta_R\sum_\theta Ef_\theta}\right)^2\\ &= \sum_{\theta_1,\theta_2,\theta_3,\theta_4}\int \eta_R Ef_{\theta_1}Ef_{\theta_2}\cdot\overline{\eta_R Ef_{\theta_3}Ef_{\tau_4}}\\ &= \sum_{\theta_1,\theta_2,\theta_3,\theta_4}\int\left(\widehat{\eta_R}*\widehat{Ef_{\theta_1}}*\widehat{Ef_{\theta_2}}\right)\cdot\left(\overline{\widehat{\eta_R}}*\overline{\widehat{Ef_{\theta_3}}}*\overline{\widehat{Ef_{\theta_4}}}\right)\\ &= \sum_{\theta_1,\theta_2,\theta_3,\theta_4}\int\left(\widehat{\eta_R}*f_{\theta_1}^*\,\mathrm{d}\sigma*f_{\theta_2}^*\,\mathrm{d}\sigma\right)\cdot\left(\overline{\widehat{\eta_R}}*\overline{f_{\theta_3}^*}\,\mathrm{d}\sigma*\overline{f_{\theta_4}^*}\,\mathrm{d}\sigma\right). \end{aligned}\tag{4.1}$$

The convolution $\widehat{\eta_R} * f^*_{\theta_1}\, d\sigma * f^*_{\theta_2}\, d\sigma$ is supported in the R^{-1} neighborhood of the Minkowski sum of the supports of $f^*_{\theta_1}$ and $f^*_{\theta_2}$; and the convolution $\overline{\widehat{\eta_R}} * \overline{f^*_{\theta_3}}\, d\sigma * \overline{f^*_{\theta_4}}\, d\sigma$ is supported in the R^{-1} neighborhood of the Minkowski sum of the supports of $f^*_{\theta_3}$ and $f^*_{\theta_4}$. The integral

$$\int \left(\widehat{\eta_R} * f^*_{\theta_1}\, d\sigma * f^*_{\theta_2}\, d\sigma\right) \cdot \left(\overline{\widehat{\eta_R}} * \overline{f^*_{\theta_3}}\, d\sigma * \overline{f^*_{\theta_4}}\, d\sigma\right) \tag{4.2}$$

is none-zero only if the supports of $\widehat{\eta_R} * f^*_{\theta_1}\, d\sigma * f^*_{\theta_2}\, d\sigma$ and $\overline{\widehat{\eta_R}} * \overline{f^*_{\theta_3}}\, d\sigma * \overline{f^*_{\theta_4}}\, d\sigma$ have nonempty intersection, and only if there exists x_i in the $R^{-\frac{1}{2}}$-neighborhood of θ_i, $i = 1, 2, 3, 4$, such that

$$\begin{cases} |x_1 + x_2 - x_3 - x_4| & \leq 2R^{-1}, \\ \left|x_1^2 + x_2^2 - x_3^2 - x_4^2\right| & \leq 2R^{-1}. \end{cases} \tag{4.3}$$

The first inequality in (4.3) implies that

$$\left|(x_1 + x_2 - x_3)^2 - x_4^2\right| \leq 8R^{-1}, \tag{4.4}$$

and we have

$$2\,|x_1 - x_3|\,|x_2 - x_3| = \left|x_1^2 + x_2^2 - x_3^2 - (x_1 + x_2 - x_3)^2\right| \leq 9R^{-1}, \tag{4.5}$$

thus

$$\min\{|x_1 - x_3|, |x_2 - x_3|\} \leq 3R^{-\frac{1}{2}}. \tag{4.6}$$

This implies that $\min\{|\theta_1 - \theta_3|, |\theta_2 - \theta_3|\} \leq 4R^{-\frac{1}{2}}$; and with the first inequality in (4.3) we conclude that either $|\theta_1 - \theta_3|, |\theta_2 - \theta_4| \leq 4R^{-\frac{1}{2}}$, or $|\theta_2 - \theta_3|, |\theta_1 - \theta_4| \leq 4R^{-\frac{1}{2}}$.

Then for fixed (θ_1, θ_2), there are $\lesssim 1$ pairs of (τ_3, τ_4), such that (4.2) is nonzero.

Now we have

(4.7)

$$\begin{aligned} &\int_{B_R} \left|\sum_{\theta} Ef_\theta\right|^4 \\ &\lesssim \sum_{\theta_1, \theta_2} \int |\eta_R Ef_{\theta_1} Ef_{\theta_2}|^2 \\ &= \sum_{\theta_1, \theta_2} \sum_{u_1, u_2 \in R^{\frac{1}{2}}\mathbb{Z}} \int_{RT_{\theta_1, u_1} \bigcap RT_{\theta_2, u_2}} \left| \eta_R \sum_{v_1, v_2 \in R^{\frac{1}{2}}\mathbb{Z}} Ef_{\theta_1, v_1} Ef_{\theta_2, v_2} \right|^2 \\ &\lesssim \sum_{\theta_1, \theta_2} \sum_{v_1, v_2, u_1, u_2 \in R^{\frac{1}{2}}\mathbb{Z}} \int_{RT_{\theta_1, u_1} \bigcap RT_{\theta_2, u_2}} |\eta_R Ef_{\theta_1, v_1} Ef_{\theta_2, v_2}|^2 \\ &\qquad \cdot (1 + |u_1 - v_1| R^{-\frac{1}{2}})^2 (1 + |u_2 - v_2| R^{-\frac{1}{2}})^2, \end{aligned}$$

and by (2.9), (4.7) is further bounded by

$$\sum_{\theta_1, \theta_2} \sum_{v_1, v_2, u_1, u_2 \in R^{\frac{1}{2}}\mathbb{Z}} \frac{R^{-1} |c_{\theta_1, v_1} c_{\theta_2, v_2}|^2 \int_{RT_{\theta_1, u_1} \bigcap RT_{\theta_2, u_2}} |\eta_R|^2}{(1 + |u_1 - v_1| R^{-\frac{1}{2}})^{10} (1 + |u_2 - v_2| R^{-\frac{1}{2}})^{10}}. \tag{4.8}$$

Taking the square of (4.8), we get

(4.9)

$$\sum_{\substack{\theta_1,\theta_2,\theta_3,\theta_4\\ v_1,v_2,u_1,u_2\\ v_3,v_4,u_3,u_4}} \frac{R^{-2}|c_{\theta_1,v_1}c_{\theta_2,v_2}c_{\theta_3,v_3}c_{\theta_4,v_4}|^2\left(\int_{RT_{\theta_1,u_1}\cap RT_{\theta_2,u_2}}|\eta_R|^2\right)\left(\int_{RT_{\theta_3,u_3}\cap RT_{\theta_4,u_4}}|\eta_R|^2\right)}{\left((1+|u_1-v_1|R^{-\frac{1}{2}})(1+|u_2-v_2|R^{-\frac{1}{2}})(1+|u_3-v_3|R^{-\frac{1}{2}})(1+|u_4-v_4|R^{-\frac{1}{2}})\right)^{10}}$$

$$\lesssim \sum_{\substack{\theta_1,\theta_2,\theta_3,\theta_4\\ v_1,v_2,u_1,u_2\\ v_3,v_4,u_3,u_4}} \frac{R^{-2}|c_{\theta_1,v_1}c_{\theta_3,v_3}c^2_{\theta_4,v_4}|^2\left(\int_{RT_{\theta_1,u_1}\cap RT_{\theta_2,u_2}}|\eta_R|^2\right)^2}{\left((1+|u_1-v_1|R^{-\frac{1}{2}})(1+|u_2-v_2|R^{-\frac{1}{2}})(1+|u_3-v_3|R^{-\frac{1}{2}})(1+|u_4-v_4|R^{-\frac{1}{2}})\right)^{10}}$$

$$+ \sum_{\substack{\theta_1,\theta_2,\theta_3,\theta_4\\ v_1,v_2,u_1,u_2\\ v_3,v_4,u_3,u_4}} \frac{R^{-2}|c_{\theta_1,v_1}c^2_{\theta_2,v_2}c_{\theta_3,v_3}|^2\left(\int_{RT_{\theta_3,u_3}\cap RT_{\theta_4,u_4}}|\eta_R|^2\right)^2}{\left((1+|u_1-v_1|R^{-\frac{1}{2}})(1+|u_2-v_2|R^{-\frac{1}{2}})(1+|u_3-v_3|R^{-\frac{1}{2}})(1+|u_4-v_4|R^{-\frac{1}{2}})\right)^{10}}$$

$$= \sum_{\substack{\theta_1,\theta_2,\theta_3,\theta_4\\ v_1,v_2,u_1,u_2\\ v_3,v_4,u_3,u_4}} \frac{2R^{-2}|c_{\theta_1,v_1}c_{\theta_3,v_3}c^2_{\theta_4,v_4}|^2\left(\int_{RT_{\theta_1,u_1}\cap RT_{\theta_2,u_2}}|\eta_R|^2\right)^2}{\left((1+|u_1-v_1|R^{-\frac{1}{2}})(1+|u_2-v_2|R^{-\frac{1}{2}})(1+|u_3-v_3|R^{-\frac{1}{2}})(1+|u_4-v_4|R^{-\frac{1}{2}})\right)^{10}}.$$

Summing over v_2, u_3, u_4, (4.9) is bounded by

$$\sum_{\substack{\theta_1,\theta_2,\theta_3,\theta_4\\ v_1,u_1,u_2\\ v_3,v_4}} \frac{R^{-2}|c_{\theta_1,v_1}c_{\theta_3,v_3}c^2_{\theta_4,v_4}|^2\left(\int_{RT_{\theta_1,u_1}\cap RT_{\theta_2,u_2}}|\eta_R|^2\right)^2}{\left(1+|u_1-v_1|R^{-\frac{1}{2}}\right)^{10}}. \tag{4.10}$$

For each θ_1, u_1, θ_2, there is

$$\sum_{u_2}\int_{RT_{\theta_1,u_1}\cap RT_{\theta_2,u_2}}|\eta_R|^2 \sim \int_{RT_{\theta_1,u_1}}|\eta_R|^2 \lesssim R^{\frac{3}{2}}, \tag{4.11}$$

by (2.4). Also, for any $\theta_1 \neq \theta_2, u_1, u_2$, there is

$$\int_{RT_{\theta_1,u_1}\cap RT_{\theta_2,u_2}}|\eta_R|^2 \lesssim \int_{RT_{\theta_1,u_1}\cap RT_{\theta_2,u_2}} 1 \lesssim R^{\frac{3}{2}}\left(|\theta_1-\theta_2|R^{\frac{1}{2}}\right)^{-1}. \tag{4.12}$$

We conclude that, for fixed θ_1, u_1, there is

$$\sum_{\theta_2,u_2}\left(\int_{RT_{\theta_1,u_1}\cap RT_{\theta_2,u_2}}|\eta_R|^2\right)^2 \lesssim R^3\sum_{\theta_2}\left(|\theta_1-\theta_2|R^{\frac{1}{2}}\right)^{-1} \sim R^3\log(R). \tag{4.13}$$

Plugging (4.13) into (4.10) and summing over u_1, we see that (4.10) is bounded by

$$\sum_{\substack{\theta_1,\theta_3,\theta_4\\ v_1,v_3,v_4}} R\log(R)\left|c_{\theta_1,v_1}c_{\theta_3,v_3}c^2_{\theta_4,v_4}\right|^2 \tag{4.14}$$

$$\leq R\log(R)M^2\left(\sum_{\theta,u}|c_{\theta,v}|^2\right)^3 \sim R\log(R)M^2\|f\|^6_{L^2}.$$

This implies that (4.8) $\lesssim_\epsilon R^{\frac{1}{2}+4\epsilon}M\|f\|^3_{L^2} = R^{\frac{1}{2}+4\epsilon}S\|f\|^4_{L^2}$, and we finish the proof. □

5. Polynomial partitioning: Proof for $p = 5$

In this section we use the approach of polynomial partitioning. It was introduced by Larry Guth and Nets Katz in [**GK15**], and further used by Larry Guth to the restriction problem (see [**Gut16a**] [**Gut16b**]). We exploit this method in this problem, and prove the following result.

PROPOSITION 5.1. *For $p = 5$, (3.2) holds for $\alpha = \frac{1}{20}$ and $\beta = \frac{1}{5}$.*

The main idea is to use a polynomial of degree D to cut $\mathbb{R}^2$ into D^2 cells, each has the same L^p norm, and bound each cell using an induction argument. As each wave packet is essentially supported in a tube, which like a line, and thus it is expected to intersect the polynomial curve $\sim D$ times. Then we expect that on average, each cell intersects $\sim D^{-1}$ of all the wave packets.

We use the following version of polynomial partitioning in our proof, which also follows Stone-Tukey ham sandwich theorem [**ST42**].

THEOREM 5.2. ;[**Gut16a**, Theorem 0.6]/ *For any nonnegative L^1 function W on $\mathbb{R}$, and degree $D \geq 1$, there is a nonzero polynomial $\mathcal{F}$ of degree at most D, whose zero set is denoted as $Z(\mathcal{F})$, and $\mathbb{R}^2 \backslash Z(\mathcal{F})$ is a union of $\sim D^2$ disjoint cells O_i, with all $\int_{O_i} W$ equal.*

Let us point out that actually the tubes are different from lines: it can intersect arbitrarily many cells. To overcome this, we consider a "strip", which is a $R^{\frac{1}{2}+\delta}$-neighborhood of of the curve $Z(\mathcal{F})$. Now we need to bound the L^p norm on the strip as well. In some cases the L^p norm on the strip makes up a significant portion of the total L^p in B_R. For each tube, when intersecting the strip, it can be either parallel or transverse to the strip. We analyze these situations separately.

The following theorem of Wongkew will be used to estimate the area of the strip.

THEOREM 5.3. [**Won93**] *Let $\mathcal{F}$ be a nonzero polynomial of degree D on $\mathbb{R}^2$, and $\mathcal{N}$ the ρ-neighborhood of $Z(\mathcal{F})$. For B_R any ball of radius R, there is*

$$\mathrm{Vol}(B_R \bigcap \mathcal{N}) \lesssim D\rho R. \tag{5.1}$$

PROOF OF PROPOSITION 5.1. In the wave packet decomposition $f = \sum_{\theta,v} f_{\theta,v} = \sum_{\theta,v} c_{\theta,v}\phi_{\theta,v}$ of size R, denote $K(f)$ to be the number of nonzero $c_{\theta,v}$. It suffices to prove the case where $K(f)$ is finite. Indeed, by Corollary 2.8, there is

$$\int_{B_R} |Ef|^5 \lesssim \int_{B_R} \left| \sum_{\theta,v:RT_{\theta,v} \bigcap B_{R^2} \neq \emptyset} Ef_{\theta,v} \right|^5 + R^{-10} \|f\|_{L^2}^5. \tag{5.2}$$

The first term has only finitely many nonzero wave packets, while the second is obviously bounded.

Now we do induction on both the radius R and $K(f)$.

In the case where $K(f) = 1$, we have $S = 1$, and (3.2) follows Theorem 1.1.

Now consider finite $K(f)$. Denote χ_{B_R} to be the indicator function of B_R. For any positive integer D, by applying Theorem 5.2 to $\chi_{B_R}|f|^5$, there exists a nonzero polynomial $\mathcal{F}_D$, of degree at most D, and $\mathbb{R}^2 \backslash Z(\mathcal{F}_D)$ is a union of $\sim D^2$ disjoint cells $\tilde{O}_i$, such that all $\int_{\tilde{O}_i \bigcap B_R} |f|^5$ equal.

Let $\delta = \frac{\epsilon}{20}$. Denote $\mathcal{N}$ to be the $R^{\frac{1}{2}+\delta}$-neighborhood of $Z(\mathcal{F}_D)$ inside B_R, and $O_i = \tilde{O}_i \bigcap B_R \backslash \mathcal{N}$. Then there is

$$\|Ef\|_{L^5(B_R)}^5 = \|Ef\|_{L^5(\mathcal{N})}^5 + \sum_{O_i} \|Ef\|_{L^5(O_i)}^5. \tag{5.3}$$

Now consider the following two situations

CASE 1. $\sum_{O_i} \|Ef\|_{L^5(O_i)}^5 > \frac{3}{4}\|Ef\|_{L^5(B_R)}^5 = \frac{3}{4}\sum_{\tilde{O}_i} \|Ef\|_{L^5(\tilde{O}_i)}^5$.
Then for half of all O_i, there is

$$\|Ef\|_{L^5(O_i)}^5 \geq \frac{1}{2}\|Ef\|_{L^5(\tilde{O}_i)}^5 \geq C_1 D^{-2}\|Ef\|_{L^5(B_R)}^5, \tag{5.4}$$

for some constant C_1.

For each θ, v, let $\mathcal{L}_{\theta,v}$ be the line $x = v + \theta t$; then $R^{1+2\delta}T_{\theta,v}$ is the $R^{\frac{1}{2}+\delta}$ neighborhood of $\mathcal{L}_{\theta,v}$. For any cell O_i, if $R^{1+2\delta}T_{\theta,v} \bigcap O_i \neq \emptyset$, we must have $\mathcal{L}_{9,v} \bigcap \tilde{O}_i \neq \emptyset$. However, since $\mathcal{F}_D$ is of degree D, $\mathcal{L}_{\theta,v}$ has at most D intersections with the curve $Z(\mathcal{L}_{\theta,v})$; thus $R^{1+2\delta}T_{\theta,v}$ intersects at most D of all O_i. In particular, $\lesssim D$ of all O_i intersect each of $R^{1+2\delta}T_{\theta,v}$.

Also, by (2.7), there is

$$\sum_{O_i} \|f_{O_i}\|_{L^2}^2 \sim \sum_{O_i} \sum_{\substack{\theta,v \\ R^{1+2\delta}T_{\theta,v} \bigcap O_i \neq \emptyset}} |c_{\theta,v}|^2 \leq D \sum_{\theta,v} \|c_{\theta,v}|^2 \sim D\|f\|_{L^2}^2, \tag{5.5}$$

then more than $\frac{3}{4}$ of all O_i satisfy that

$$\|f_{O_i}\|_{L^2}^2 \leq C_2 D^{-1}\|f\|_{L^2}^2, \tag{5.6}$$

for some constant C_2.

Now we can find a cell O_{i_o}, which does not intersect every $R^{1+2\delta}T_{\theta,v}$, and satisfies

$$\|Ef\|_{L^5(O_{i_o})}^5 \geq C_1 D^{-2}\|Ef\|_{L^5(B_R)}^5, \quad \|f_{O_{i_o}}\|_{L^2}^2 \leq C_2 D^{-1}\|f\|_{L^2}^2. \tag{5.7}$$

We denote

$$f_{O_{i_o}} = \sum_{\theta,v:\ R^{1+2\delta}T_{\theta,v} \bigcap O_{i_o} \neq \emptyset} f_{\theta,v}. \tag{5.8}$$

Then there is

$$\int_{O_{i_o}} |Ef|^5 = \int_{O_{i_o}} \left|\sum_{\theta,v} Ef_{\theta,v}\right|^5 \lesssim \int_{O_{i_o}} |Ef_{O_{i_o}}|^5 + \int_{O_{i_o}} |E(f - f_{O_{i_o}})|^5. \tag{5.9}$$

Since $f - f_{O_{i_o}}$ is the sum of all $f_{\theta,v}$ where each $R^{1+2\delta}T_{\theta,v} \bigcap O_{i_o} = \emptyset$, by Corollary 2.8 there is

$$\|Ef\|_{L^5(O_{i_o})} = \left(\int_{O_{i_o}} |Ef|^5\right)^{\frac{1}{5}} \lesssim C_\delta \|Ef_{O_{i_o}}\|_{L^5(B_R)} + R^{-2}\|f\|_{L^2}, \tag{5.10}$$

where C_δ is a constant relying on δ.

Using the induction hypothesis for $f_{O_{i_o}}$, we obtain

(5.11)

$$\begin{aligned}
&\|Ef\|_{L^5(B_R)} \le C_1^{-\frac{1}{5}} D^{\frac{2}{5}} \|Ef\|_{L^5(O_{i_o})} \\
\le& C_1^{-\frac{1}{5}} D^{\frac{2}{5}} \left(C_\delta \|Ef_{O_{i_o}}\|_{L^5(B_R)} + R^{-2}\|f\|_{L^2}\right) \\
\le& C_1^{-\frac{1}{5}} D^{\frac{2}{5}} \left(C_\delta C\left(\frac{1}{20}, \frac{1}{5}, 5, \epsilon\right) R^{\frac{1}{20}+\epsilon} M^{\frac{1}{5}-\epsilon} \|f_{O_{i_o}}\|_{L^2}^{\frac{4}{5}+\epsilon} + R^{-2}\|f\|_{L^2}\right) \\
\le& C_1^{-\frac{1}{5}} C_2^{\frac{2}{5}+\frac{\epsilon}{2}} C_\delta C\left(\frac{1}{20}, \frac{1}{5}, 5, \epsilon\right) R^{\frac{1}{20}+\epsilon} M^{\frac{1}{5}-\epsilon} D^{-\frac{\epsilon}{2}} \|f\|_{L^2}^{\frac{4}{5}+\epsilon} + C_1^{-\frac{1}{5}} D^{\frac{2}{5}} R^{-2}\|f\|_{L^2}.
\end{aligned}$$

Now one can take D large enough (relying on ϵ, δ and C_o), such that when R large enough (relying on D), there is

$$\begin{aligned}
\|Ef\|_{L^5(B_R)} &\le C\left(\frac{1}{20}, \frac{1}{5}, \epsilon, 5\right) R^{\frac{1}{20}+\epsilon} M^{\frac{1}{5}-\epsilon} \|f\|_{L^2}^{\frac{4}{5}+\epsilon} \\
&= C\left(\frac{1}{20}, \frac{1}{5}, 5, \epsilon\right) R^{\frac{1}{20}+\epsilon} S^{\frac{1}{5}-\epsilon} \|f\|_{L^2}.
\end{aligned} \tag{5.12}$$

CASE 2. $\|Ef\|_{L^5(\mathcal{N})}^5 > \frac{1}{4}\|Ef\|_{L^5(B_R)}^5$.

For any θ, v, consider the tube $2R^{\frac{1}{2}+\delta}T_{\theta,v}$. The boundary of the tube consists of two lines, and the curve $Z(\mathcal{F}_D$ can cross it $\lesssim D$ times; and also $Z(\mathcal{F}_D$ can cross the boundary of B_R for $\lesssim D$ times. This means that $2R^{\frac{1}{2}+\delta}T_{\theta,v} \bigcap Z(\mathcal{F}_D) \bigcap B_r$ consists of $\lesssim D$ segments. Denote them as $Z_1, \cdots, Z_r$; and for each $1 \le i \le r$, denote $\mathcal{N}_i$ to be the $R^{\frac{1}{2}+\delta}$-neighborhood of Z_i.

For any point $a \in R^{\frac{1}{2}+\delta}T_{\theta,v} \bigcap \mathcal{N}$, there is $b \in Z(\mathcal{F}_D)$, such that $|a-b| \le R^{\frac{1}{2}+\delta}$. This means that $b \in 2R^{\frac{1}{2}+\delta}T_{\theta,v}$, $b \in Z_i$ for some $1 \le i \le r$, and then $a \in \mathcal{N}_i$. Thus there is

$$R^{\frac{1}{2}+\delta}T_{\theta,v} \bigcap \mathcal{N} \subset \bigcup_{1\le i\le r} \mathcal{N}_i. \tag{5.13}$$

Let G be a constant (to be determined later). For any θ, v and a corresponding set $\mathcal{N}_i$, if $\mathcal{N}_i$ has diameter $> \frac{R}{G}$, we say that $\mathcal{N}_i$ is *tangent* to $\mathcal{N}$; otherwise it is *transverse* to $\mathcal{N}$. Let $\Omega_{\theta,v,tang}$ be the collection of all tangent sets $\mathcal{N}_i$, and $\Omega_{\theta,v,trans}$ be the collection of all transverse sets $\mathcal{N}_i$.

Now divide B_R into $\sim G^2$ squares, each of size $\frac{R}{G} \times \frac{R}{G}$, and denote the collection of these squares as $\mathbb{S}$. For each $Q \in \mathbb{S}$, define $f_{trans}, f_{non} : [0,1] \to \mathbb{C}$ as following:

$$f_{Q,trans} = \sum_{\substack{\theta,v:\\ \exists \mathcal{N}_i \in \Omega_{\theta,v,trans},\\ \mathcal{N}_i \bigcap Q \neq \emptyset}} f_{\theta,v}, \quad f_{Q,non} = \sum_{\substack{\theta,v:\\ \forall \mathcal{N}_i \in \Omega_{\theta,v,trans},\\ \mathcal{N}_i \bigcap Q = \emptyset}} f_{\theta,v}. \tag{5.14}$$

Obviously, for each Q there is $f = f_{Q,trans} + f_{Q,non}$, and

$$\|Ef\|_{L^5(\mathcal{N})}^5 \lesssim \sum_{Q\in\mathbb{S}} \|Ef_{Q,trans}\|_{L^5(\mathcal{N}\bigcap Q)}^5 + \sum_{Q\in\mathbb{S}} \|Ef_{Q,non}\|_{L^5(\mathcal{N}\bigcap Q)}^5. \tag{5.15}$$

First consider $\sum_{Q\in\mathbb{S}} \|Ef_{Q,trans}\|_{L^5(\mathcal{N}\bigcap Q)}^5 \le \sum_Q \|Ef_{Q,trans}\|_{L^5(Q)}^5$. We can cover any Q by a ball of radius $\frac{R}{G}$. By Proposition 2.9 we can choose a wave packet composition of size R/G, and use the induction hypothesis for the smaller ball. Then we have

(5.16)
$$\|Ef_{Q,trans}\|_{L^5(Q)}^5 \leq C\left(\frac{1}{20},\frac{1}{5},5,\epsilon\right)^5\left(\frac{R}{G}\right)^{\frac{1}{4}+5\epsilon}\left(MG^{\frac{1}{4}}\right)^{1-5\epsilon}\|f_{Q,trans}\|_{L^2}^{4+5\epsilon}.$$

For each θ, v, $|\Omega_{\theta,v,trans}| \lesssim D$, and each $\mathcal{N}_i \in \Omega_{\theta,v,trans}$ can be covered by ~ 1 squares. Then by Proposition 2.7 there is

(5.17)
$$\sum_Q \|f_{Q,trans}\|_{L^2}^2 \lesssim \sum_Q \sum_{\substack{\theta,v\\ \exists \mathcal{N}_i\in\Omega_{\theta,v,trans},\mathcal{N}_i\bigcap Q\neq\emptyset}} |c_{\theta,v}|^2 \lesssim D\sum_{\theta,v}|c_{\theta,v}|^2 \sim D\|f\|_{L^2}^2,$$

then $\sum_Q \|f_{Q,trans}\|_{L^2}^{4+5\epsilon} \lesssim D^{2+\frac{5}{2}\epsilon}\|f\|_{L^2}^{4+5\epsilon}$.

By taking G large enough (relying on ϵ,δ,D), we have

(5.18)
$$\begin{aligned}\|Ef_{Q,trans}\|_{L^5(B_R)}^5 &= \sum_Q \|Ef_{Q,trans}\|_{L^5(Q)}^5\\ &\leq \frac{1}{8}C\left(\frac{1}{20},\frac{1}{5},5,\epsilon\right)^5 R^{\frac{1}{4}+5\epsilon}M^{1-5\epsilon}\|f\|_{L^2}^{4+5\epsilon}.\end{aligned}$$

Now consider $\sum_Q \|Ef_{Q,non}\|_{L^5(\mathcal{N}\bigcap Q)}^5$. Define $g_{tang}, g_o : B_R \to \mathbb{C}$ as following: for any $Q\in\mathbb{S}$ and point $q\in Q$, let

(5.19)
$$\begin{aligned}g_{tang}(q) &= \sum_{\substack{\theta,v:\\ \forall\mathcal{N}_i\in\Omega_{\theta,v,trans},\mathcal{N}_i\bigcap Q=\emptyset\\ \exists\mathcal{N}_j\in\Omega_{\theta,v,tang},q\in\mathcal{N}_j}} Ef_{\theta,v}(q),\\ g_o(q) &= \sum_{\substack{\theta,v:\\ \forall\mathcal{N}_i\in\Omega_{\theta,v,trans},\mathcal{N}_i\bigcap Q=\emptyset\\ \forall\mathcal{N}_j\in\Omega_{\theta,v,tang},q\notin\mathcal{N}_j}} Ef_{\theta,v}(q).\end{aligned}$$

Then for any $q\in Q$, $Ef_{Q,non}(q) = g_{tang}(q) + g_o(q)$, and we have

(5.20)
$$\sum_Q \|Ef_{Q,non}\|_{L^5(\mathcal{N}\bigcap Q)}^5 \lesssim \|g_{tang}\|_{L^5(\mathcal{N})}^5 + \|g_o\|_{L^5(\mathcal{N})}^5.$$

For any $q\in\mathcal{N}$, if q is not in any $\mathcal{N}_i\in\Omega_{\theta,v,trans}\bigcup\Omega_{\theta,v,tang}$, from (5.13), $q\notin R^{\frac{1}{2}+\delta}T_{\theta,v}$. From Corollary 2.8, this implies that $|g_o(q)|^5 \lesssim_\delta R^{-100}\|f\|_{L^2}^5$ and

(5.21)
$$\|g_o\|_{L^5(\mathcal{N})}^5 \lesssim_\delta R^{-50}\|f\|_{L^2}^5.$$

For $\|g_{tang}\|_{L^5(\mathcal{N})}^5$, we show that for each point $q\in\mathcal{N}$, it is contained in $\lesssim D^2R^{2\delta}$ sets in $\bigcup_{\theta,v}\Omega_{\theta,v,tang}$.

Suppose that there are W sets in $\bigcup_{\theta,v}\Omega_{\theta,v,tang}$ that contains q. For any θ, v and $\mathcal{N}_i\in\Omega_{\theta,v,tang}$, $q\in\mathcal{N}_i$, there are two points $a,b\in\mathcal{N}_i$ with $|a-b|\geq R/G$. Then there is $c\in\{a,b\}$ such that $|c-q|\geq R/2G$. This implies that there are points $c',q'\in Z(\mathcal{F}_D)$, such that the segment of $Z(\mathcal{F}_D)$ between c',q' is in $2R^{\frac{1}{2}+\delta}T_{\theta,v}$, and $|c-c'|,|q-q'|\leq R^{\frac{1}{2}+\delta}$. Let $P_{\theta,v,i}$ be the segment of $Z(\mathcal{F}_D)$ between c' and q'.

Denote Y_1 and Y_2 to be the circles centered at q with radius $R/3G$ and $R^{\frac{1}{2}+\delta}$. Then each $P_{\theta,v,i}$ intersects both Y_1 and Y_2. As $Z(\mathcal{F}_D)$ intersects either Y_1 and Y_2 at $\lesssim D$ points, there are $\gtrsim W/D^2$ different $P_{\theta,v,i}$, that intersect Y_1 and Y_2 at the same point, respectively. Let the two points be $y_1 \in Y_1$ and $y_2 \in Y_2$, and the segments be $P_{\theta_1,v_1,i_1}, \cdots, P_{\theta_k,v_k,i_k}$, for $k \gtrsim W/D^2$. Since for any fixed θ, v and $i \neq j$, $P_{\theta,v,i}$ does not overlap with $P_{\theta,v,j}$, the pairs $(\theta_1, v_1), \cdots, (\theta_k, v_k)$ are mutually different. Now there are k tubes $R^{\frac{1}{2}+\delta}T_{\theta_1,v_1}, \cdots, R^{\frac{1}{2}+\delta}T_{\theta_k,v_k}$, each containing y_1, y_2. Note that as $|y_1 - y_2| \sim R/G$, for any $R^{\frac{1}{2}+\delta}T_{\theta,v}$ containing both of them, θ can take only GR^δ values, and v can take only R^δ values. Then we have that $W/D^2 \lesssim k \leq GR^{2\delta}$, and $W \lesssim D^2GR^{2\delta}$.

By Theorem 5.3, we conclude that

(5.22)

$$\|g_{tang}\|^5_{L^5(\mathcal{N})} = \int_{\mathcal{N}} |g_{tang}|^5 \lesssim (D^2GR^{2\delta})^4 \int_{\mathcal{N}} \sum_{\theta,v} R^{-\frac{5}{4}} |c_{\theta,v}|^5 \lesssim (D^2G)^4 R^{\frac{1}{4}+9\delta} M \|f\|^4_{L^2}.$$

Finally, add (5.21) and (5.22) together, and take $C\left(\frac{1}{20}, \frac{1}{5}, 5, \epsilon\right)$, relying on G, D, ϵ, such that when R large enough. Then there is

$$\sum_{Q \in \mathbb{S}} \|Ef_{Q,non}\|^5_{L^5(\mathcal{N} \bigcap Q)} \leq \frac{1}{8} C\left(\frac{1}{20}, \frac{1}{5}, 5, \epsilon\right)^5 R^{\frac{1}{4}+5\epsilon} M^{1-5\epsilon} \|f\|^{4+5\epsilon}_{L^2}, \tag{5.23}$$

and

(5.24)

$$\begin{aligned}\|Ef\|^5_{L^5(B_R)} \leq 4\|Ef\|^5_{L^p(\mathcal{N})} &\leq 4 \sum_{Q \in \mathbb{S}} \|Ef_{Q,trans}\|^5_{L^5(\mathcal{N} \bigcap Q)} + \|Ef_{Q,non}\|^5_{L^5(\mathcal{N} \bigcap Q)} \\ &\leq C\left(\frac{1}{20}, \frac{1}{5}, 5, \epsilon\right)^5 R^{\frac{1}{4}+5\epsilon} M^{1-5\epsilon} \|f\|^{4+5\epsilon}_{L^2},\end{aligned}$$

and taking $\frac{1}{5}$th power in both sides finishes the proof. □

6. Using l^2 decoupling theorem: Proof for $p = 6$

Now we directly apply the l^2 Decoupling Theorem to give another bound for $p = 6$. The theorem was proved in recent years by Bourgain and Demeter for compact surfaces with positive definite second fundamental form. Some more explanations on the ideas in the proof can be found at [**BD16**].

In $1 + 1$ dimension and parabola, it can be stated as follows.

THEOREM 6.1. [**BD15**, Theorem 1.1] *For $\delta > 0$, let $\mathbb{S}$ be a set of segments of P such that finitely covers P, and each segment in $\mathbb{S}$ is of length $\sim \delta^{\frac{1}{2}}$. For any smooth function $g : \mathbb{R}^2 \to \mathbb{C}$, if $\widehat{g}$ is supported in the δ-neighborhood of P, and we write*

$$g = \sum_{\iota \in \mathbb{S}} g_\iota, \tag{6.1}$$

where each $\widehat{g}_\iota$ is supported in the δ-neighborhood of ι ; then for any $\epsilon > 0$, there is

$$\|g\|_{L^6} \lesssim_\epsilon \delta^{-\epsilon} \left(\sum_{\iota \in \mathbb{S}} \|g_\iota\|^2_{L^6}\right)^{\frac{1}{2}}. \tag{6.2}$$

We directly use Theorem 6.1 to obtain another sharp bound for $p = 6$.

PROPOSITION 6.2. *For $p = 6$, (3.2) holds for $\alpha = \frac{1}{6}$ and $\beta = \frac{2}{3}$.*

PROOF. Consider $Ef \cdot \eta_R = \sum_\theta Ef_\theta \cdot \eta_R$: each $\widehat{Ef_\theta \cdot \eta_R} = \widehat{Ef_\theta} * \widehat{\eta_R} = \widehat{f_\theta} * \widehat{\eta_R}$, and is supported in the $O(R^{-1})$ neighborhood of a segment of P, where the segment is the $\sim R^{-\frac{1}{2}}$ neighborhood of (θ, θ^2) on P. Also $\widehat{Ef \cdot \eta_R} = \widehat{Ef_\theta} * \widehat{\eta_R}$ is supported in the $O(R^{-1})$ neighborhood of P. By Theorem 6.1 we conclude that

$$\|Ef\|_{L^6(B_R)} \lesssim \|Ef \cdot \eta_R\|_{L^6} \lesssim_\epsilon R^\epsilon \left(\sum_\theta \|Ef_\theta \cdot \eta_R\|_{L^6}^2 \right)^{\frac{1}{2}}. \tag{6.3}$$

For each θ, there is

$$\|Ef_\theta \cdot \eta_R\|_{L^6}^6 = \int |Ef_\theta \cdot \eta_R|^6 \leq \sum_{u \in R^{\frac{1}{2}}\mathbb{Z}} \int_{RT_{\theta,u}} |Ef_\theta \cdot \eta_R|^6 \tag{6.4}$$
$$\lesssim \sum_{u,v \in R^{\frac{1}{2}}\mathbb{Z}} \int_{RT_{\theta,u}} |Ef_{\theta,v} \cdot \eta_R|^6 (1 + |u-v|R^{-\frac{1}{2}})^6.$$

In (2.9) we take $N = 100$, and get

$$\begin{aligned} &\|Ef_\theta \cdot \eta_R\|_{L^6}^6 \\ &\lesssim \sum_{u,v \in R^{\frac{1}{2}}\mathbb{Z}} R^{-\frac{3}{2}} (1 + |u-v|R^{-\frac{1}{2}})^6 \int_{RT_{\theta,u}} \left| \left(1 + |x - v - \theta t| R^{-\frac{1}{2}}\right)^{-100} \eta_R c_{\theta,v} \right|^6 \\ &\lesssim \sum_{u,v \in R^{\frac{1}{2}}\mathbb{Z}} R^{-\frac{3}{2}} (1 + |u-v|R^{-\frac{1}{2}})^{-2} \int_{RT_{\theta,u}} |\eta_R c_{\theta,v}|^6 . \end{aligned} \tag{6.5}$$

Use (2.4), taking $N = 10$, and (6.5) is bounded by

$$\sum_{u,v \in R^{\frac{1}{2}}\mathbb{Z}} R^{-\frac{3}{2}} (1 + |u-v|R^{-\frac{1}{2}})^{-2} \int_{RT_{\theta,u}} |c_{\theta,v}|^6 (1 + |t|/R)^{-10} \tag{6.6}$$
$$\lesssim \sum_{u,v \in R^{\frac{1}{2}}\mathbb{Z}} (1 + |u-v|R^{-\frac{1}{2}})^{-2} |c_{\theta,v}|^6,$$

and is bounded by $\sum_{v \in R^{\frac{1}{2}}\mathbb{Z}} |c_{\theta,v}|^6 \leq M^4 \sum_{v \in R^{\frac{1}{2}}\mathbb{Z}} |c_{\theta,v}|^2$, by summing over u.

Finally, there is

$$\|Ef\|_{L^6(B_R)} \lesssim_\epsilon R^\epsilon M^{\frac{2}{3}} \left(\sum_\theta \left(\sum_v |c_{\theta,v}|^2 \right)^{\frac{1}{3}} \right)^{\frac{1}{2}} \lesssim R^\epsilon M^{\frac{2}{3}} \left(R^{\frac{1}{3}} \left(\sum_{\theta,v} |c_{\theta,v}|^2 \right)^{\frac{1}{3}} \right)^{\frac{1}{2}}, \tag{6.7}$$

and it is $\sim R^{\frac{1}{6}+\epsilon} S^{\frac{2}{3}} \|f\|_{L^2}$. □

7. Examples

In this section we give examples, which provide necessary conditions for (3.2) to hold.

The example of Many Wave Packets is constructed to make S arbitrarily small, showing that there must be $\beta \leq 1$ for (3.2) to hold. The next two scenarios are constructed to maximize the left hand side of (3.2), by overlapping wave packets. In the Bundle Scenario, we take a large area inside B_R with equal number of overlaps inside the area; in Figure 1 this corresponds to the two planes intersect at the line XF. In the Star Scenario, we make every nonzero wave packet pass through the origin, thus make large number of overlaps in a small area; in Figure 1 this corresponds to the two planes intersect at the line YF. It is not surprising that the setting of the Star Scenario is larger for large p, since there $|Ef|$ highly concentrates in a small area and is small elsewhere; while the Bundle Scenario is larger for small p, since there $|Ef|$ is "equally large" in a nontrivial proportion of B_R.

7.1. Many wave packets. We start with a trivial example. Take any $U \in \left(0, R^{-\frac{1}{2}}\right)$, let $f : [-1,1] \to \mathbb{C}$ be a smooth function, satisfying $f(\omega) = 1$ for $|\omega| < \frac{U}{2}$, $|f| < 1$, and f is supported in $[-U, U]$.

PROPOSITION 7.1. *To make* (3.2) *hold, there must be* $\beta \leq 1$.

PROOF. Under the wave packet decomposition (of size R) in Proposition 2.7, there is $f_0 = f$ and any $f_\theta = 0$, for $\theta \neq 0$. By direct computation or Theorem 2.1, there is $|Ef_0(x,0)| \lesssim U(1+|x|U)^{-N}$ for any $N > 0$, and $|Ef_0(x,t)| \sim U$ for $|x| \lesssim U^{-1}$ and $|t| \lesssim 1$. Then each $|c_{0,v}| \lesssim R^{\frac{1}{4}}U$, $M \lesssim R^{\frac{1}{4}}U$ and $\|Ef_0\|_{L^p(B_R)} \gtrsim R^{\frac{1}{p}}U$. Also $\|f\|_{L^2} \sim U^{\frac{1}{2}}$.

Then (3.2) becomes

$$R^{\frac{1}{p}}U \lesssim_\epsilon R^{\alpha+\epsilon}\left(R^{\frac{1}{4}}U\right)^{\beta-\epsilon} U^{\frac{1-\beta+\epsilon}{2}}. \tag{7.1}$$

Since U can be taken arbitrarily small, there must be $1 \geq \beta - \epsilon + \frac{1-\beta+\epsilon}{2} = \frac{1}{2} + \frac{\beta}{2} - \frac{\epsilon}{2}$, for any $\epsilon > 0$. This implies that $\beta \leq 1$. ☐

7.2. Bundle scenario. Then consider the configuration where the wave packets take consecutive $\sim N$ directions (or θ in the decomposition), and in every direction there are $\sim N$ locations (or v in the decomposition) centered at the origin. More specifically, for any $R > 0$, and positive integer $N < R^{\frac{1}{2}}$, consider the following function

$$f_{R,N} : [-1,1] \to \mathbb{C}, \quad f_{R,N}(\omega) = \sum_{n=-N}^{N} \Theta_{R,N}(\omega - nR^{-\frac{1}{2}}), \tag{7.2}$$

where $\Theta_{R,N} : [-1,1] \to \mathbb{R}$ is the smooth function, such that $|\Theta_{R,N}| \leq 1$, $\Theta_{R,N}(\omega) = 1$ for $\omega \in \left[-\frac{1}{2}R^{-\frac{1}{2}}N^{-1}, \frac{1}{2}R^{-\frac{1}{2}}N^{-1}\right]$, and $\Theta_{R,N}$ is supported in

$$\left[-R^{-\frac{1}{2}}N^{-1}, R^{-\frac{1}{2}}N^{-1}\right].$$

PROPOSITION 7.2. *To make* (3.2) *hold, there must be* $4p\alpha + p \geq 6$ *and* $2p\alpha - p\beta + p \geq 4$.

PROOF. We can do a wave packet decomposition of $f_{R,N}$ in the following way. For $\theta = nR^{-\frac{1}{2}}$, $-N \leq n \leq N$, denote $f_\theta : [-1,1] \to \mathbb{C}$, $f_\theta(\omega) = \Theta_{R,N}(\omega - \theta)$. Then each f_θ is supported in an interval of length $2R^{-\frac{1}{2}}$ centered at θ; and we compute Ef_θ as

$$\int e^{i(\omega^2 t+\omega x)} f_\theta \, d\omega = e^{i\left(n^2R^{-1}t + nR^{-\frac{1}{2}}x\right)} \int e^{i\left(\omega^2 t + \omega\left(2nR^{-\frac{1}{2}}t + x\right)\right)} \Theta_{R,N}(\omega)\, d\omega. \tag{7.3}$$

By Theorem 2.1, the absolute value of (7.3) is $\sim R^{-\frac{1}{2}}N^{-1}$, when $\left|2nR^{-\frac{1}{2}}t+x\right| < R^{\frac{1}{2}}N$ and $(x,t)\in B_R$, and fast decays outside this area. This means that we can split each f_θ into $2N$ wave packets of equal size, thus there is

$$S=\frac{\left(\frac{1}{2N}\int|\Theta_{R,N}(\omega)|^2\,\mathrm{d}\omega\right)^{\frac{1}{2}}}{\|f_{R,N}\|_{L^2}}\sim N^{-1}. \tag{7.4}$$

The L^2-norm of $f_{R,N}$ can be easily computed as

$$\|f_{R,N}\|_{L^2}=\left(\int|f_{R,N}(\omega)|^2d\omega\right)^{\frac{1}{2}}=\left((2N+1)2R^{-\frac{1}{2}}N^{-1}\right)^{\frac{1}{2}}\sim R^{-\frac{1}{4}}. \tag{7.5}$$

Now we show that for $2\le p\le 6$ and any $\epsilon>0$, there is

$$\|Ef_{R,N}\|_{L^p(B_R)}\gtrsim_{\epsilon,p} R^{\frac{3}{2p}-\frac{1}{2}-\epsilon}N^{\frac{1}{p}-\frac{1}{2}}. \tag{7.6}$$

For $p=2$, since Fourier transform is unitary, and the L^2 norm is conserved, there is

$$\|Ef_{R,N}\|^2_{L^2(B_R)}\le\|Ef_{R,N}\|^2_{L^2(\mathbb{R}\times[-R,R])}=2R\|Ef_{R,N}|_{t=0}\|^2_{L^2_x}=2R\|f\|^2_{L^2}\sim R^{\frac{1}{2}}. \tag{7.7}$$

For $p=6$, as seen in the proof of Proposition 6.2, there is

$$\|Ef_{R,N}\|_{L^6(B_R)}\lesssim_\epsilon R^\epsilon\left(\sum_\theta\|Ef_\theta\cdot\eta_R\|^2_{L^6}\right)^{\frac{1}{2}}. \tag{7.8}$$

For each θ, we have that

$$\|Ef_\theta\cdot\eta_R\|_{L^6}\sim\left(R^{-3}N^{-6}\cdot NR^{\frac{3}{2}}\right)^{\frac{1}{6}}=R^{-\frac{1}{4}}N^{-\frac{5}{6}}, \tag{7.9}$$

and thus

$$\|Ef_{R,N}\|_{L^6(B_R)}\lesssim_\epsilon R^{-\frac{1}{4}+\epsilon}N^{-\frac{1}{3}}. \tag{7.10}$$

For $p=4$, let $\delta=\frac{1}{10}$. Then $\eta_{R^{1-\delta}}$ fast decays outside $B_{R^{1-\delta}}$, and we have

$$\left\|Ef_{R,N}\cdot\eta^{\frac{1}{2}}_{R^{1-\delta}}\right\|^4_{L^4}\lesssim\|Ef_{R,N}\|^4_{L^4(B_R)}+R^{-100}\|f_{R,N}\|^4_{L^2}. \tag{7.11}$$

By Plancherel Theorem we have that

$$\begin{aligned}\left\|Ef_{R,N}\cdot\eta^{\frac{1}{2}}_{R^{1-\delta}}\right\|^4_{L^4}&=\int\left|\eta_{R^{1-\delta}}\sum_\theta Ef_\theta\right|^4\\&=\sum_{\theta_1,\theta_2,\theta_3,\theta_4}\int|\eta_{R^{1-\delta}}|^2Ef_{\theta_1}Ef_{\theta_2}\overline{Ef_{\theta_3}Ef_{\theta_4}}\\&=\sum_{\theta_1,\theta_2,\theta_3,\theta_4}\int\left(\widehat{\eta}_{R^{1-\delta}}*f^*_{\theta_1}\,\mathrm{d}\sigma*f^*_{\theta_2}\,\mathrm{d}\sigma\right)\overline{\left(\widehat{\eta}_{R^{1-\delta}}*f^*_{\theta_3}\,\mathrm{d}\sigma*f^*_{\theta_4}\,\mathrm{d}\sigma\right)},\end{aligned} \tag{7.12}$$

and for any θ_1,θ_2, the convolution $\widehat{\eta}_{R^{1-\delta}}*f^*_{\theta_1}\,\mathrm{d}\sigma*f^*_{\theta_1}\,\mathrm{d}\sigma$ is real and nonnegative in the Minkowski sum of their support, and vanishes outside. Then for any $\theta_1,\theta_2,\theta_3$, θ_4, the integral

$$\int|\eta_{R^{1-\delta}}|^2Ef_{\theta_1}Ef_{\theta_2}\overline{Ef_{\theta_3}Ef_{\theta_4}} \tag{7.13}$$

is nonnegative. Then there is

(7.14)

$$\left\|\eta_{R^{1-\delta}}^{\frac{1}{2}} E f_{R,N}\right\|_{L^4}^4 \gtrsim \sum_{\theta_1,\theta_2} \int |\eta_{R^{1-\delta}}|^2 |Ef_{\theta_1}|^2 |Ef_{\theta_2}|^2 = \int |\eta_{R^{1-\delta}}|^2 \left(\sum_\theta |Ef_\theta|^2\right)^2$$

$$\gtrsim \int_{\mathcal{A}} \left(\sum_\theta |Ef_\theta|^2\right)^2 \sim R^{\frac{3}{2}} N \cdot R^{-2} N^{-2} = R^{-\frac{1}{2}} N^{-1},$$

where $\mathcal{A}$ is the area defined by $|x| \le \frac{1}{2} R^{\frac{1}{2}} N$ and $|t| \le \frac{R}{4}$, where each $|Ef_\theta| \sim R^{-\frac{1}{2}} N^{-1}$, and $|\eta_{R^{1-\delta}}| \sim 1$. Thus when R large enough, there is

$$\|Ef_{R,N}\|_{L^4(B_R)}^4 \gtrsim R^{-\frac{1}{2}} N^{-1}. \tag{7.15}$$

Finally, by Hölder's inequality, for any $2 \le p < 4$ there is

$$\|Ef_{R,N}\|_{L^p(B_R)} \ge \frac{\|Ef_{R,N}\|_{L^4(B_R)}^{\frac{12}{p}-2}}{\|Ef_{R,N}\|_{L^6(B_R)}^{\frac{12}{p}-3}} \gtrsim_\epsilon R^{\frac{3}{2p}-\frac{1}{2}-(\frac{12}{p}-3)\epsilon} N^{\frac{1}{p}-\frac{1}{2}}, \tag{7.16}$$

and for any $4 < p \le 6$ there is

$$\|Ef_{R,N}\|_{L^p(B_R)} \ge \frac{\|Ef_{R,N}\|_{L^4(B_R)}^{2-\frac{4}{p}}}{\|Ef_{R,N}\|_{L^2(B_R)}^{2-\frac{4}{p}}} \gtrsim R^{\frac{3}{2p}-\frac{1}{2}} N^{\frac{1}{p}-\frac{1}{2}}. \tag{7.17}$$

And these imply (7.6).

Finally, if (3.2) is true for some p, α, β, there is

$$R^{\frac{3}{2p}-\frac{1}{2}-\epsilon} N^{\frac{1}{p}-\frac{1}{2}} \lesssim_\epsilon R^{\alpha+\epsilon} N^{-\beta+\epsilon} R^{-\frac{1}{4}}. \tag{7.18}$$

By taking $N = 1$ and $N = R^{\frac{1}{2}}$, we get $4p\alpha + p \ge 6$ and $2p\alpha - p\beta + p \ge 4$, respectively. □

7.3. Star scenario. Let us consider another configuration, where for consecutive N directions, there is exactly one wave packet at each direction, passing through the origin.

Namely, for any $R > 0$, and positive integer $N < R^{\frac{1}{2}}$, consider the following function

$$f_{R,N} : [-1,1] \to \mathbb{C}, \quad f_{R,N}(\omega) = \sum_{n=-N}^{N} \Phi_{R,N}(\omega - nR^{-\frac{1}{2}}), \tag{7.19}$$

where $\Phi_{R,N} : [-1,1] \to \mathbb{R}$ is the smooth function, such that $|\Phi_{R,N}| \le 1$, $\Phi_{R,N}(\omega) = 1$ for $\omega \in \left[-\frac{1}{2}R^{-\frac{1}{2}}, \frac{1}{2}R^{-\frac{1}{2}}\right]$, and $\Phi_{R,N}$ is supported in $\left[-R^{-\frac{1}{2}}, R^{-\frac{1}{2}}\right]$.

We also require that $\Phi_{R,N}$ is an even function, and for any $\omega \in \left[0, \frac{1}{2}R^{-\frac{1}{2}}\right]$, there is $\Phi_{R,N}(\omega) = \Phi_{R,N}\left(\frac{1}{2}R^{-\frac{1}{2}} - \omega\right)$. Then $f_{R,N}(\omega) = 1$ for any

$$\omega \in \left[-\left(N + \frac{1}{2}\right) R^{-\frac{1}{2}}, \left(N + \frac{1}{2}\right) R^{-\frac{1}{2}}\right],$$

and $f_{R,N}$ is real and supported in $\left[-(N+1) R^{-\frac{1}{2}}, (N+1) R^{-\frac{1}{2}}\right]$.

PROPOSITION 7.3. *To make* (3.2) *hold, there must be* $4p\alpha+p \geq 6$ *and* $4\alpha-\beta \geq 0$.

PROOF. The wave packet decomposition of $f_{R,N}$ is as following: for $\theta = nR^{-\frac{1}{2}}$, $-N \leq n \leq N$, denote $f_\theta : [-1,1] \to \mathbb{C}$, $f_\theta(\omega) = \Phi_{R,N}(\omega - \theta)$. Then each f_θ is supported in an interval of length $2R^{-\frac{1}{2}}$ centered at θ. We compute Ef_θ as

$$\int \mathrm{e}^{\mathbf{i}(\omega^2 t+\omega x)} f_\theta \, \mathrm{d}\omega = \mathrm{e}^{\mathbf{i}\left(n^2R^{-1}t+nR^{-\frac{1}{2}}x\right)} \int \mathrm{e}^{\mathbf{i}\left(\omega^2 t+\omega\left(2nR^{-\frac{1}{2}}t+x\right)\right)} \Phi_{R,N}(\omega) \, \mathrm{d}\omega. \tag{7.20}$$

The absolute value of (7.20) $\sim R^{-\frac{1}{2}}$ inside the area defined by $\left|2nR^{-\frac{1}{2}}t + x\right| < 2R^{\frac{1}{2}}$ and $|t| < R$; and by Theorem 2.1, it also fast decays outside. This means that each f_θ corresponds to ~ 1 wave packets, of equal size, in wave packet decomposition. Thus we have $S \sim N^{-\frac{1}{2}}$.

As for the L^2-norm of $f_{R,N}$, there is

$$\|f_{R,N}\|_{L^2} \sim \left(\int_{-(N+\frac{1}{2})R^{-\frac{1}{2}}}^{(N+\frac{1}{2})R^{-\frac{1}{2}}} 1 \, \mathrm{d}\omega\right)^{\frac{1}{2}} \sim R^{-\frac{1}{4}}N^{\frac{1}{2}}. \tag{7.21}$$

Then it remains to estimate $\|Ef_{R,N}\|_{L^p(B_R)}$ for $2 \leq p \leq 6$. Let $\mathcal{C}$ be the area defined by $|t| \leq \frac{1}{10}RN^{-2}$, and $|x| \leq \frac{1}{10}R^{\frac{1}{2}}N^{-1}$. Then for any $(x,t) \in \mathcal{C}$, there is

$$|Ef_{R,N}(x,t)| = \left|\int \mathrm{e}^{\mathbf{i}(\omega^2 t+\omega x)} f_{R,N} \, \mathrm{d}\omega\right| \sim R^{\frac{1}{2}}N, \tag{7.22}$$

since $f_{R,N}$ is real and supported in $\left[-(N+1)R^{-\frac{1}{2}}, (N+1)R^{-\frac{1}{2}}\right]$. We conclude that

$$\|Ef_{R,N}\|_{L^p(B_R)} \geq \|Ef_{R,N}\|_{L^p(\mathcal{C})} \gtrsim \left(R^{\frac{3}{2}}N^{-3} \cdot N^p R^{\frac{p}{2}}\right)^{\frac{1}{p}} \geq R^{\frac{3}{2p}-\frac{1}{2}}N^{1-\frac{3}{p}}. \tag{7.23}$$

To make (3.2) hold, there must be

$$R^{\frac{3}{2p}-\frac{1}{2}}N^{1-\frac{3}{p}} \lesssim_\epsilon R^{\alpha+\epsilon}N^{\frac{\epsilon-\beta}{2}}R^{-\frac{1}{4}}N^{\frac{1}{2}}, \tag{7.24}$$

and via taking $N = 1$ and $N = R^{\frac{1}{2}}$ we conclude $4p\alpha + p \geq 6$ and $4\alpha - \beta \geq 0$, respectively. □

8. Conjecture

We make the conjecture that (3.4) is sufficient for (3.2) to be true. It suffices to consider the follow case.

CONJECTURE 8.1. *The inequality* (3.2) *holds for* $p = \frac{14}{3}$, $\alpha = \frac{1}{14}$ *and* $\beta = \frac{2}{7}$.

This corresponds to point F in Figure 1. Again, by Hölder's inequality and Lemma 3.2, this implies that all possible p, α, β are given by (3.4), and the remaining cases can be settled.

Acknowledgments

This work was conducted at the 2016 Summer Program in Undergraduate Research (SPUR) of the MIT Department of Mathematics, where the second author was an undergraduate and the first author was the graduate mentor. We would like to thank Professor Larry Guth for proposing this problem, and Proffessors David Jerison and Ankur Moitra for many useful discussions and for suggesting the direction of this work.

References

[BD15] Jean Bourgain and Ciprian Demeter, *The proof of the l^2 decoupling conjecture*, Ann. of Math. (2) **182** (2015), no. 1, 351–389, DOI 10.4007/annals.2015.182.1.9. MR3374964

[BD16] Jean Bourgain and Ciprian Demeter, *A study guide for the l^2 decoupling theorem*, Chinese Ann. Math. Ser. B **38** (2017), no. 1, 173–200, DOI 10.1007/s11401-016-1066-1. MR3592159

[Bou91] Jean Bourgain, *Some new estimates on oscillatory integrals*, Essays on Fourier analysis in honor of Elias M. Stein (Princeton, NJ, 1991), Princeton Math. Ser., vol. 42, Princeton Univ. Press, Princeton, NJ, 1995, pp. 83–112. MR1315543

[Eva10] Lawrence C. Evans, *Partial differential equations*, 2nd ed., Graduate Studies in Mathematics, vol. 19, American Mathematical Society, Providence, RI, 2010, DOI 10.1090/gsm/019. MR2597943

[GK15] Larry Guth and Nets Hawk Katz, *On the Erdős distinct distances problem in the plane*, Ann. of Math. (2) **181** (2015), no. 1, 155–190, DOI 10.4007/annals.2015.181.1.2. MR3272924

[Gut16a] Larry Guth, *A restriction estimate using polynomial partitioning*, J. Amer. Math. Soc. **29** (2016), no. 2, 371–413, DOI 10.1090/jams827. MR3454378

[Gut16b] Larry Guth, *Restriction estimates using polynomial partitioning II*, Acta Math. **221** (2018), no. 1, 81–142, DOI 10.4310/ACTA.2018.v221.n1.a3. MR3877019

[SM93] Elias M. Stein, *Harmonic analysis: real-variable methods, orthogonality, and oscillatory integrals*, Princeton Mathematical Series, vol. 43, Princeton University Press, Princeton, NJ, 1993. With the assistance of Timothy S. Murphy; Monographs in Harmonic Analysis, III. MR1232192

[ST42] Arthur H. Stone and John W. Tukey, *Generalized "sandwich" theorems*, Duke Math. J. **9** (1942), 356–359. MR7036

[Ste86] Elias M. Stein, *Oscillatory integrals in Fourier analysis*, Beijing lectures in harmonic analysis (Beijing, 1984), Ann. of Math. Stud., vol. 112, Princeton Univ. Press, Princeton, NJ, 1986, pp. 307–355. MR864375

[Str77] Robert S. Strichartz, *Restrictions of Fourier transforms to quadratic surfaces and decay of solutions of wave equations*, Duke Math. J. **44** (1977), no. 3, 705–714. MR512086

[Tao03] Terence Tao, *A sharp bilinear restrictions estimate for paraboloids*, Geom. Funct. Anal. **13** (2003), no. 6, 1359–1384, DOI 10.1007/s00039-003-0449-0. MR2033842

[Tao04] Terence Tao, *Some recent progress on the restriction conjecture*, Fourier analysis and convexity, Appl. Numer. Harmon. Anal., Birkhäuser Boston, Boston, MA, 2004, pp. 217–243, DOI 10.1198/106186003321335099. MR2087245

[Tao06] Terence Tao, *Nonlinear dispersive equations*, CBMS Regional Conference Series in Mathematics, vol. 106, Published for the Conference Board of the Mathematical Sciences, Washington, DC; by the American Mathematical Society, Providence, RI, 2006. Local and global analysis, DOI 10.1090/cbms/106. MR2233925

[Tom75] Peter A. Tomas, *A restriction theorem for the Fourier transform*, Bull. Amer. Math. Soc. **81** (1975), 477–478, DOI 10.1090/S0002-9904-1975-13790-6. MR358216

[TVV98] Terence Tao, Ana Vargas, and Luis Vega, *A bilinear approach to the restriction and Kakeya conjectures*, J. Amer. Math. Soc. **11** (1998), no. 4, 967–1000, DOI 10.1090/S0894-0347-98-00278-1. MR1625056

[Wol03] Thomas H. Wolff, *Lectures on harmonic analysis*, University Lecture Series, vol. 29, American Mathematical Society, Providence, RI, 2003. With a foreword by Charles

Fefferman and a preface by Izabella Łaba; Edited by Łaba and Carol Shubin, DOI 10.1090/ulect/029. MR2003254

[Won93] Richard Wongkew, *Volumes of tubular neighbourhoods of real algebraic varieties*, Pacific J. Math. **159** (1993), no. 1, 177–184. MR1211391

Department of Mathematics, Massachusetts Institute of Technology, Cambridge, Massachusetts 02139

Current address: Department of Mathematics, University of California, Los Angeles, Los Angeles, CA 90095

Email address: hongwang@math.ucla.edu

Department of Mathematics, Massachusetts Institute of Technology, Cambridge, Massachusetts 02139

Current address: Department of Statistics, University of California, Berkeley, Berkeley, CA 94720

Email address: lfzhang@berkeley.edu

Selected Published Titles in This Series

792 **Shaoming Guo, Zane Kun Li, and Brian Street, Editors,** Recent Developments in Harmonic Analysis and its Applications, 2024

790 **Peter Crooks and Alexandru I. Suciu, Editors,** Compactifications, Configurations, and Cohomology, 2023

789 **I. S. Krasil′shchik, A. B. Sossinsky, and A. M. Verbovetsky, Editors,** The Diverse World of PDEs, 2023

788 **I. S. Krasil′shchik, A. B. Sossinsky, and A. M. Verbovetsky, Editors,** The Diverse World of PDEs, 2023

787 **Hemen Dutta, Editor,** Mathematical Modelling, 2023

786 **Hemen Dutta, Editor,** Mathematical Modelling, 2023

785 **A. Leroy and S. K. Jain, Editors,** Algebra and Coding Theory, 2023

784 **Dinh-Liem Nguyen, Loc Hoang Nguyen, and Thi-Phong Nguyen, Editors,** Advances in Inverse Problems for Partial Differential Equations, 2023

783 **Alla Detinko, Michael Kapovich, Alex Kontorovich, Peter Sarnak, and Richard Schwartz, Editors,** Computational Aspects of Discrete Subgroups of Lie Groups, 2023

782 **Galina Filipuk, Alberto Lastra, and Sławomir Michalik, Editors,** Recent Trends in Formal and Analytic Solutions of Diff. Equations, 2023

781 **Emanuel Indrei, Diego Marcon, and Levon Nurbekyan, Editors,** Geometric and Functional Inequalities and Recent Topics in Nonlinear PDEs, 2023

780 **Erik Koelink, Stefan Kolb, Nicolai Reshetikhin, and Bart Vlaar, Editors,** Hypergeometry, Integrability and Lie Theory, 2022

779 **Samuele Anni, Valentijn Karemaker, and Elisa Lorenzo García, Editors,** Arithmetic, Geometry, Cryptography, and Coding Theory 2021, 2022

778 **Carlos Galindo, Alejandro Melle Hernández, Julio José Moyano-Fernández, and Wilson A. Zúñiga-Galindo, Editors,** p-Adic Analysis, Arithmetic and Singularities, 2022

777 **Bang-Yen Chen, Nicholas D. Brubaker, Takashi Sakai, Bogdan D. Suceavă, Makiko Sumi Tanaka, Hiroshi Tamaru, and Mihaela B. Vajiac, Editors,** Differential Geometry and Global Analysis, 2022

776 **Aaron Wootton, S. Allen Broughton, and Jennifer Paulhus, Editors,** Automorphisms of Riemann Surfaces, Subgroups of Mapping Class Groups and Related Topics, 2022

775 **Fernando Galaz-García, Cecilia González-Tokman, and Juan Carlos Pardo Millán, Editors,** Mexican Mathematicians in the World, 2021

774 **Randall J. Swift, Alan Krinik, Jennifer M. Switkes, and Jason H. Park, Editors,** Stochastic Processes and Functional Analysis, 2021

773 **Nicholas R. Baeth, Thiago H. Freitas, Graham J. Leuschke, and Victor H. Jorge Pérez, Editors,** Commutative Algebra, 2021

772 **Anatoly M. Vershik, Victor M. Buchstaber, and Andrey V. Malyutin, Editors,** Topology, Geometry, and Dynamics, 2021

771 **Nicolás Andruskiewitsch, Gongxiang Liu, Susan Montgomery, and Yinhuo Zhang, Editors,** Hopf Algebras, Tensor Categories and Related Topics, 2021

770 **Stéphane Ballet, Gaetan Bisson, and Irene Bouw, Editors,** Arithmetic, Geometry, Cryptography and Coding Theory, 2021

769 **Kiyoshi Igusa, Alex Martsinkovsky, and Gordana Todorov, Editors,** Representations of Algebras, Geometry and Physics, 2021

768 **Dražen Adamović, Andrej Dujella, Antun Milas, and Pavle Pandžić, Editors,** Lie Groups, Number Theory, and Vertex Algebras, 2021

For a complete list of titles in this series, visit the
AMS Bookstore at **www.ams.org/bookstore/conmseries/**.